Biotechnology: From Theories to Practice

Biotechnology: From Theories to Practice

Editor: Jordan Bryant

R CALLISTO
REFERENCE

www.callistoreference.com

Callisto Reference,
118-35 Queens Blvd., Suite 400,
Forest Hills, NY 11375, USA

Visit us on the World Wide Web at:
www.callistoreference.com

ISBN: 978-1-64116-261-6 (Hardback)

Cataloging-in-Publication Data

Biotechnology : from theories to practice / edited by Jordan Bryant.
   p. cm.
Includes bibliographical references and index.
ISBN 978-1-64116-261-6
1. Biotechnology. 2. Genetic engineering. I. Bryant, Jordan.
TP248.2 .B56 2020
660.6--dc23

Table of Contents

Permissions

List of Contributors

Index

Preface

Biotechnology is an area of biology which involves the use of living organisms to create products or processes for human use. Human society has benefitted tremendously from biotechnology, as evident in the cultivation of plants, domestication of animals and manufacture of biopharmaceutics. However, the technological bases for biotechnology are the breakthroughs in genetic engineering and molecular biology in the late 20th century. The understanding of the structure of DNA and the techniques to manipulate it have led to the emergence of transgenic animals, genetically modified crops, the Human Genome Project, and cutting-edge research in bioinformatics. Cloning and stem cell technology are some of its modern tools. Biotechnology has extended beyond its domain of application in health care and agriculture, and stepped into environmental remediation and industry. There has been rapid progress in this field and its applications are finding their way across multiple industries. Different approaches, evaluations, methodologies and advanced studies on biotechnology have been included in this book. It is a vital tool for all researching and studying this field.

After months of intensive research and writing, this book is the end result of all who devoted their time and efforts in the initiation and progress of this book. It will surely be a source of reference in enhancing the required knowledge of the new developments in the area. During the course of developing this book, certain measures such as accuracy, authenticity and research focused analytical studies were given preference in order to produce a comprehensive book in the area of study.

This book would not have been possible without the efforts of the authors and the publisher. I extend my sincere thanks to them. Secondly, I express my gratitude to my family and well-wishers. And most importantly, I thank my students for constantly expressing their willingness and curiosity in enhancing their knowledge in the field, which encourages me to take up further research projects for the advancement of the area.

Editor

Chemical Profiling and Antimicrobial Properties of Phyto-Active Extracts from *Terminalia glaucescens* Stem against Water Microbial Contaminants

Adeyemi Ojutalayo Adeeyo[1,2,*], John Odiyo[2] and Kehinde Odelade[1]

[1]*Department of Pure and Applied Biology, Ladoke Akintola University of Technology, P.M.B 4000, Ogbomoso, Oyo State, Nigeria*
[2]*School of Environmental Sciences, University of Venda, Private Bag X5050, Thohoyandou 0950, South Africa*

Abstract:

Background:

The present study was designed to evaluate the phytochemicals of *Terminalia glaucescens* stem extracts and test their antimicrobial potency against water microbial contaminants reported to be multidrug resistant.

Method:

Dry stem powder was extracted with ethanol, ethyl acetate and chloroform. These fractions were then examined for antimicrobial activity by using disc diffusion assay against typical clinical bacteria and fungal isolates which have been reported as water contaminants. The microbial strains were exposed to five different concentrations of extracts: 500 mg/ml, 250 mg/ml, 100 mg/ml, 50 mg/ml and 25 mg/ml.

Result:

It was observed in this study that increase in the concentration of extracts correlated with microbial growth inhibition. In-vitro phytochemical screening of plant extracts revealed the presence of alkaloid, flavonoid, saponin, terpenoid, steroid and anthraquinones. Ethanolic extract performs better than ethyl acetate and chloroform extracts, as it recorded the highest zone of inhibition of 20.5 mm against *Streptococcus pneumoniae* while ethyl acetate and chloroform recorded 17.50 mm each against *Streptococcus pneumoniae* and *Bacillus cereus*, respectively. Ethanolic extract also showed the highest antifungal activity against *Trichoderma* sp. and *Aspergillus niger*. The antibacterial and antifungal activities of active extracts were observed in the increasing order Ethanol>Chloroform≥ethyl acetate with respect to the maximum zone of inhibition. Activity of crude extract from ethanol, when further compared with commercial antibiotics (Gentamicin, Streptomycin and Nystatin), was significantly higher.

Conclusion:

This plant crude extracts could therefore serve as potential source of new biocides with application in water research and other biotechnological fields.

Keywords: Antimicrobials, *In-vitro*, Phytochemical, *Terminalia glaucescens*, Chemical profiling, Contaminants.

1. INTRODUCTION

Fungi and bacteria are of veterinary and human significance. *Bacillus cereus* has been involved in foodborne intoxication [1]. *Escherichia coli* and *Pseudomonas aeruginosa* cause ailments like mastitis, abortions and upper

* Address correspondence to this author at the School of Environmental Sciences, University of Venda, Private Bag X5050, Thohoyandou 0950, South Africa; E-mail: firstrebby@gmail.com

respiratory complications. *Streptococcus* spp is a pathogenic bacteria usually found in the intestines of birds [1, 2]. Fungi like *Aspergillus niger* has been reported to cause lung diseases, aspergillosis and otomycosis. *Aspergillus flavus* is a livestock and human pathogen associated with aspergillosis of the lungs and sometimes causes corneal, otomycotic and naso-orbital infections. They produce significant quantity of aflatoxin [3, 4]. *Candida albicans* is reported to cause asthma [5], vaginitis and yeast mastitis. These sets of organisms have been described to be multi-drug resistant and established as recent contaminants of water bodies [6 - 10]. Research in recent years have revealed multidrug resistance in different bacterial species [6, 8]. Some emerging fungal water contaminants have also been reported [7]. Some of these resistant microbial spp are currently being found in water bodies [9,10]. Reports exist to show the continuous isolation of pathogenic microorganisms especially, antibiotic resistant strains from water and wastewater. Sisti [11] and Sidhu [12], reported a high incidence of bacteria spp from influent and effluent of urban waste and water purification plants. In addition, bacteria spp isolated from sewage have also been found in fresh water and drinking water even after chlorination. Health and environmental implications of these organisms, necessitate a search for novel lead compounds that could be used against such microbes. The pitfalls of chemical coagulants and disinfectants in water such as alum and chlorine also necessitate studies to further develop biomaterials as green and low cost treatment alternative for global treatment and management amidst growing global water crises [5]. It is also incumbent to curb and treat this water contamination bearing in mind that the total water volume on earth cannot be increased. Interest in plants with antimicrobial properties has been revitalized as a consequence of antimicrobial resistance attributed to indiscriminate use of commercial drugs [13], undesirable side effects of synthetic antimicrobials with respect to the emergence of previously uncommon diseases [14, 15], limited effective life span and high cost of synthetic antimicrobials [16]. The efficacy of synthetic antimicrobial agents has been abridged due to the continuing emergence of drug resistant organisms and the adaptations by microbial pathogens to commonly used antimicrobials [17]. While hundreds of plant species had been tested for antimicrobial properties, the vast majority of plants remained unassessed [18 - 20]. In spite of the presence of numerous approaches to drug discovery, plants still remain the leading reservoir of natural medicines [21] and represent a starting point for antimicrobial compounds discovery [22].

Terminalia glaucescens is a key component of the plants used in the formulation of the "wonder cure" concoction used in the management of tuberculosis in Nigeria. The efficacy of the plant extract on *Mycobacterium tuberculosis* was established by Adeleye *et al.* [23]. *Terminalia* species are used as traditional medication in the treatment of dysentery, and also found use in the last stages of management of AIDS [24, 25]. It is anticipated that plant secondary metabolic extracts with target sites other than those used by antibiotics will be effective against drug resistant pathogens [26]. These extracts are of great importance to scientists working on infectious diseases because they signify a possible source of unique antibiotic prototypes [6, 16]. Approximately 80% of the populaces in developing nations use traditional medical specialties for their wellness and care [27]. The genus *Terminalia* is the second largest genus of the *Combretaceae* after *Combretum*, with about 200 species. These plants spread in tropical areas of the globe with the greatest genetic diversity in Southeast Asia [28]. Genus *Terminalia* gets its name from Latin terminus, as the leaves appear at the tips of the shoots [29]. *Terminalia* species range from shrubs to huge deciduous forest trees. Mostly they are very large trees reaching in height up to 75 m tall [30]. *Terminalia* spp are widely used in traditional medicine in a number of continents in the world for the treatment of diseases including, abdominal disorders, bacterial infections, colds, diarrhea, dysentery, sore throats, conjunctivitis, fever, heart diseases, hookworm, gastric ulcers, headaches, hypertension, jaundice, edema, pneumonia, leprosy, nosebleed, and skin diseases [31]. The fruits of both *T. chebula* and *T. bellerica* are vital constituents of "triphala", a popular Ayurvedic formulation that possesses several activities in the Indian traditional medicine [32]. *T. chebula* fruit possesses an amazing power of healing and is called the "King of Medicine" in Tibet as it is used for the treatment of various illnesses [33, 34]. The Bark of *T. arjuna* is used as cardioprotective and anti-hyperlipidemia in folklore treatment [35]. In Africa, *T. mollis* is used to treat gonorrhea, malaria, diarrhea, and HIV management, while *T. brachystemma* is used for the treatment of shistosomiasis and gastrointestinal complaints [36].

Phytochemicals, are often secondary metabolites existing in smaller amounts in higher plants which include terpenoids, tannins, alkaloids, steroids, flavonoids and many others [37]. Different studies and researchers have used different extraction procedures and investigated medicinal importance in several plants. Tom *et al.* [38] and others [39 - 42] have equally investigated the medicinal importance of different *Terminalia* spp using different extraction methods. A current expectation in this field of inquiry is the hunt for emerging plants and plant materials with novel biotechnological applications such as the application of *Cola nitida* and *D. eriocarpum* in water purification and nanoparticle synthesis as paint additives [43, 44]. Table **1** consists of some biotechnological applications of plant extracts.

Table 1. Phytobiotechnological application prospects of plant extracts.

Application areas	Plant	Reference
Bio-fertilizers	Tea-seed (Camellia sp) Ascophyllum nodusum	Andresen and Cedergreen, 2010; Norrie and Keathley, 2006. [70, 71]
Artificial insemination	Curcurbita pepo Khaya senegalensis	Yongabi, 2005 [72]
Bio-shampoo	Accacia concinna (Willd.) DC, Averrhoa bilimbi Linn. and Tamarindus indica Linn.)	Rassami and Soonwera, 2013 [73]
Drink flavours	Tetrapleura tetraptera	Korankye, 2010 [74]
Food colourant	Tefashia sp	Korankye, 2010 [74]
Fumigants	Syzygium aromaticum	Rahuman et al., 2011 [75]
Insect repellants, larvicides insecticides and pesticides	Mesua ferra L, Tephrosia vogeli, Petivera alliacea.	Olaitan and Abiodun, 2011; Singha et al., 2011 [76, 77]
Livestock ethno-therapy	Carica papaya Zingiber officinale	Yongabi, 2005 [72]
Nanoparticle production and paint additives	Cola nitida	Lateef et al. 2015; Lateef and Adeeyo, 2015 [43, 46]
Water purification	Moringa oleifera, Jatropha curcas, Guar gum and Dicerocaryum eriocarpum	Pritchard et al., 2009 Odiyo et al., 2017 [44, 78]

The present study reports the phytochemical screening and antimicrobial activities of Terminalia glaucescens extracts as potential biocide against emerging microbial water contaminants.

2. MATERIALS AND METHODS

2.1. Materials

The chemicals used were obtained from Sigma–Aldrich, Germany through Lab Trade Chemicals Limited Nigeria and are of Analytical grades. Filter papers were from Whatman, GE Healthcare companies, China. The bacterial strains were obtained from Ladoke Akintola Univerity of Technology (LAUTECH) Teaching Hospital, and Obafemi Awolowo Univeristy (OAU), Nigeria.

2.2. Plant Sample

Newly harvested Terminalia glaucescens were obtained from Oja Igbo market, Ogbomoso North local government, Ogbomoso, Oyo State, Nigeria. Plant samples were moved to the lab immediately in polyethylene bags for preparation in order to avoid the decomposition of bioactive compounds.

2.3. Preparation of Samples

The plant stems were carefully washed in order to get rid of contaminants and impurities and were, shredded into smaller pieces, dried at 25±2 °C, ground into powder, filtered into fine particles and then extracted.

2.4. Extraction Procedure for Dry Leaf Powder Samples

50 g of dried and powdered Terminalia glaucescens stem was extracted in 150 ml of chloroform, ethanol and ethyl acetate for a duration of 48 hrs. The solvents were separated using sterile muslin cloth and sieved through sterile Whatmann filter paper (No. 01). The resultant extract in solvents was evaporated to dryness in a rotary evaporator and used as crude extract. The dry crude extracts were used for antimicrobial and phytochemical studies.

2.5. Preliminary Phytochemicals Screening

The phytochemical screening of the extracts was investigated according to the methods of Harborne [45] with little modifications as follows;

2.5.1. Saponins

The presence of saponin was detected by froth test. Exactly 500 mg of extract was measured into a 250 ml conical flask with 25 ml of sterile distilled water and heated for 5 minutes. This was then sieved and after which 2.5 ml of the filtrate was added to 10 ml of sterile distilled water in a test tube. The mixture was then vigorously shaken for about 30 seconds and left for 30 minutes. Honeycomb froth showed the presence of saponins.

2.5.2. Alkaloids

Alkaloid test was carried out by measuring 2 ml of chloroform into a test tube to which few drops of Wagners reagent and 1 ml of the crude filtrate solution were added. A reddish brown precipitate formation indicated the presence of alkaloids.

2.5.3. Tannins

To investigate the presence of tannins, to a portion of crude filtrate solution in distilled water, about 4 drops of 20% ferric chloride solution were carefully added. The formation of green, blue or blue-black colour indicated the presence of tannins.

2.5.4. Reducing Sugars

To 0.5 ml of extracts 1 ml of water and few drops of Fehling's solution were added. The mixture was heated over water-bath and the formation of brick red precipitate confirmed the presence of reducing sugars.

2.5.5. Terpenoids

The presence of terpenoids was confirmed by adding 1 ml of crude extract solution to 2 ml chloroform followed by the addition of acetic anhydride (1 ml). One milliliter of concentrated sulphuric acid was carefully introduced to the solution. The formation of red violet colour mixture showed the presence of terpenoids.

2.5.6. Flavonoids

Few drops of diluted sodium hydroxide solution was added to 0.5 ml extract. A deep yellow colour mixture which became colourless upon the addition of few drops of diluted H_2SO_4 acid proved the presence of flavonoids.

2.5.7. Steroids

About 0.5 ml of acetic anhydride and few drops of concentrated H_2SO_4 were added to 1 ml of extract filtrate. A blue-green precipitate shows the presence of steroids.

2.5.8. Anthraquinones

To 10 ml of chloroform, 1 ml of the crude filtrate solution was added. A yellowish green precipitate observed in the mixture indicated the presence of anthraquinones.

2.5.9. Phlobatannins

To 1 ml of extract solution, few drops of 1% HCl were added. Reddish brown colouration was observed in the presence of phlobatannins.

2.6. Antimicrobial Activity Study

Extracts preparation: Crude extract of the *T. glaucescens,* was prepared as different concentrations in different solvents through serial dilution in appropriate solvents (ethanol, ethyl acetate and chloroform) to give (500 mg/ml, 250 mg/ml, 100 mg/ml, 50 mg/ml and 25 mg/ml designated as C_5, C_4, C_3, C_2 and C_1, respectively) and was used for the antimicrobial screening. The antimicrobial potency test was carried out using the agar disc diffusion method [46]. Whatman No 1 filter paper was used in the preparation of 6.0 mm diameter discs embedded with extracts at various concentrations and used in the sensitivity test. Negative controls (C) were prepared by using the same solvents employed to dissolve the samples. Inhibition zones were measured and compared with the standard synthetic antibiotics as reference.

Antimicrobial Sensitivity Test: Test organism was swabbed evenly on the surface of the agar plate using sterile swab sticks. Impregnated paper discs containing the plant extract at different concentrations were then arranged radially and pressed slightly and firmly on the inoculated agar surface to ensure even contact. The plates were then incubated at 37 °C for 24 hours on Nutrient Agar (for bacteria isolates) while the fungal plates (on Potato Dextrose Agar) were incubated at 25 ±2 °C for 48 hrs. The level of sensitivity was defined by assessing the diameter (in millimeter) of the visible zone of inhibition of microbial growth produced by the dispersion of the infusion. Each method in this experiment was repeated three times.

Test organisms: The bacterial isolates used in this study include *Klebsiella* spp, *E.coli*, *Pseudomonas auriginosa*, *Streptococcus pneumoniae* and *Bacillus aureus*. While the fungal isolates were *Aspergillus flavus*, *Aspergillus niger*, *Trichoderma spp* and *Candida spp*, and the cultures were maintained on nutrient agar and potato dextrose agar, respectively.

Statistical Analysis: The Statistical Package for Social Scientists (SPSS, version 19.0) was used for the analysis of the data obtained. Two way ANOVA test was applied to set the degree of significance of the crude extracts at different concentrations. Also, comparison of the different solvent extractions (i.e. ethanol, ethyl acetate and chloroform) were made statistically; the general antimicrobial effects of the extracts were compared with the standard antibiotics and antifungal disc, respectively.

3. RESULTS

The results obtained from the antimicrobial activity of *T. glaucescens* on the test organisms are presented in Tables **2** to **5**. The data were analyzed statistically and the significance level was obtained at $p<0.05$. Values are shown as mean ± standard deviation (SD) for each immersion. Values with (*) are significantly higher than the control at $p < 0.05$.

Table 2. Diameter Zones of Inhibition (mm) of the Crude Extract of *T. glaucescens* Against Tested Bacterial Isolates.

Zones of inhibition of different extracts						
		Ethanol (mm)		Ethyl Acetate(mm)		Chloroform(mm)	
P. aeruginosa	C	8.50	± 0.707	7.50	± 0.707	7.00	±0.000
	C_1	9.50	± 0.707	8.50	± 0.707	7.50	± 0.707
	C_2	11.50	± 0.707*	9.00	± 1.414	8.50	± 0.707
	C_3	11.50	± 0.707*	11.50	± 0.707*	8.50	± 0.707
	C_4	13.50	± 0.707*	12.00	± 1.414*	9.50	± 0.707
	C_5	14.50	± 0.707*	13.50	± 0.707*	10.50	± 0.707*
Klebsiella sp	C	9.00	±1.414	7.50	± 0.707	7.00	±0.000
	C_1	8.50	±2.121	7.50	± 0.707	7.50	± 0.707
	C_2	9.50	± 0.707	8.50	± 0.707	8.00	±0.000
	C_3	11.50	± 0.707	10.00	±0.000	8.50	± 0.707
	C_4	12.00	±1.414*	10.50	± 0.707*	8.50	± 0.707
	C_5	13.50	± 0.707*	11.50	± 0.707*	10.00	±0.000*
E. coli	C	8.50	± 0.707	7.00	±0.000	7.50	± 0.707
	C_1	10.50	± 0.707	8.50	± 0.707	8.00	±01.414
	C_2	14.00	±1.414*	9.50	± 0.707	8.50	± 0.707
	C_3	15.00	±1.414*	11.00	±0.000*	9.50	± 0.707
	C_4	18.00	±1.414*	12.50	± 0.707*	11.50	± 0.707
	C_5	20.00	±1.414*	13.00	±1.414*	13.50	±02.121*
S. pneumonia	C	8.50	± 0.707	7.00	±0.000	7.50	± 0.707
	C_1	12.00	±1.414	8.50	± 0.707	8.00	±0.000
	C_2	14.00	±1.414*	9.50	± 0.707	7.50	± 0.707
	C_3	17.00	±1.414*	12.00	±1.414*	8.50	± 0.707
	C_4	19.00	±0.000*	14.50	± 0.707*	10.50	± 0.707*
	C_5	20.50	± 0.707*	17.50	± 0.707*	13.50	± 0.707*
B. cereus	C	8.50	± 0.707	7.00	±0.000	8.00	±0.000
	C_1	12.00	±1.414	9.00	±1.414	10.50	± 0.707
	C_2	15.00	±1.414*	12.50	± 0.707*	13.00	±1.414*
	C_3	16.00	±1.414*	13.50	± 0.707*	14.00	±1.414*
	C_4	18.50	± 0.707*	13.50	±2.121*	16.00	±1.414*
	C_5	19.00	±1.414*	16.00	±1.414*	17.50	± 0.707*

Values are expressed as mean ± SD for n=2 for each concentration. Values with (*) are significantly higher than the control at p < 0.05. C-Control, C_1-25ml, C_2-50ml, C_3-100ml, C_4-250ml, C_5-500ml

Table 3. Diameter Zones of Inhibition (mm) of the Crude Extract of *T. glaucescens* Against Tested Fungal Isolates.

		Zones of inhibition of different extracts					
		Ethanol (mm)		Ethyl Acetate(mm)		Chloroform(mm)	
A. flavus	C	8.50	± 0.707	7.50	± 0.707	7.00	±0.000
	C_1	12.00	±2.828	9.50	± 0.707	9.50	±2.121
	C_2	13.50	±2.121	12.00	±1.414*	10.50	± 0.707
	C_3	17.00	±1.414*	13.00	±1.414*	14.50	± 2.121*
	C_4	17.50	± 0.707*	14.50	± 0.707*	15.50	± 2.121*
	C_5	19.00	±1.414*	15.50	± 0.707*	17.00	± 1.414*
A. niger	C	7.50	± 0.707	7.00	± 0.000	7.00	± 0.000
	C_1	11.00	±1.414	8.50	± 0.707	10.00	± 1.414
	C_2	14.00	±1.414*	11.00	± 1.414*	13.00	± 1.414*
	C_3	17.50	± 0.707*	13.50	± 0.707*	15.50	± 0.707*
	C_4	18.50	± 0.707*	13.00	± 1.414*	16.50	± 0.707*
	C_5	19.00	± 0.000*	15.50	± 0.707*	17.50	± 0.707*
Trichoderma sp	C	8.50	± 0.707	7.00	± 0.000	7.50	± 0.707
	C_1	9.00	± 1.414	8.50	± 0.707	9.50	± 0.707
	C_2	10.00	± 1.414	8.00	± 1.414	9.00	± 1.414
	C_3	11.00	± 1.414	8.50	± 0.707	10.00	± 1.414
	C_4	14.00	± 1.414*	11.00	± 1.414*	12.50	± 0.707*
	C_5	19.00	± 1.414*	12.50	± 0.707*	16.00	± 1.414*
Candida sp	C	8.00	± 1.414	C	± 0.000	7.50	± 0.707
	C_1	11.00	± 1.414	9.00	± 1.414	10.00	± 1.414
	C_2	13.00	± 1.414	9.50	± 0.707	11.00	± 1.414
	C_3	12.00	± 1.414	10.00	± 1.414	11.00	± 1.414
	C_4	13.50	± 0.707*	12.00	± 0.000*	11.00	± 0.000
	C_5	15.00	± 1.414*	12.00	± 1.414*	13.00	± 1.414*

Values are expressed as mean ± SD for n=2 for each concentration. Values with (*) are significantly higher than the control at p < 0.05. C-Control, C_1-25ml, C_2-50ml, C_3-100ml, C_4-250ml, C_5-500ml

Table 4. Comparative activity of ethanolic extract against Bacteria and fungi isolates.

Bacteria	Highest zone of inhibition (mm)
Ethanol	20.5
Ethyl acetate	17.5
Chloroform	17.5
Fungi	
Ethanol	19.0
Ethyl acetate	15.5
Chloroform	17.5

Table 5. Diameter Zones of Inhibition (mm) of the ethanolic extract of *T. glaucescens* and the standard.

		Ethanol	
P. aeruginosa	STR	11.50	± 0.707
	C_1	9.50	± 0.707
	C_2	11.50	± 0.707
	C_3	11.50	± 0.707
	C_4	13.50	± 0.707
	C_5	14.50	± 0.707*

(Table 5) contd.....

		Ethanol	
Klebsiella sp	STR	10.50	± 0.707
	C_1	8.50	±2.121
	C_2	9.50	± 0.707
	C_3	11.50	± 0.707
	C_4	12.00	± 1.414
	C_5	13.50	± 0.707
E.coli	STR	10.00	± 0.000
	C_1	10.50	± 0.707
	C_2	14.00	± 1.414
	C_3	15.00	± 1.414*
	C_4	18.00	± 1.414*
	C_5	20.00	± 1.414*
S. pneumonia	GEN	13.50	± 0.707
	C_1	12.00	± 1.414
	C_2	14.00	± 1.414
	C_3	17.00	± 1.414
	C_4	19.00	± 0.000
	C_5	20.50	± 0.707*
B. cereus	GEN	13.00	± 0.000
	C_1	12.00	± 1.414
	C_2	15.00	± 1.414
	C_3	16.00	± 1.414
	C_4	18.50	± 0.707*
	C_5	19.00	± 1.414*
A. flavus	NIS	15.00	± 0.000
	C_1	12.00	± 2.828
	C_2	13.50	± 2.121
	C_3	17.00	± 1.414
	C_4	17.50	± 0.707
	C_5	19.00	± 1.414
A. niger	NIS	16.50	± 0.707
	C_1	11.00	± 1.414
	C_2	14.00	± 1.414
	C_3	17.50	± 0.707
	C_4	18.50	± 0.707
	C_5	19.00	± 0.000
Trichoderma sp	NIS	14.00	± 0.000
	C_1	9.00	± 1.414
	C_2	10.00	± 1.414
	C_3	11.00	± 1.414
	C_4	14.00	± 1.414
	C_5	19.00	± 1.414
Candida sp	NIS	14.00	± 0.000
	C_1	11.00	± 1.414
	C_2	13.00	± 1.414
	C_3	12.00	± 1.414
	C_4	13.50	± 0.707
	C_5	15.00	± 1.414

Values are expressed as mean ± SD for n=2 for each concentration. Values with (*) are significantly higher than the standard at p < 0.05. STR-Streptomycin, GEN- Gentamycin, NIS- Nystatin

3.1. Antimicrobial Effect of Crude Extracts of *Terminalia glaucescens*

The antimicrobial activities of ethanol, chloroform and ethyl acetate extract of *T. glaucescens* with respect to their zones of inhibition (mm) on the test organisms are shown in Tables **2** to **4**. All the extracts showed potential antimicrobial activities against the test organisms. The antimicrobial activity of the extracts increased with increase in extract concentrations.

Ethanolic extract: Ethanolic extract (C_1 – C_5) effectively inhibited majority of the test organisms and was comparatively better in antimicrobial performance than the control. The zone of inhibition against test organisms ranged from 8.5 mm at C_1 – to 20.5 mm at C_5. The extract shows inhibition against test fungi with zones of inhibition ranging from 9.0 mm at C_1 – 19.0 mm at C_5 while zones of inhibition against bacteria isolates tested ranged between 8.50 mm at C_1 - 20.5 mm at C_5. Ethanolic extract at concentration C_5 (500 mg/ml) showed consistently higher zones of inhibition against *S. pneumonia* (20.5 mm), *E.coli* (20.0 mm), *B. cereus*, *A. niger*, *Trichoderma* sp. and *A. flavus* (19.0 mm) respectively, while the lowest zone of inhibition at this concentration was recorded against *Klebsiella* sp with a diameter of 13.5 mm which is, however, higher than the performance of the control with a record range of 8.50 mm – 9.00 mm.

Ethyl acetate extract: Ethyl acetate extract like ethanolic extract effectively inhibited the growth of all tested bacteria and fungi and performed better than the control except for activity against *Klebsiella* sp where the performances were comparatively the same with the control and recorded as 7.50 mm. The range of inhibition of the extract was from 7.5 mm at C_1 – 17.5 mm at C_5. Ethyl acetate extract recorded zone of inhibition against test fungal isolates in the range of 8.0 mm at C_2 – 15.5 mm at C_5 while zones of inhibition against bacteria isolates tested ranged from 7.50 mm at C_1 - 17.5 mm at C_5. The highest zone of inhibition recorded for ethyl acetate was observed at 500 mg/ml against *S. pneumonia* with a diameter of 17.5 mm while the lowest zone of inhibition at the same concentration was recorded against *Klebsiella* sp with a diameter of 11.5 mm.

Chloroform extract: In a similar manner as Ethanol and Ethyl acetate extracts, the chloroform extract exhibited antibacterial and antifungal activities against the test isolates with zones of inhibition against bacteria isolates ranging from 7.50 mm at C_1 - 17.5 mm at C_5. It was also comparatively better in antimicrobial performance than the control. The extract exhibited inhibition against tested fungal isolates in the range of 9.0 mm at C_2 - 17.5 mm at C_5. The highest zone of inhibition was also recorded at 500 mg/ml against *B. cereus and A. niger* with a diameter of 17.5 mm in each case, while the lowest zone of inhibition for the same concentration was recorded against *Klebsiella sp* with a diameter of 10.0 mm.

Comparative activities of extracts: Ethanolic extract performs better than ethyl acetate and chloroform extracts as it recorded the highest zone of inhibition of 20.5 mm against *Streptococcus pneumoniae* while ethyl acetate and chloroform recorded 17.50 mm each against *Streptococcus pneumoniae* and *Bacillus cereus*, respectively (Table 4). The extracts of *Terminalia glaucescens* seem to exhibit a good antibacterial activity against *Streptococcus pneumoniae* and *Bacillus* sp and lesser activity against *Klebsiella* sp. The extract exhibited higher antifungal activity against *Trichoderma* sp. and *Aspergillus niger* (Table **4**).

Crude extract of ethanol at the highest concentration C_5 tested was a more potent antimicrobial agent when further compared with standard (Gentamicin, Streptomycin and Nystatin). Ethanolic extract recorded higher zones of inhibition against *Pseudomonas aeruginosa*, *Klebsiella* sp and *E. coli* as 14.5 mm, 13.5 mm and 20. 0 mm at C_5 while Streptomycin recorded 11.5 mm, 10.5 mm and 10.0 mm, respectively, against the same set of organisms (Table 5). The extract recorded highest zones of inhibition against *Streptococcus pnuemoniae* and *Bacillus cereus* as 20.5 mm and 19.0 mm at C_5 while the record of Gentamycin against the organisms were 13.5 and 13.0 mm, respectively. The ethanolic extract equally exerted zones of inhibition that were comparatively higher than that of nystatin against the tested fungi, but were not statistically different from the values obtained for nystatin ($P<0.05$). Higher activities were recorded against *Aspergillus flavus*, *Aspergillus niger* and *Trichoderma* sp; 19.0 mm at concentration C_5, while the least was recorded as 15.0 mm against *Candida* sp at the same concentration. The antimicrobial activity of ethanol was therefore notable against all tested organisms and commercial antibiotics at C_5 ($P<0.05$). The results are presented in Table **5**. Plates (**1** and **2**) illustrates reported characteristic antibacterial (against *E.coli*) and antifungal (against *A. flavus*) activities of ethanolic extracts.

Chemical Profiling and Antimicrobial Properties of Phyto-Active Extracts from Terminalia glaucescens...

9

Plate (1). The Zones of Inhibition of Ethanol Crude Extract of *T. glaucescens* on *E. coli*
C – Control consisting of solvent without crude extract,
25-500 – Amount of crude extract dissolve in 1 ml of solvent.

Plate (2). The Zones of Inhibition of Ethanol Crude Extract of *T. glaucescens* on *A. flavus*
C – Control consisting of solvent without crude extract
25-500 – Amount of crude extract dissolve in 1 ml of solvent.

3.2. Phytochemical Screening of *Terminalia glaucescens*

The results obtained from the qualitative phytochemical screening of *Terminalia glaucescens* are presented in Table 6. In ethanol, saponins, tannins, alkaloids, flavonoids, terpenoids, steroids and anthraquinones were present while reducing sugars and phlobatannins were absent. In ethyl acetate and chloroform, saponins, tannins, alkaloids, terpenoids, steroids and anthraquinones were present while Reducing sugars, Phlobatannins and Flavonoids were absent.

Table 6. Phytochemical Analysis Results of *T. glaucescens*.

	Ethanol	Ethyl Acetate	Chloroform
Saponins	+	+	+
Tannins	+	+	+
Reducing Sugars	-	-	-
Alkaloids	+	+	+
Flavonoids	+	-	-
Terpenoids	+	+	+
Phlobatannins	-	-	-
Steroids	+	+	+
Anthraquinones	+	+	+

+ = Present,
- = Absent

4. DISCUSSION

4.1. Comparative Activities of Plant Extracts and Commercial Antibiotics

Tables **2** to **5** present the summary of the antibacterial and antifungal activities of the various extracts of *Terminalia glaucescens* and that of selected commercial antibiotics against different fungal and bacterial strains tested. Antimicrobial activities were found in all the plant extracts and with remarkable zones of inhibitions. The antibacterial and antifungal activities of active extracts were observed in the increasing order of Ethanol>Chloroform≥ethyl acetate with respect to the maximum zone of inhibition. Ethanolic extract compared favourably against commercial streptomycin, gentamycin and nystatin which were used as standards with significant statistical differences. The zones of inhibition reported in this study against bacterial isolates tested ranged from 7.5 to 20.5 mm while that recorded against fungal isolates ranged from 8.0 to 19.0 mm. The extent of antimicrobial activity of extract based on the zone of inhibition has been described as low (12-18 mm), moderate (19-22 mm) and strong (23-38 mm) [47], we may therefore infer that ethanolic extract of *T. glaucescens* which exhibited inhibitions in a range of 18.0-20.0 mm against most of the microbial strain tested has considerable antimicrobial activity. Previous screening by earlier researchers had demonstrated the antimicrobial efficacy of variuos plant extracts including *Holarrhenea antidyssentrica* [48]; *Tapinthus senssilifolius* [49]; *Rauelfia tetraphylla* and *Physalis minima* [50]; *Achillea santolina, Salvia dominica* and *Salvia officinalis* [51]; *Psidium guajava* and *Mangifera indica* [52] and *Salicornia brachiata* [53] against different bacteria and fungal isolates. Viji *et al.* [17], have demonstrated that ethanolic and chloroform extracts can potentially be used against *E. coli, Pseudomonas aeruginosa, Bacillus* sp and *Klebsiella* sp. which were used in this study.

4.2. Phytochemicals

Plants are known to contain a number of phytochemicals such as flavonoids, saponins, tannins and other phenolic compounds that have antimicrobial activities [54 - 56]. The results of phytochemical screening in *T. glaucescens* reveal the presence of saponins, tannins, alkaloids, terpenoids, steroids and anthraquinones in all extracts investigated. Reducing sugars and phlobatannins were absent. Flavonoids was however present in ethanolic extract but not in ethyl acetate and chloroform extracts. The presence of antifungal and antimicrobial substances in higher plants is well established as they have provided a source of inspiration for novel drug compounds [17]. This suggests that the antimicrobial activities of the plant under investigation may be as a result of the phytochemicals present.

4.3. Antimicrobial Responses and Mechanisms

It was noted that all the extracts exhibited antibiosis against gram positive and gram negative bacteria as well as the tested fungal isolates (broad spectrum activities). Noteworthy, is the activity of the ethanolic extract at 500 mg/ml against *Streptococcus pneumoniae* (20.5 mm), *E.coli* (20 mm) and B. cereus (19.0 mm). The values of zones of inhibitions show that ethanolic extract performed better than ethyl acetate and chloroform extracts. The susceptibility of the tested microbes to the extracts varies across the different solvents of extraction. This suggests that ethanolic extracts of *T. glaucescens* was more effective than ethyl acetate and chloroform extract. The variation in responses of the various microbial strains to individual extracts may be attributed to the nature of the microbial cells and their genetic diversity [57]. The chemistry of the extracting solvent and subsequent bio-active components extracted by various solvents may also explain the difference in antimicrobial potency of the extract in different solvents [16], which may affect overall activity [57]. Resistance could be due to the permeability barrier provided by the cell wall or to the membrane accumulation mechanism [58].

Variation in phytochemical components of the various extracts may result from variance in the chemistry of the extracting solvent which selectively affect extraction of various bio-active metabolites. The mechanisms of antimicrobial actions of these compounds may be via cell membranes perturbations [59] and may involve diverse molecular modes, such as binding and increasing the permeability of cell wall and membrane component [60]. These induce membrane destabilization, leakage of cytoplasmic contents, loss of membrane potential, change of membrane permeability, lipid distribution, the entry of the peptide and blocking of anionic cell components or the triggering of autolytic enzymes and the final death of the microbial cell [61]. This finding is similar to that of Dahot [62] who reported that plant extract from *M. oleifera* had antimicrobial activity against *E. coli*, *S. aureus* and *B. subtilis*. Dahot [62], however, reported resistance from *Aspergillus niger* and *Aspergillus flavus* from which notable susceptibility was recorded in this study. The curative advantage of such a plant used in this finding is that consumers including animals tend to consume the plant material in large quantities and in high concentrations. This suggests the ability of the plant to meet the required physiological levels to inhibit the pathogenic growth *in situ* and point to the potential of *T. glaucescens* as biocides to be explored in various fields. The differences in ethanolic extract activity from that of ethyl acetate and chloroform extracts as well as the similarities in the activities of ethyl acetate and chloroform extracts follow the pattern observed in the similarities and differences in phytochemical compositions of the extracts (Table 6). This may suggest the positive contributions of these phytochemicals in the antimicrobial activities of the extracts.

There exist reports to show that the mechanism of water purification by plant material is associated with their flocculating, coagulative and disinfecting properties [63]. Polysaccharides as well as protein associated phyto-chemicals have been implicated in the purification process. The phytochemicals may form flocs which settles slowly while sweeping out suspended impurities in water. Phytochemicals with their net charge; either positive, negative or neutral are thought to combine with active sides of colloids and impurities; such interaction produces a bridging effect, binding impurities and phytochemicals together into a large particle which settles under the action of gravity. The disinfecting properties of lemon and moringa extracts in water have been partly associated with ability to alter the pH of water, making it unsuitable for some living contaminants [5,64,65] while the effectiveness in coagulation and colour reduction is a function of particles size and concentration [66].

While some of the current chemical compounds like alum and chlorine used in disinfecting water has been tagged a precursor for cancer, as it forms tetrahalomethane compounds and lead to hormone mimics as well as generating dementia in young and elderly, reports exist to show that plant based technology are at different stages of development for water purification. The technology will be very simple, non-toxic and with no major machinery nor specialized labour needed [5,67].

In this study, *T. glaucescens* extracts had a dose dependent bactericidal properties against all bacterial strains which are mostly known to be multi-resistant [6, 8]. According to several authors, these bacteria are generally less sensitive to the activity of plant extracts [68, 69]. The ability of the extracts to antagonise and exhibit broad spectrum antibiosis is therefore noteworthy. The plant ethanolic extracts with optimal antimicrobial performance can therefore be recommended for further research in development of potent antimicrobial agents against multidrug and emerging resistant microbes.

CONCLUSION

The present study which was aimed at establishing the antimicrobial efficacy of different crude extracts of *T. glaucescens* showed that the crude extracts from the plant possessed potent activity against the employed bacteria and fungi. Similarly, phytochemical screening showed that the antimicrobial activities of the crude extracts of the plant may depend on the presence of phytochemicals such as saponins, tannins, alkaloids, flavonoids, terpenoids, steroids and anthraquinones. This plant crude extracts could serve as potential sources of new antimicrobials and for green and eco-friendly water treatment technology development.

RECOMMENDATIONS

Further research is needed towards isolation and identification of active metabolites present in the extracts which could be adopted for different biotechnological uses.

HUMAN AND ANIMAL RIGHTS

No Animals/Humans were used for studies of this research.

ACKNOWLEDGEMENTS

Declared None.

REFERENCES

[1]   Granum PE, Lund T. *Bacillus cereus* and its food poisoning toxins. FEMS Microbiol Lett 1997; 157(2): 223-8.
        [http://dx.doi.org/10.1111/j.1574-6968.1997.tb12776.x] [PMID: 9435100]

[2]   Fraser CM. The Merck Veterinary Manual. 6th ed. New Jersey: Merck 1986.

[3]   Samson RA, Houbraken J, Summerbell RC, Flannigan B, Miller JD. Common and important species of fungi and actinomycetes in door environment.Micro-organisms on home and indoor work environments. New York: Taylor and Francis 2001; pp. 287-92.

[4]   Klich MA. *Aspergillus flavus*: The major producer of aflatoxin. Mol Plant Pathol 2007; 8(6): 713-22.
        [http://dx.doi.org/10.1111/j.1364-3703.2007.00436.x] [PMID: 20507532]

[5]   Yongabi KA. Biocoagulants for Water and Waste Water Purification: A Review International Review of Chemical Engineering 2010; 2: 444-58.

[6]   Afolayan AJ. Extracts from the shoots of *Arcotis arctotoides* inhibit the growth of bacteria and fungi. Pharm Biol 2003; 41: 22-5.
        [http://dx.doi.org/10.1076/phbi.41.1.22.14692]

[7]   Pfaller MA, Diekema DJ. Rare and emerging opportunistic fungal pathogens: Concern for resistance beyond *Candida albicans* and *Aspergillus fumigatus*. J Clin Microbiol 2004; 42(10): 4419-31.
        [http://dx.doi.org/10.1128/JCM.42.10.4419-4431.2004] [PMID: 15472288]

[8]   Boussaada O, Ammar S, Saidana D, *et al.* Chemical composition and antimicrobial activity of volatile components from capitula and aerial parts of *Rhaponticum acaule* DC growing wild in Tunisia. Microbiol Res 2008; 163(1): 87-95.
        [http://dx.doi.org/10.1016/j.micres.2007.02.010] [PMID: 17482441]

[9]   Chouhan S. Enumeration of standard plate count bacteria in raw water supplies. Journal of Environmental Science. Toxicol Food Technol 2015; 9: 2319-4012.

[10]   Nandita D, Uchechukwu S, Tomilola DA. Physicochemical and microbiological assessment of Lagos lagoon water, Lagos Nigeria. J Pharmacy Biol Sci 2015; 10: 2319-7676.

[11]   Sisti M, Albano A, Brandi G. Bactericidal effect of chlorine on motile *Aeromonas* spp. in drinking water supplies and influence of temperature on disinfection efficacy. Lett Appl Microbiol 1998; 26(5): 347-51.
        [http://dx.doi.org/10.1046/j.1472-765X.1998.00346.x] [PMID: 9674163]

[12]   Sidhu J, Gibbs RA, Ho GE, Unkovich I. Selection of *Salmonella typhimurium* as an indicator for pathogen regrowth potential in composted biosolids. Lett Appl Microbiol 1999; 29(5): 303-7.
        [http://dx.doi.org/10.1046/j.1365-2672.1999.00626.x] [PMID: 10664970]

[13]   Aliero AA, Afolayan AJ. Antimicrobial activity of *Solanum tomentosum*. Afr J Biotechnol 2006; 5: 369-72.

[14]   Marchese A, Schito GC. Resistance patterns of lower respiratory tract pathogens in Europe. Int J Antimicrob Agents 2000; 16(Suppl. 1): S25-9.
        [http://dx.doi.org/10.1016/S0924-8579(00)00302-2] [PMID: 11137405]

[15]   Poole K. Overcoming antimicrobial resistance by targeting resistance mechanisms. J Pharm Pharmacol 2001; 53(3): 283-94.
        [http://dx.doi.org/10.1211/0022357011775514] [PMID: 11291743]

[16]   Busani M, Julius MP, Voster M. Antimicrobial activities of *Moringa oleifera* Lam leaf extracts. Afr J Biotechnol 2012; 11: 2797-802.

[17]   Viji M, Sathiya M, Murugesan S. Phytochemical analysis and antibacterial activity of medicinal plant *Cardiospermum helicacabum* linn. Pharmacologyonline 2010; 2: 445-56.

[18]   Balandrin MF, Klocke JA, Wurtele ES, Bollinger WH. Natural plant chemicals: Sources of industrial and medicinal materials. Science 1985; 228(4704): 1154-60.
[http://dx.doi.org/10.1126/science.3890182] [PMID: 3890182]

[19]   Erdogrul OT. Antimicrobial activities of some plant extracts used in folklore medicine. Pharm Biol 2002; 40: 269-73.
[http://dx.doi.org/10.1076/phbi.40.4.269.8474]

[20]   Parek J, Karathia N, Chandra S. Screening of some traditionally used medicinal plants for potential antibacterial activity. Indian J Pharm Sci 2006; 68: 832-4.
[http://dx.doi.org/10.4103/0250-474X.31031]

[21]   Mahomed IM, Ojewole JA. Anticonvulsant activity of Harpagophytum procumbens DC [Pedaliaceae] secondary root aqueous extract in mice. Brain Res Bull 2006; 69(1): 57-62.
[http://dx.doi.org/10.1016/j.brainresbull.2005.10.010] [PMID: 16464685]

[22]   Cseke IJ, Kirakosyan A, Kaufman PB, Warber SL, Duke JA, Brielmann HL. Natural product from plants CRC Press. United States of America: Taylor & Francis Group, LLC 2006.

[23]   Adeleye IA, Onubogu CC, Ayolabi CI, Isawumi AO, Nshiogu ME. Screening of crude extracts of twelve medicinal plants and "Wonder –Cure" concoction used in Nigeria unorthodox medicine for activity against mycobacterium tuberculosis isolated from tuberculosis patients sputum. Afr J Biotechnol 2008; 7: 3182-7.

[24]   Koudou J, Roblot G, Wylde R. Tannin constituents of *Terminalia glaucescens*. Planta Med 1995; 61(5): 490-1.
[http://dx.doi.org/10.1055/s-2006-958153] [PMID: 7480219]

[25]   Rahman AU, Choudhary MI. Biodiversity as a source of new pharmacophores: A new theory of memory. Pure Appl Chem 2005; 77: 75-81.

[26]   Ahmad I, Beg AZ. Antimicrobial and phytochemical studies on 45 Indian medicinal plants against multi-drug resistant human pathogens. J Ethnopharmacol 2001; 74(2): 113-23.
[http://dx.doi.org/10.1016/S0378-8741(00)00335-4] [PMID: 11167029]

[27]   Kim H, Park SW, Park JM, Moon KH, Lee CK. Screening and isolation of antibiotic resistance inhibitors from herb material Resistant Inhibition of 21 Korean plants. Nat Prod Sci 2005; 1: 50-4.

[28]   de Morais Lima GR, de Sales IR, Caldas Filho MR, *et al.* Bioactivities of the genus Combretum (Combretaceae): A review. Molecules 2012; 17(8): 9142-206.
[http://dx.doi.org/10.3390/molecules17089142] [PMID: 22858840]

[29]   Saxena VG, Mishra G, Saxena A, Vishwakarma KK. A comparative study on quantitative estimation of tannins in Terminalia chebula, *Terminalia belerica, Terminalia arjuna* and *Saraca indica* using spectrophotometer. Asian J Pharm Clin Res 2013; 6: 148-9.

[30]   Stace CA. Combretaceae: Flowering plants. Eudicots 2007; 9: 67-82.

[31]   Eloff JN, Katerere DR, McGaw LJ. The biological activity and chemistry of the southern African Combretaceae. J Ethnopharmacol 2008; 119(3): 686-99.
[http://dx.doi.org/10.1016/j.jep.2008.07.051] [PMID: 18805474]

[32]   Yang MH, Vasquez Y, Ali Z, Khan IA, Khan SI. Constituents from Terminalia species increase PPARα and PPARγ levels and stimulate glucose uptake without enhancing adipocyte differentiation. J Ethnopharmacol 2013; 149(2): 490-8.
[http://dx.doi.org/10.1016/j.jep.2013.07.003] [PMID: 23850833]

[33]   Bag A, Bhattacharyya SK, Chattopadhyay RR. The development of *Terminalia chebula* Retz. (Combretaceae) in clinical research. Asian Pac J Trop Biomed 2013; 3(3): 244-52.
[http://dx.doi.org/10.1016/S2221-1691(13)60059-3] [PMID: 23620847]

[34]   Pellati F, Bruni R, Righi D, *et al.* Metabolite profiling of polyphenols in a *Terminalia chebula Retzius* ayurvedic decoction and evaluation of its chemopreventive activity. J Ethnopharmacol 2013; 147(2): 277-85.
[http://dx.doi.org/10.1016/j.jep.2013.02.025] [PMID: 23506992]

[35]   Dixit D, Dixit AK, Lad H, Gupta D, Bhatnagar D. Radioprotective effect of *Terminalia Chebula* Retzius extract against irradiation-induced oxidative stress. Biomed Aging Pathol 2013; 3: 83-8.
[http://dx.doi.org/10.1016/j.biomag.2012.10.008]

[36]   Liu M, Katerere DR, Gray AI, Seidel V. Phytochemical and antifungal studies on *Terminalia mollis* and *Terminalia brachystemma*. Fitoterapia 2009; 80(6): 369-73.
[http://dx.doi.org/10.1016/j.fitote.2009.05.006] [PMID: 19446614]

[37]   Nonita PP, Mylene MU. Antioxidant and cytotoxic activities and phytochemical screening of four Philippine medicinal plants. J Med Plants Res 2010; 4: 407-14.

[38]   Tom EN, Demougeot C, Mtopi OB, *et al.* The aqueous extract of *Terminalia superba* (Combretaceae) prevents glucose-induced hypertension in rats. J Ethnopharmacol 2011; 133(2): 828-33.
[http://dx.doi.org/10.1016/j.jep.2010.11.016] [PMID: 21075190]

[39]   Kathirvel A, Sujatha V. *In vitro* assessment of antioxidant and antibacterial properties of *Terminalia chebula* Retz. leaves. Asian Pac J Trop Biomed 2012; 2: S788-95.
         [http://dx.doi.org/10.1016/S2221-1691(12)60314-1]

[40]   Fyhrquist P, Laakso I, Garcia-Marco S, Julkunen-Tiitto R, Hiltunen R. Antimycobacterial activity of ellagitannin and ellagic acid derivate rich crude extracts and fractions of five selected species of *Terminalia* used for treatment of infectious diseases in African traditional medicine. S Afr J Bot 2014; 90: 1-16.
         [http://dx.doi.org/10.1016/j.sajb.2013.08.018]

[41]   Tom EN, Girard-Thernier C, Martin H, *et al.* Treatment with an extract of *Terminalia superba* Engler & Diels decreases blood pressure and improves endothelial function in spontaneously hypertensive rats. J Ethnopharmacol 2014; 151(1): 372-9.
         [http://dx.doi.org/10.1016/j.jep.2013.10.057] [PMID: 24212074]

[42]   Fahmy NM, Al-Sayed E, Abdel-Daim MM, Karonen M, Singab A. Protective effect of *Terminalia muelleri* against carbon tetrachloride-induced hepato– nephro toxicity in mice and characterization of its bioactive constituents. Pharm Biol 2015; 1-11.
         [PMID: 25894213]

[43]   Lateef A, Azeez MA, Asafa TB, *et al.* Biogenic synthesis of silver nanoparticles using pod extract of *Cola nitida*: antibacterial, antioxidant activities and application as additive in paint. J Taibah University Sci 2016; 10: 551-62.
         [http://dx.doi.org/10.1016/j.jtusci.2015.10.010]

[44]   Odiyo JO, Bassey OJ, Ochieng A, Chimuka L. Coagulation efficiency of *Dicerocaryum eriocarpum* (DE) plant. Water SA 2017; 43
         [http://dx.doi.org/10.4314/wsa.v43i1.01]

[45]   Harborne JB. Phytochemical Methods, A Guide to Modern Techniques of Plant analysis. 2nd ed. London: Chapman and Hall 1998; pp. 54-84.

[46]   Lateef A, Adeeyo AO. Green synthesis and antibacterial activities of silver nanoparticles using extracellular laccase of *Lentinus edodes*. Not Sci Biol 2015; 7: 405-11.
         [http://dx.doi.org/10.15835/nsb.7.4.9643]

[47]   Ahmad I, Zaiba-Beg AZ, Mehmood Z. Antimicrobial potency of selected medicinal plants with special interest in activity against phytopathogenic fungi. Indian Vet med J 1999; 23: 299-306.

[48]   Kavitha D, Shilpa PN, Devaraj SN. Antibacterial and antidiarrhoeal effects of alkaloids of Holarrhena antidysenterica WALL. Indian J Exp Biol 2004; 42(6): 589-94.
         [PMID: 15260110]

[49]   Tarfa FD, Obodozie OO, Mshelia E, Ibrahim K, Temple VJ. Evaluation of phytochemical and antimicrobial properties of leaf extract of *Tapinanthus sessilifolius* (P. Beauv) van Tiegh. Indian J Exp Biol 2004; 42(3): 326-9.
         [PMID: 15233306]

[50]   Shariff MS, Sudarshana S, Umesha P, Hariprasad S. Antimicrobial activity of *Rauvolfia tetraphylla* and *Physalis minima* leaf and callus extracts. Afr J Biotechnol 2006; 5: 946-50.

[51]   Hassawi D, Kharma A. Antimicrobial activity of medicinal plants against *Candida albicans*. J Biol Sci 2006; 6: 104-9.

[52]   Akinpelu DA, Onakoya TM. Antimicrobial activities of medicinal plants used in folklore remedies in south-western Africa. Afr J Biotechnol 2006; 5: 1078-81.

[53]   Manikandan T, Neelakandan T, Usha RG. Antibacterial activity of *Salicornia brachiata*, a halophyte. J Phytol 2009; 1: 441-3.

[54]   Sato Y, Shibata H, Arai T, *et al.* Variation in synergistic activity by flavone and its related compounds on the increased susceptibility of various strains of methicillin-resistant *Staphylococcus aureus* to β-lactam antibiotics. Int J Antimicrob Agents 2004; 24(3): 226-33.
         [http://dx.doi.org/10.1016/j.ijantimicag.2004.02.028] [PMID: 15325425]

[55]   Cushnie TP, Lamb AJ. Antimicrobial activity of flavonoids. Int J Antimicrob Agents 2005; 26(5): 343-56.
         [http://dx.doi.org/10.1016/j.ijantimicag.2005.09.002] [PMID: 16323269]

[56]   Mboto CI, Eja ME, Adegoke AA, *et al.* Phytochemical properties and antimicrobial activities of combined effect of extracts of the leaves of *Garcinia Kola, Vernonia amygdalina* and honey on some medically important microorganisms. Afr J Microbiol Res 2009; 3: 557-9.

[57]   Aiyegoro OA, Afolayan AJ, Okoh AI. Interactions of antibiotics and extracts of *Helichrysum pedunculatum* against bacteria implicated in wound infections. Folia Microbiol (Praha) 2010; 55(2): 176-80.
         [http://dx.doi.org/10.1007/s12223-010-0026-5] [PMID: 20490761]

[58]   Adwan K, Abu-Hasan N. Gentamicin resistance in clinical strains of *Enterobacteriaceae* associated with reduced gentamicin uptake. Folia Microbiol (Praha) 1998; 43(4): 438-40.
         [http://dx.doi.org/10.1007/BF02818588] [PMID: 9821296]

[59]   Esimone CO, Iroha IR, Ibezim EC, Okeh CO, Okpana EM. *In vitro* evaluation of the interaction between tea extracts and penicillin G against *Staphylococcus aureus*. Afr J Biotechnol 2006; 5: 1082-6.

[60]   Chuang PH, Lee CW, Chou JY, Murugan M, Shieh BJ, Chen HM. Anti-fungal activity of crude extracts and essential oil of *Moringa oleifera* Lam. Bioresour Technol 2007; 98(1): 232-6.
         [http://dx.doi.org/10.1016/j.biortech.2005.11.003] [PMID: 16406607]

[61]   Zasloff M. Antimicrobial peptides of multicellular organisms. Nature 2002; 415(6870): 389-95.
         [http://dx.doi.org/10.1038/415389a] [PMID: 11807545]

[62]   Dahot MU. Antimicrobial activity of small protein of *Moringa oleifera* leaves. J Islam Acad Sci 1998; 11: 27-32.

[63]   Burkill HM. The useful plants of West Africa. 92nd ed. Kew: Families A-D, Royal Botanical Gardens 1985; pp. 101-20.

[64]   Dalsgaard A, Reichert P, Mortensen HF, Sandström A, Kofoed P-E, Larsen JL. Application of lime (*Citrus aurantifolia*) juice to drinking water and food as a cholera-preventive measure. J Food Prot 1997; 60: 1329-33.
        [http://dx.doi.org/10.4315/0362-028X-60.11.1329]

[65]   Hindi NK, Chabuck ZA. Antimicrobial activity of different aqueous lemon extracts. J Appl Pharm Sci 2013; 3: 074-8.

[66]   Jahn SA. Traditional water clarification methods using scientific observation to maximise efficiency. Waterlines 1984; 2: 27-8.
        [http://dx.doi.org/10.3362/0262-8104.1984.010]

[67]   Kebreab AG. Moringa seed and Pumice as alternative natural materials for drinking water treatment, PhD thesis, KTH, Sweden, ISRNKTHLWR/Phd 1013-SE 2004. 2004; 156.

[68]   Pintore G, Usai M, Juliano C, *et al.* Chemical composition and antimicrobial activity of *Rosmarinus officinalis* L. oils from Sardina and Corsica. Flavour Fragrance J 2002; 17: 15-9.
        [http://dx.doi.org/10.1002/ffj.1022]

[69]   Wilkinson JM, Hipwell M, Ryan T, Cavanagh HM. Bioactivity of *Backhousia citriodora*: Antibacterial and antifungal activity. J Agric Food Chem 2003; 51(1): 76-81.
        [http://dx.doi.org/10.1021/jf0258003] [PMID: 12502388]

[70]   Andresen M, Cedergreen N. Plant growth is stimulated by tea-seed extract: A new natural growth regulator. HortSci 2010; 45: 1848-53.

[71]   Norrie J, Keathley JP. Benefits of *Escophyllum nodusum* marine plant extract application to 'Thompson seedless' grape production. Acta Hort 2006; 243-50.

[72]   Yongabi KA. The role of medicinal plants in environmental biotechnology and integrated Biosystems E-seminar Proceedings of International Organisation for Biotechnology and Bioengineering. 8-22.

[73]   Rassami W, Soonwera M. *In vitro* pediculicidal activity of herbal shampoo base on Thai local plants against head louse (Pediculus humanus capitis De Geer). Parasitol Res 2013; 112(4): 1411-6.
        [http://dx.doi.org/10.1007/s00436-013-3292-8] [PMID: 23334727]

[74]   Korankye O. Extraction and application of plant dyes to serve as colourants for food and textiles. 2010.

[75]   Bagavan A, Rahuman AA, Kamaraj C, *et al.* Contact and fumigant toxicity of hexane flower bud extract of Syzygium aromaticum and its compounds against Pediculus humanus capitis (Phthiraptera: Pediculidae). Parasitol Res 2011; 109(5): 1329-40.
        [http://dx.doi.org/10.1007/s00436-011-2425-1] [PMID: 21541752]

[76]   Olaitan AF, Abiodun AT. Comparative toxicity of botanical and synthetic insecticides against major field insects pests of cowpea (Vigna unquiculata (L) Walp). J Nat Prod Plant Resour 2011; 1: 86-95.

[77]   Singha S, Adhikari U, Chandra G. Smoke repellency and mosquito larvicidal potentiality of *Mesua ferra* L. leaf extract against filarial vector *Culex quinquefasciatus* Say. Asian Pac J Trop Biomed 2011; 1: S119-23.
        [http://dx.doi.org/10.1016/S2221-1691(11)60137-8]

[78]   Pritchard M, Mkandawire T, Edmondson A, O'Neill JG, Kululanga G. Potential of using plant extracts for purification of shallow well water in Malawi. Phys Chem Earth 2009; 34: 799-805.
        [http://dx.doi.org/10.1016/j.pce.2009.07.001]

Optimization of Lipase Production in Solid-State Fermentation by *Rhizopus Arrhizus* in Nutrient Medium Containing Agroindustrial Wastes

Georgi Dobrev[1], Hristina Strinska[1], Anelia Hambarliiska[1], Boriana Zhekova[1,*] and Valentina Dobreva[2]

[1]*Department of Biochemistry and Molecular Biology, University of Food Technologies, 26 Maritza Blvd., 4002 Plovdiv, Bulgaria*
[2]*Department of Engineering Ecology, University of Food Technologies, 26 Maritza Blvd., 4002 Plovdiv, Bulgaria*

Abstract:

Background:

Rhizopus arrhizus is a potential microorganism for lipase production. Solid-state fermentation is used for microbial biosynthesis of enzymes, due to advantages, such as high productivity, utilization of abundant and low-cost raw materials, and production of enzymes with different catalytic properties.

Objective:

The objective of the research is optimization of the conditions for lipase production in solid-state fermentation by *Rhizopus arrhizus* in a nutrient medium, containing agroindustrial wastes.

Method:

Biosynthesis of lipase in solid-state fermentation by *Rhizopus arrhizus* was investigated. The effect of different solid substrates, additional carbon and nitrogen source, particles size and moisture content of the medium on enzyme production was studied. Response surface methodology was applied for determination of the optimal values of moisture content and tryptone concentration. A procedure for efficient lipase extraction from the fermented solids was developed.

Results:

Highest lipase activity was achieved when wheat bran was used as a solid substrate. The addition of 1% (w/w) glucose and 5% (w/w) tryptone to the solid medium significantly increased lipase activity. The structure of the solid medium including particles size and moisture content significantly influenced lipase production. A mathematical model for the effect of moisture content and tryptone concentration on lipase activity was developed. Highest enzyme activity was achieved at 66% moisture and 5% (w/w) tryptone. The addition of the non-ionic surfactant Disponyl NP 3070 in the eluent for enzyme extraction from the fermented solids increased lipase activity about three folds.

Conclusion:

After optimization of the solid-state fermentation the achieved 1021.80 U/g lipase activity from *Rhizopus arrhizus* was higher and comparable with the activity of lipases, produced by other fungal strains. The optimization of the conditions and the use of low cost components in solid-state fermentation makes the process economicaly effective for production of lipase from the investigated strain *Rhizopus arrhizus*.

* Address correspondence to this author at the Department of Biochemistry and Molecular Biology, University of Food Technologies, 26 Maritza Blvd., 4002 Plovdiv, Bulgaria; E-mail: zhekova_b@yahoo.com

Keywords: Lipase, Solid-state fermentation, *Rhizopus arrhizus*, Response surface methodology, Agroindustrial wastes, Biosynthesis.

1. INTRODUCTION

Lipases (triacylglycerol acylhydrolases, EC 3.1.1.3) are one of the most important classes of industrial enzymes. They take a second place after proteases and carbohydrases in world enzyme market and have a share of about 5% [1]. Lipases hydrolyse triglycerides into diglycerides, monoglycerides, glycerol and fatty acids. They are able to catalyse not only hydrolysis but also synthesis reaction (esterification and transesterification) in media with low water content [2].

Enzymes of industrial interest traditionally have been produced by Submerged Fermentation (SmF). Solid-State Fermentation (SSF) is defined as the fermentation process on moist solid substrate in the absence or near absence of free water [3]. However, SSF was used for enzymes production, due to its advantages, such as high productivity, generation of high-quality products, and use of abundant and low-cost raw materials such as agro industrial wastes. The use of complex matrix of agro industrial wastes as a culture medium also induces the production of different hydrolytic enzymes such as amylases and proteases in the same fermentation batch, or even pools of lipases with different catalytic properties [4]. Diaz *et al.* found that lipase produced by SSF of *Rhizopus homothallicus* was significantly thermostable, and the optimal operating temperature was 10°C higher in comparison to the enzyme produced in SmF [5].

SSF encounters problems related to mass and heat transfer phenomena associated with solid substrates. The use of natural solid substrates can hinder downstream processes, especially when extracting lipophilic enzymes such as lipases. Fermented solids have been used as naturally immobilized biocatalysts for synthesis reactions in lyophilized or dried form. This approach can lead to lower costs of enzyme preparations since no extraction and purification steps are carried out [3].

For lipase production in SSF by fungi strains, various wastes from the agricultural industry are used. Silva *et al.* used babassu cake with 6.25% molasses in the nutrient medium to produce lipase from *Penicillium simplicissimum* at 65% moisture [4]. Awan *et al.* used almond meal examining the effect of the additional carbon source and the mineral composition of the medium on lipase biosynthesis from *Rhizopus oligosporus* [6].

For optimization of the composition of the nutrient medium and cultivation conditions, many authors used mathematical experimental designs. Fallony *et al.* conducted an experimental design and developed a mathematical model to optimize the composition of the nutrient medium and the conditions for SSF [7]. A similar study was conducted by Godoy *et al.* which by the Plackett-Burman design found the optimal values of several factors influencing lipase biosynthesis in SSF of *Penicilium simplicissimum* - particle size, initial moisture, initial pH, inoculum concentration and molasses concentration [8].

An important step in lipase production by SSF is the development of methodology for enzyme extraction. Silva *et al.* carrying out an experimental design found that the significant factors influencing lipase extraction obtained in SSF of *Penicillium simplicissimum* were pH, Tween 80 concentration, temperature and buffer molarity [4].

The aim of the research is optimization of the conditions for lipase production in SSF by *R. arrhizus* in a nutrient medium, containing agroindustrial wastes as solid substrates.

2. MATERIAL AND METHODS

2.1. Microorganism

The studied *Rhizopus arrhizus* strain used in this study was provided by Biovet® Peshtera. It was grown in the following medium, g/l: malt extract 10.00; yeast extract 4.00; glucose 4.00; agar-agar 20.00. pH was adjusted to 7.0. The strain was cultivated at 28°C for 14 days and stored at 4°C.

2.2. SSF and Media Preparation

SSF was carried out in 500 ml Erlenmeyer flasks. The flasks contained minimal nutrient medium of 10 g solid substrate and salt solution (g/l): $NH_4H_2PO_4$ 6.5, $(NH_4)_2C_2O_4$ 0.90, $MgSO_4$ 0.95, KCl 0.95. pH of the salt solution was adjusted to 7.0, and the final moisture content was adjusted to the required value with the salt solution before autoclaving. After sterilization at 121°C for 30 min, the flasks were inoculated with 5 ml inoculum with 10^7 spores/ml and incubated at 30°C for 168 h.

2.2.1. Solid Substrates for Lipase Biosynthesis in SSF

The following solid substrates were studied for lipase biosynthesis: corn flour, wheat bran, rice bran, wheat flour, sunflower meal, sunflower cake, oat bran. SSF was performed with each of the substrates as described above.

2.2.2. Effect of Additional Carbon Source on Lipase Biosynthesis in SSF

Glucose, sucrose, starch and olive oil were tested as additional carbon sources for lipase biosynthesis. The solid substrate was supplemented with 1% (w/w) additional carbon source and SSF was performed as described above. The effect of glucose concentration on lipase production was studied in the range of 1-5% (w/w).

2.2.3. Effect of Additional Organic Nitrogen Source on Lipase Biosynthesis in SSF

The effect of the additional organic nitrogen source on lipase biosynthesis was studied by addition of 1% (w/w) nitrogen sources such as peptone, tryptone, yeast extract, soybean meal, cotton flour and fish meal. The effect of tryptone concentration was investigated in the range of 1-10% (w/w).

2.2.4. Effect of Particles Size of Solid Substrate on Lipase Biosynthesis in SSF

SSF with different particles size of wheat bran was performed. Particles with size >1.0 mm, particles of size 0.2-1.0 mm and particles of size <0.2 mm were investigated.

2.2.5. Effect of Moisture Content on Lipase Biosynthesis in SSF

SSF was performed as described above with different moisture content in the nutrient medium (50-90%), which was achieved by addition of different amount of the salt solution.

2.3. Optimization of Lipase Production in SSF by Response Surface Methodology

Response surface methodology and 2^2 Optimal Composite Design (OCD) with "star points" around the center point was used to determine the optimum concentrations of moisture and tryptone for lipase biosynthesis. The distance from the design center to a factorial point was a = ±1 [9, 10]. The quadratic regression model was expressed as follows:

$$Y = b_0 + \sum_{i=1}^{k} b_i . x_i + \sum_{i=1}^{k} b_{ii} . x_i^2 + \sum_{i=1}^{k-1} \sum_{j=2}^{k} b_{ij} . x_i . x_j \tag{1}$$

where Y is the response variable, b the regression coefficients of the model and x the coded levels of the independent variables.

SygmaPlot software from Systat Software, Inc. was used for regression and graphical analysis. The independent variables participating in the 2^2 OCD and their values are presented in Table 1.

Table 1. Levels of the independent variables.

Independent variable	-1	0	+1
Moisture (x_1), %(w/w)	50	60	70
Tryptone (x_2), %(w/w)	3	5	7

2.4. Extraction of Lipase from the Fermented Solids

Following SSF, extraction of lipase from the fermented solids was performed with 50 ml eluent 1% Tween 80 for 30 minutes with constant agitation. Different salts and commercial surfactants in concentration of 0.5 and 1.0% were also tested as eluents. The solids were removed by filtration, the filtrates were centrifuged and tested for lipase activity. The results are presented in units lipase activity obtained from 1g of solid substrate.

2.5. Determination of Lipase Activity

For lipase activity determination the method described by Babu *et al.* [2], Saifuddin *et al.* [11], and Dobrev *et al.* [12] was adapted. The substrate was 30 mg ρ-nitrophenyl palmitate solution in 10 ml isopropanol, 90ml 0.05M Tris-HCl buffer with pH 7.2, 0.4g Triton X-100 and 0.1g gum Arabic. A reaction mixture consisting of 2.4 ml of the substrate solution and 0.1ml enzyme was incubated at 35°C for 30 min. The enzyme reaction was stopped with 1.0 mL

0.5 M EDTA with pH 8.0 and the absorbance at 405 nm was measured. One unit (U) of lipase activity was defined as the amount of enzyme forming 1 µmol ρ-nitrophenol for 1 min at 35°C and pH 7.2.

2.6. Assays

Moisture, ash, fat, protein and starch content of wheat bran were determined by ICC standard methods No. 109/1, 104/1, 136, 105/2 and 122/1 [10, 13]. Total sugars were determined by the method of Dubois [14].

3. RESULTS AND DISCUSSION

3.1. Effect of Solid Substrate on Lipase Biosynthesis in SSF

The choice of solid substrate is of great importance to SSF, because in addition to being a source of nutrients, it forms the structure of the culture medium and determines the possibility of mass and heat transfer. The results for lipase biosynthesis on different solid substrates are shown in Fig. (1).

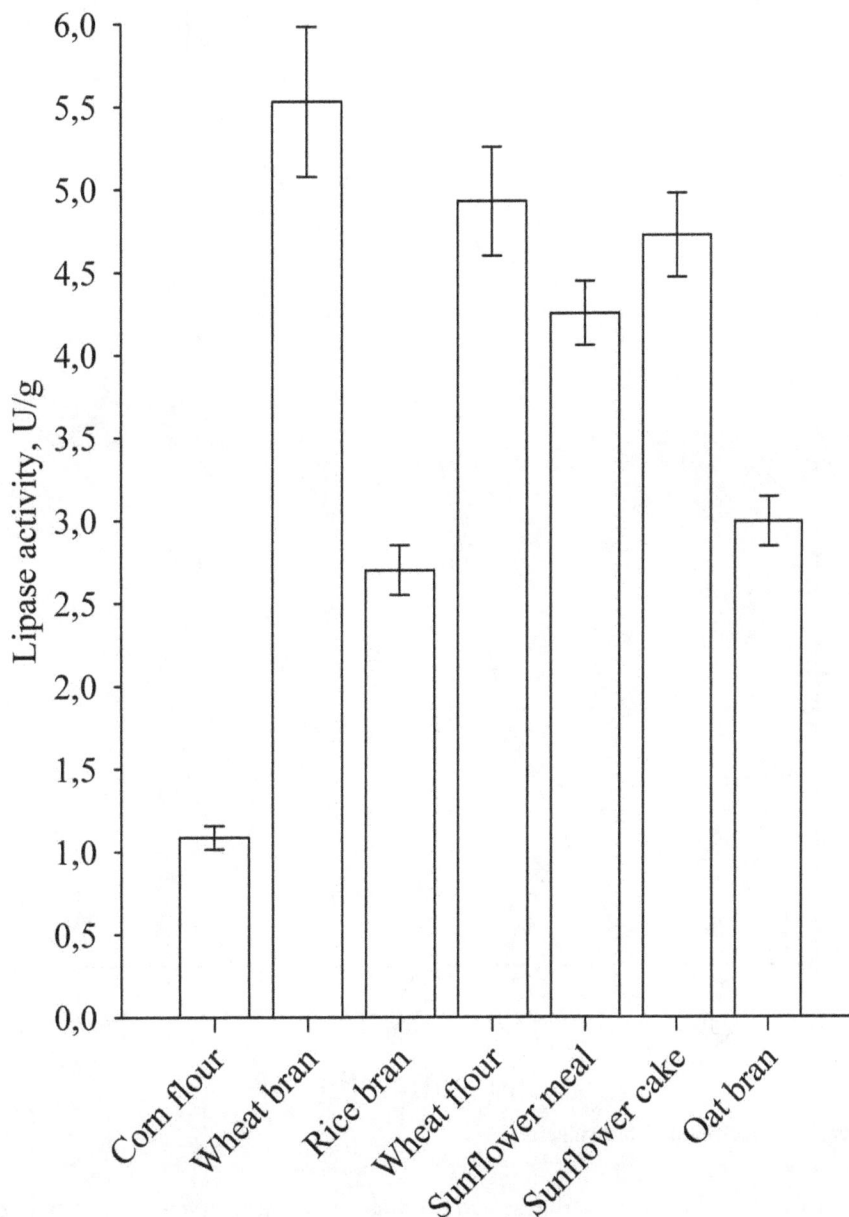

Fig. (1). Lipase biosynthesis on different solid substrates.

Highest lipase activity was noticed when wheat bran was used as a solid substrate. Significant lipase activity was also achieved with the use of other wastes from the agro and food industry, such as sunflower meal and sunflower cake, as well as oat bran and rice bran. The low price and availability of wheat bran make them a preferred solid substrate for carrying SSF [15].

The effect of fermentation time on lipase biosynthesis, when wheat bran was used as solid substrate is presented in Fig. (2).

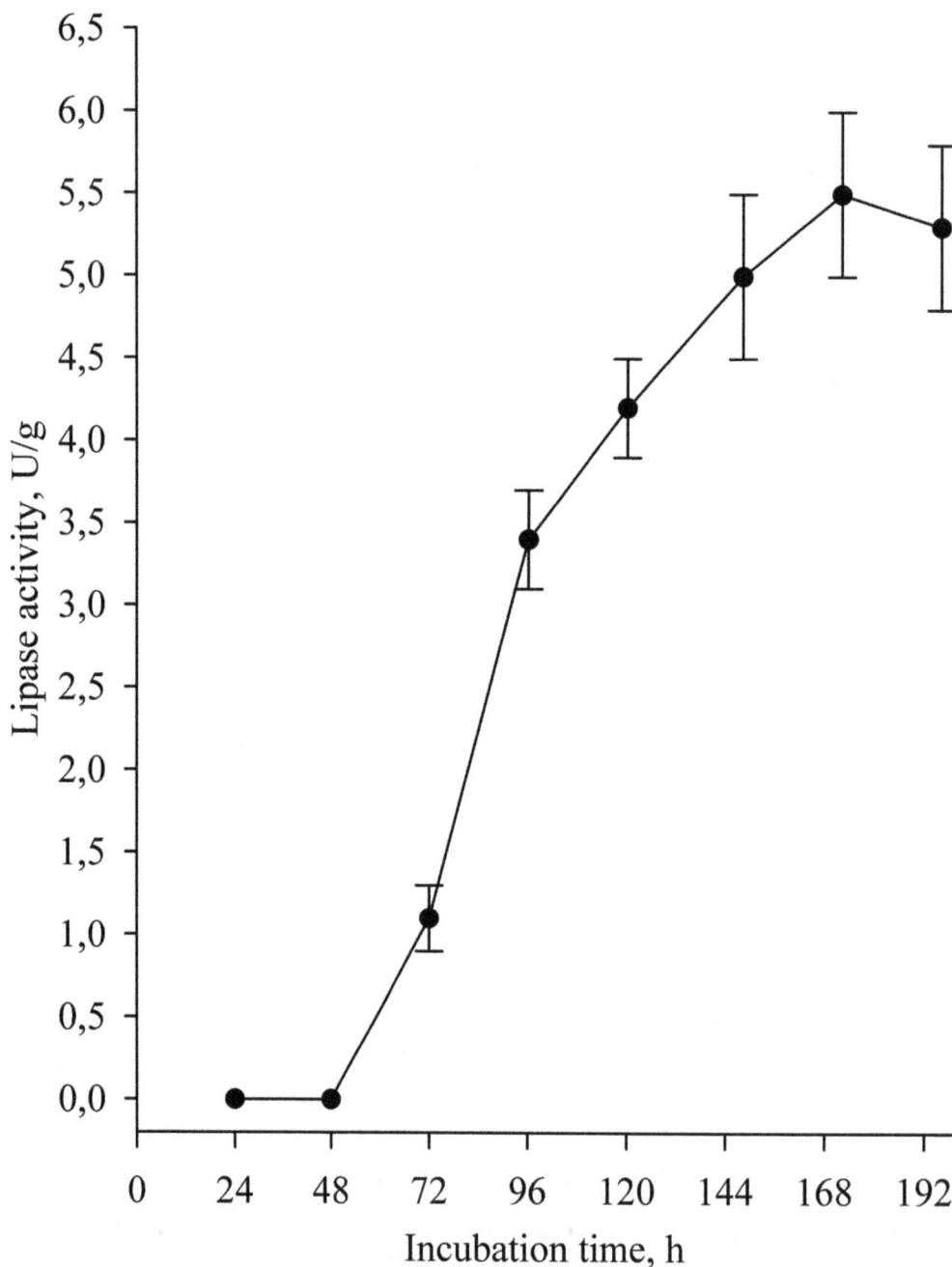

Fig. (2). Effect of fermentation time on lipase biosynthesis.

After 3 days of cultivation, a sharp peak of lipase activity was observed, with maximal activity being achieved at 168h, after which the activity was retained.

3.2. Effect of Additional Carbon Source on Lipase Biosynthesis in SSF

The effect of the additional carbon source to the medium of wheat bran on lipase biosynthesis is presented in Fig. (3). Significantly higher lipase activity was achieved by the use of glucose as an additional carbon source at a concentration of 1% (w/w) compared to the addition of starch and olive oil. The results obtained differ from the results of Awan *et al*. that found that the addition of olive oil and Tween 80 significantly increased lipase production by *Rhizopus oligosporus* compared to glucose addition [6].

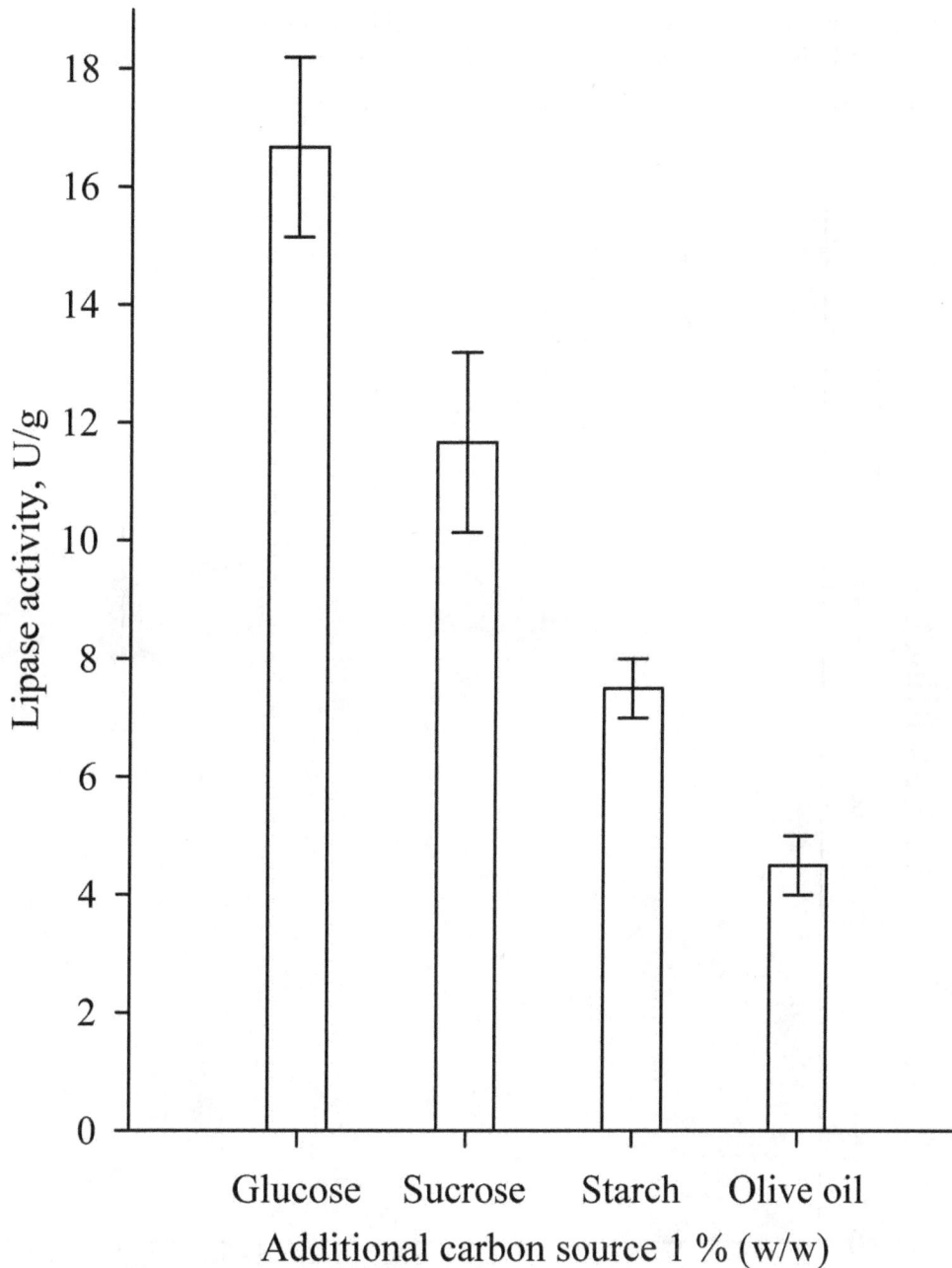

Fig. (3). Effect of additional carbon source on lipase biosynthesis.

For explanation of the results the chemical composition of the solid substrate wheat bran was determined (Table **2**).

Table 2. Composition of wheat bran.

Solid Substrate	Moisture, %	Protein, %	Total Sugars, %	Starch, %	Fat, %	Ash, %
Wheat Bran*	12.22±0.23	15.02±0.18	78.45±1.8	25.03±0.56	2.48±0.08	3.82±0.06

* Results are expressed as percent of dry material

The chemical composition of wheat bran varies significantly according to the milling process and the nutritional components are typically in wide range. Proteins content is about 9.6-18.6%, total carbohydrate content is in the range of 60.0-75.0%, and starch varies in the range of 9.10-38.9% of dry matter [16]. The starch content of the used wheat bran was significantly high (25.03%). It should be noted that the use of starch as an additional carbon source in the nutrient medium significantly reduced lipase activity. This was probably due to the fine particle size and the high starch water holding capacity, which created problems in mass transfer and hindered strain development.

The influence of glucose concentration on lipase activity is presented in Fig. (**4**).

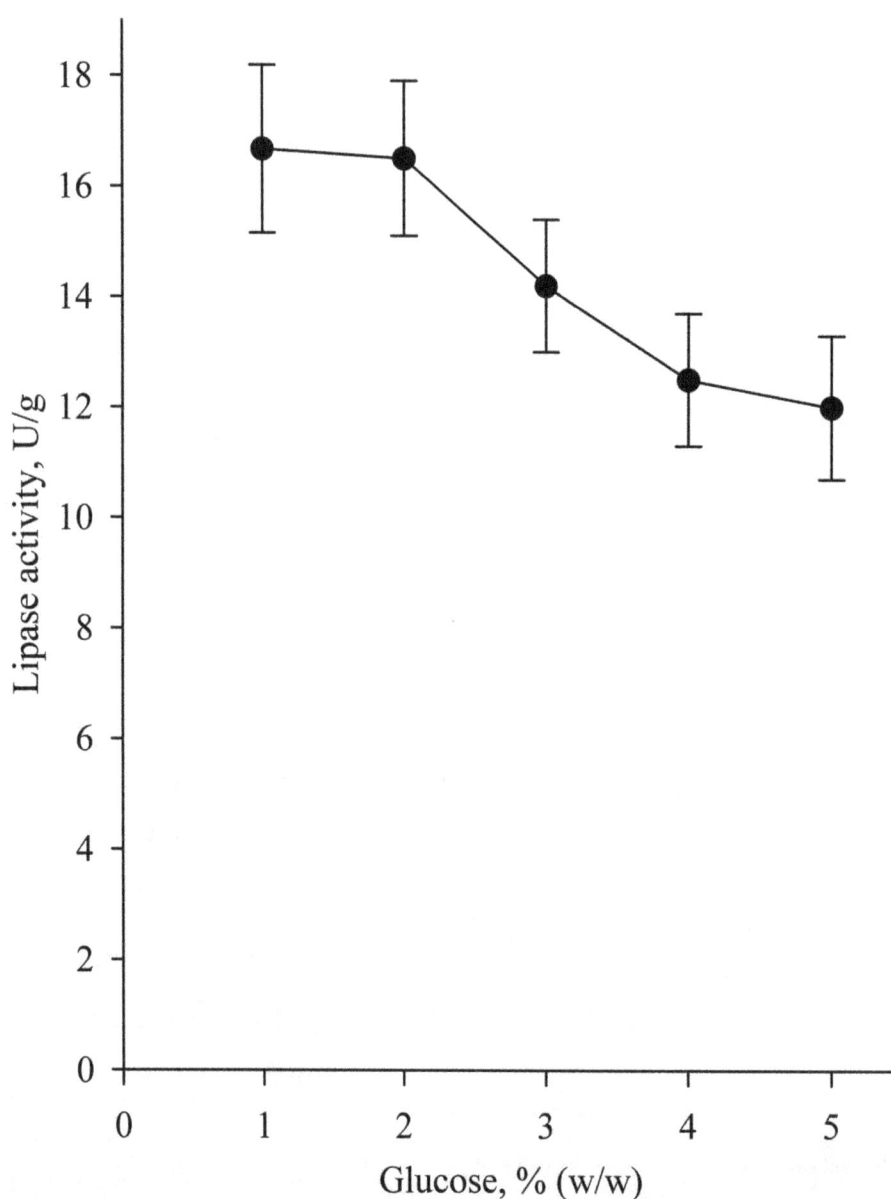

Fig. (4). Effect of glucose concentration on lipase biosynthesis.

The results showed that the addition of glucose in concentrations higher than 2% (w/w) reduced lipase activity.

3.3. Effect of Additional Organic Nitrogen Source on Lipase Biosynthesis

The addition of 2% (w/w) tryptone, peptone and yeast extract as an additional source of organic nitrogen significantly increased lipase production (Fig. **5**). Highest lipase activity was achieved when tryptone was used as an additional organic nitrogen source. The use of soybean meal, cotton flour and fishmeal led to a decrease in lipase activity. Similar results were obtained by Sun *et al.* for the supplementation of peptone to wheat bran and wheat flour solid substrates for SSF production of lipase by *Rhizopus chinensis*. The authors reported also an increase in lipase activity with the addition of yeast and beef extract [17].

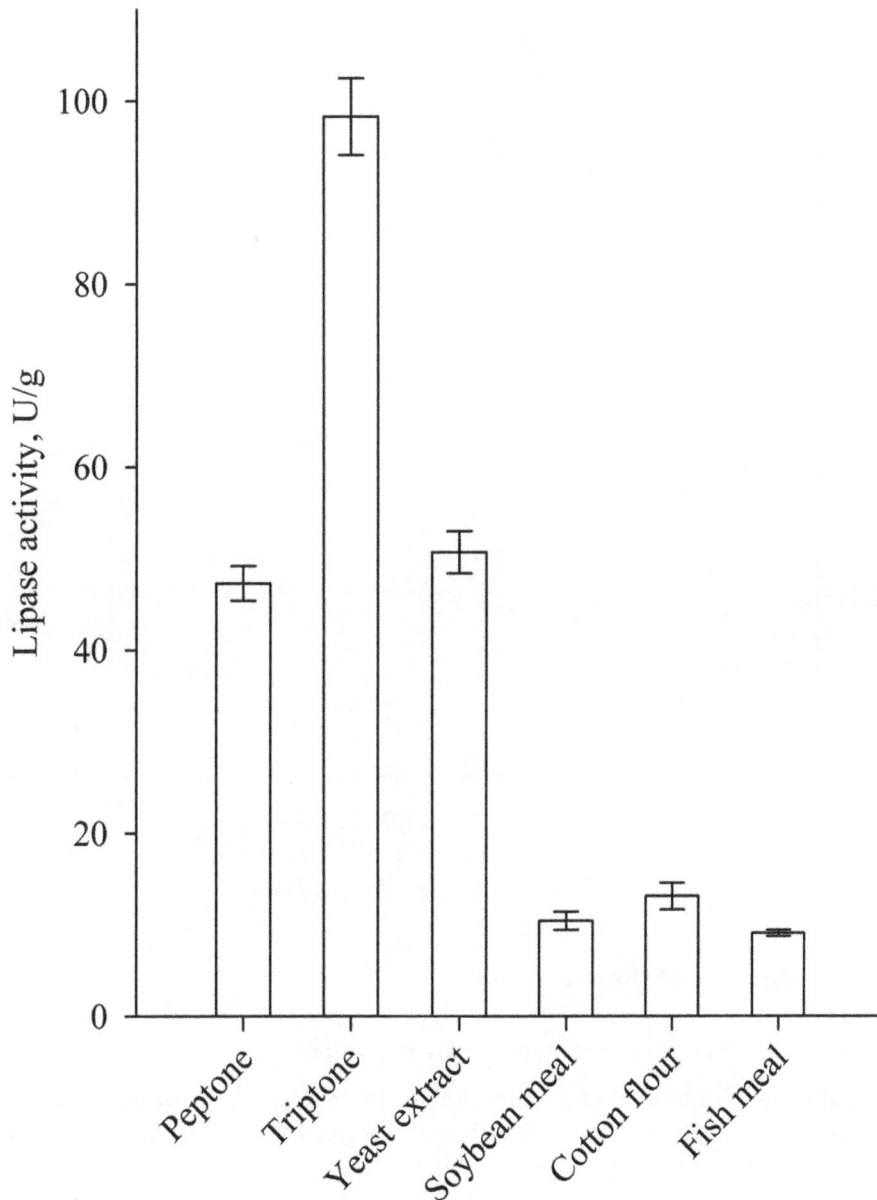

Fig. (5). Effect of additional organic nitrogen on lipase biosynthesis.

The highest lipase activity (about 270U/g) was achieved, when tryptone was added to the nutrient medium in concentration of 5.0% (w/w) (Fig. **6**).

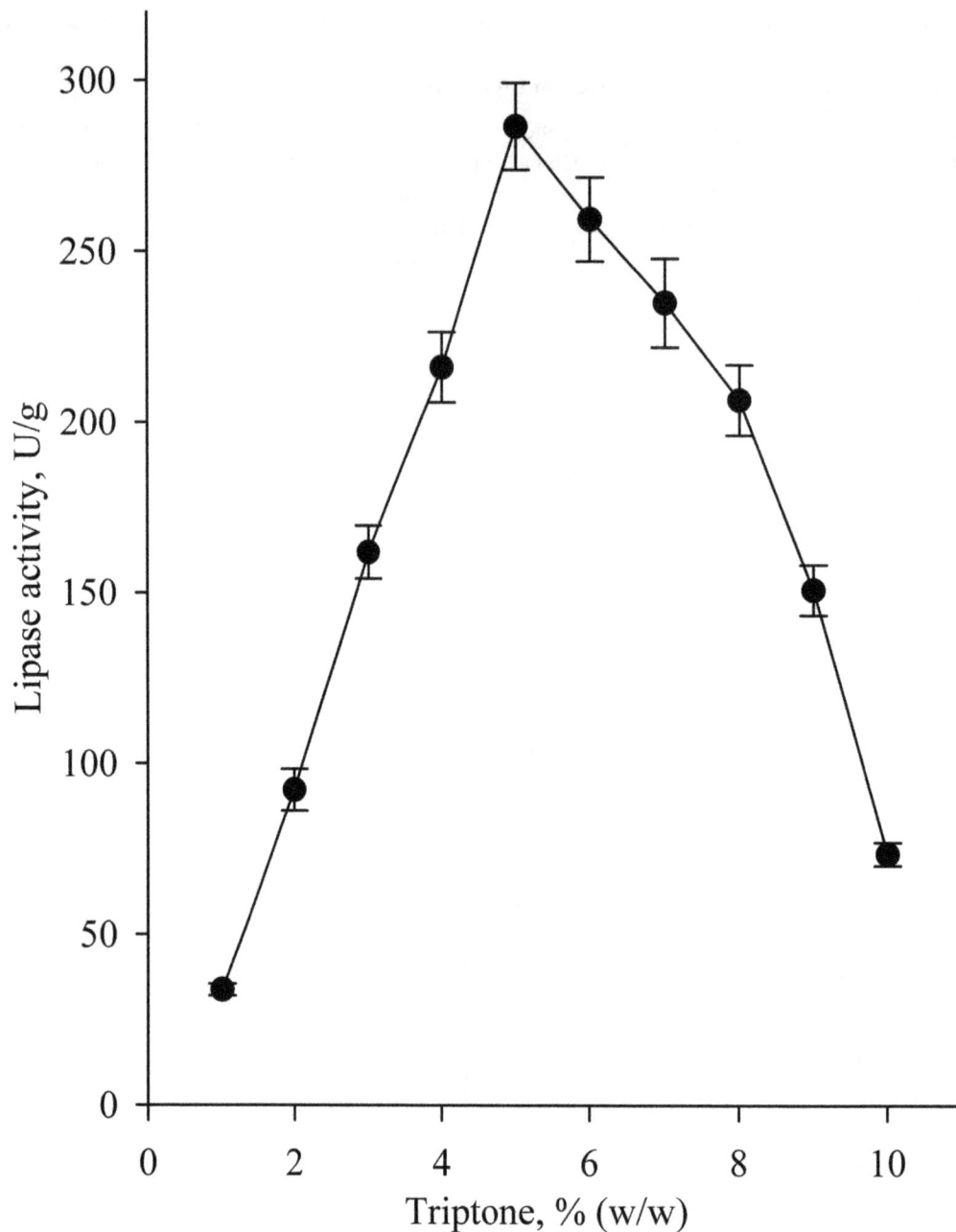

Fig. (6). Effect of tryptone concentration on lipase biosynthesis.

3.4. Effect of Particles Size and Moisture Content on Lipase Biosynthesis

The particles size of the solid substrate has a significant impact on the structure of the fermentation medium and hence on the possibility for mass transfer in the solid medium and on the growth of the strain. The effect of the size of wheat bran particles on lipase production is presented in Table **3**.

Table 3. Effect of wheat bran particles size on lipase biosynthesis.

Particles Size	Lipase Activity, U/g
Reference	288.53
>1 mm	249.88
0.2-1 mm	191.75
<0.2 mm	56.72

With particle size reduction, mass transfer in the solid medium was most likely to be impaired, which significantly reduced lipase activity.

Another factor that significantly influences lipase biosynthesis is the moisture of the medium. Typically, the optimum moisture for growth of fungi strains in SSF is in the range of 50-70% w/w [18 - 20]. Highest lipase activity of 437.25 U/g was achieved at 60% moisture (Fig. 7). Moisture higher than 70% led to a decrease in lipase activity, due to a change in the structure of the medium resulting in porosity decrease, which hinders mass transfer and the strain development. Ferraz *et al.* have received similar results and determined that the optimal moisture for lipase production by *Sporobolomyces ruberrimus* was 60% [18].

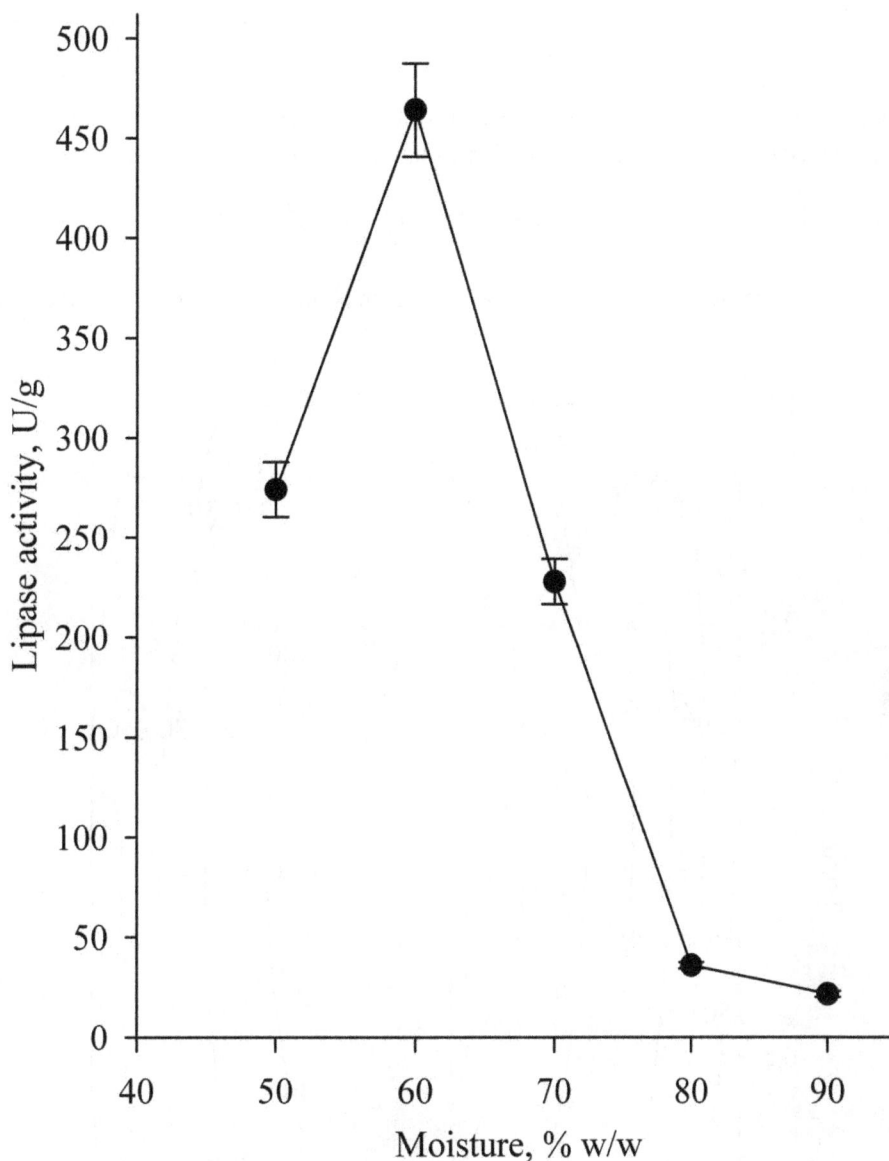

Fig. (7). Effect of moisture content of solid medium on lipase biosynthesis.

3.5. Optimization of Lipase Production by Response Surface Methodology

Of the studied factors, tryptone concentration and moisture content of the solid medium had the most significant effect on lipase biosynthesis. For these reasons, a mathematical experimental design was conducted for deveplopment of mathematical model to study the co-influence of both factors on lipase biosynthesis. The matrix of 2^2 OCD is shown in Table **4**.

Table 4. OCD 2^2 for lipase biosynthesis in SSF.

N°	Coded Levels		Lipase Activity, U/g	Predicted Lipase Activity, U/g
	Moisture (x_1)	Tryptone (x_2)		
1	-1	-1	74.68	74.84
2	0	-1	400.00	392.27
3	+1	-1	167.74	175.31
4	-1	0	135.96	141.57
5	0	0	437.25	440.84
6	+1	0	214.90	205.71
7	-1	+1	110.00	104.24
8	0	+1	381.20	385.34
9	+1	+1	130.41	132.04

The coefficients with a P-value <0.05 had a significant effect on lipase activity (Table **5**). The mathematical model, describing the effect of moisture and tryptone concentration on lipase biosynthesis in SSF was developed as follows:

$$Y = 440.84 + 32.069\,x_1 - 18.167 x_1 x_2 - 267.201 x_1^2 - 52.034 x_2^2 \tag{2}$$

By studying the response surface (Fig. **8**) it was found that the maximum lipase activity predicted by equation (2) was 441.81 U/g.

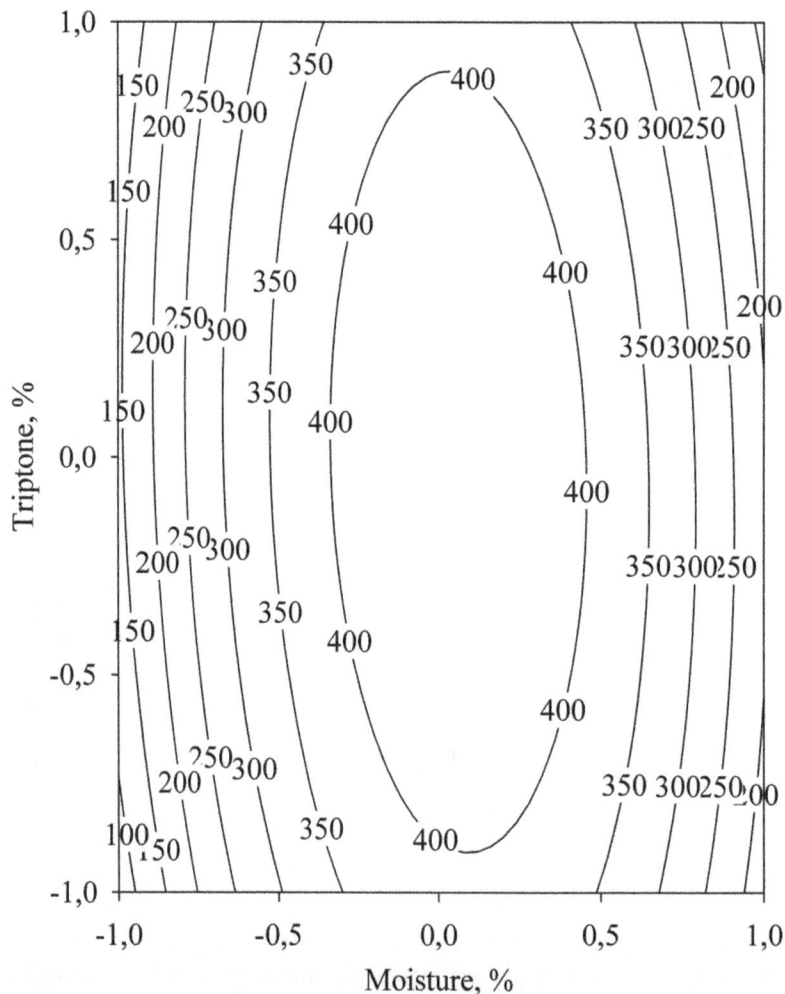

Fig. (8). Response surface of the effect of moisture and tryptone concentration on lipase biosynthesis in SSF.

Table 5. Coefficients of the model.

Coeficient	Value	P-value
Intercept	440.84	1.06E-05
x_1	32.06883	0.004273
x_2	-3.46817	0.457192
$x_1 x_2$	-18.1668	0.035725
x_1^2	-267.201	4.05E-05
x_2^2	-52.0335	0.005157

The optimal levels of the studied independent variables were 66% moisture (coded value 0.06) and 5.0% tryptone (coded value 0.01).

With the optimum values of moisture and tryptone concentration, 5 consecutive SSF experiments were performed, and the resulting lipase activity was 418.88 U/g ± 40.51. The achieved lipase activity was close to the value predicted by the model, which verified the adequacy of the model.

3.6. Lipase Extraction from the Fermented Solids

One of the processes that has a significant impact on enzyme yield in SSF is the enzyme extraction procedure from the fermented solids. Most often, to improve the extraction of lipase produced by SSF, nonionic surfactants are used, which do not denature the enzymes in contrast to most anionic and cationic surfactants. The results for the effect of different eluents for enzyme extraction on lipase activity is shown in Fig. (**9**). Highest lipase activity was achieved at 1.0% of Disponil NP 3070, which is a non-ionic coemulsifier. The lipase activity was 1021.80 U/g and it was almost 3 times higher than the activity of the reference sample with 1% Tween 80. Similar results were reported by other authors. Silva *et al.* found that the addition of nonionic surfactant Tween 80 increased lipase activity in the extract by 2.5 fold [4]. Rodrigez *et al.* found that the use of 1% Triton X-100 increased nearly 10 times the activity of lipase in the extract after SSF [21].

Lipase biosynthesis by *Rhizopus arrhizus* in SSF was compared with the process by other fungal strains grown on agricultural wastes (Table **6**). The enzyme activity of lipase obtained after optimization of SSF conditions and extraction procedure (1021.80 U/g) was comparable and even higher than the values cited in the literature.

Table 6. Comparison of lipase production in SSF by fungal strains.

Microorganism	Solid Substrate	Moisture,%	Conditions	Extraction	Lipase Activity, U/g	References
Rhizopus arrhizus	Wheat bran + tryptone	66	168 h, 30°C	1% Agnique	1021.80	In this work
Penicillium simplicissimum	Babassu cake + molasses	65	72 h, 30°C	0.1% Tween 80	85.7	[4]
Rhizopus oligosporus	Almond meal + Tween 80	45	48 h, 30°C	Phosphate buffer pH 7	81.22	[6]
Aspergillus sp.	Soybean meal + rice husk	60	12 days, 30°C	0.2 M pH 7.0 Phosphate buffer	25.07 U	[19]
Rhizopus homothallicus	Sugar-cane bagasse	75	12 h, 40 °C	20mM Tris–HCl buffer (pH 8), with 0.5% Triton X-100 and 2mM benzamidine	1500	[5]
Aspergillus niger	Wheat bran	65	7 days, 30°C	100 mL of distilled water	9.14	[7]
Penicillium simplicissimum	Babassu cake	70	72 h, 30°C	Phosphate buffer (100 mM, pH 7.0	19.6	[20]
Penicillium sp	Soybean meal	55ml/100g	48 h, 27°C,	0.1M Phosphate buffer, pH 7.0	200	[22]
Rhizopus microsporus	Sugarcane bagasse+ wheat bran	84	18 h, 40°C	-	262	[15]
Rhizopus homothallicus	Sugarcane bagasse	75	12 h, 40°C	1% Triton-X-100	1224	[21]
Aspergillus versicolor CJS-98	Jatropha seed cake	40	5 days, 25°C	Chilled distilled water	1079.47	[23]

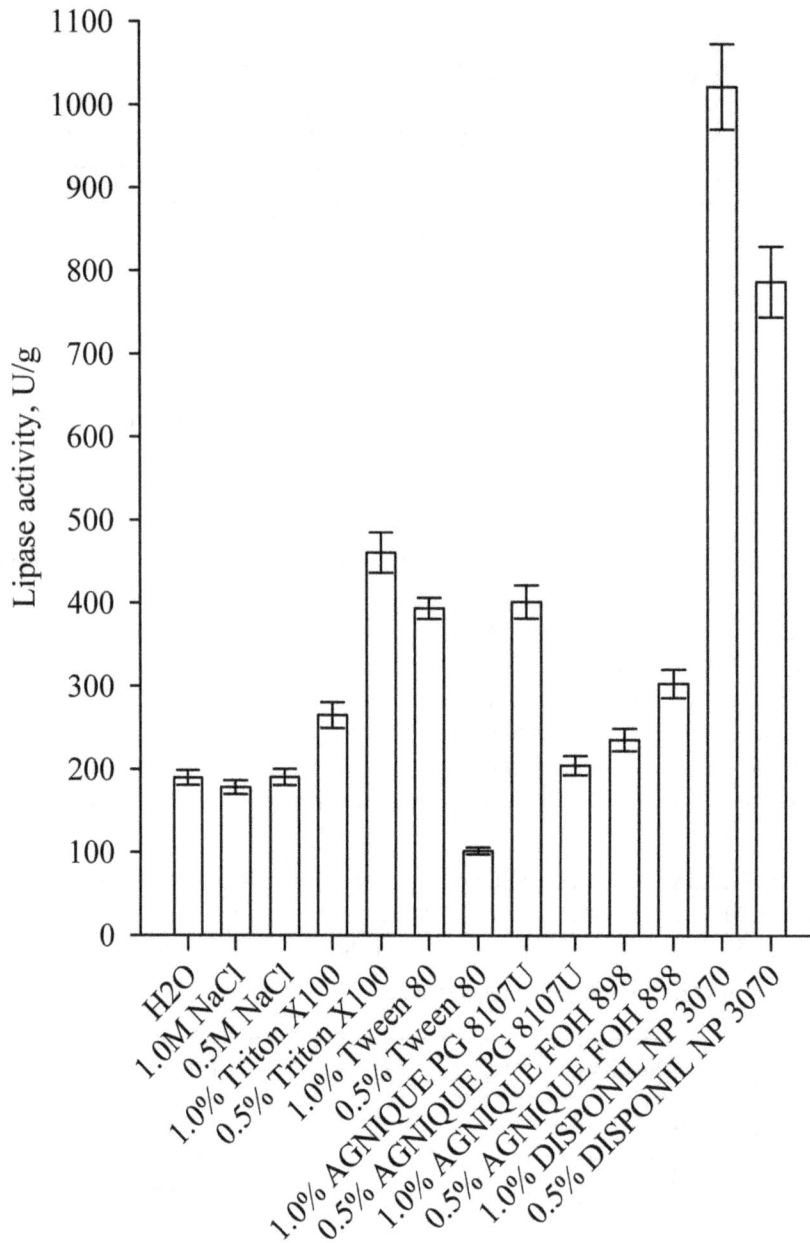

Fig. (9). Effect of different eluents on lipase extraction from the solid medium.

CONCLUSION

The optimal conditions for production of lipase in SSF by *Rhizopus arrhizus* were determined. Maximum lipase activity was achieved when the strain was grown on wheat bran as solid substrate. Addition of glucose at a concentration of 1% (w/w) as an additional carbon source and tryptone at a concentration of 5% (w/w) as a source of organic nitrogen, significantly stimulated lipase biosynthesis. Using response surface methodology, the optimum moisture of the medium was determined to be 66% and the optimum tryptone concentration was 5% (w/w). As a result of the optimization, 418.88 U/g lipase activity was achieved. The procedure for enzyme extraction from the fermented solids had a significant effect on lipase yield. The addition of the non-ionic surfactant Disponyl NP 3070 in the eluent increased lipase activity about three folds. The achieved activity of 1021.80 U/g was higher and comparable with the activity of lipases, produced by other fungal strains. The optimization of the conditions and the use of low cost components in SSF makes the process economicaly effective for production of lipase from the investigated strain *Rhizopus arrhizus*.

HUMAN AND ANIMAL RIGHTS

No Animals/Humans were used for studies in this research.

ACKNOWLEDGEMENTS

The research was financially supported by Bulgarian National Science Fund (BNSF), Project 17/25-2017.

REFERENCES

[1]   Jooyandeh H, Amarjeet K, Minhas KS. Lipases in dairy industry: A review. J Food Sci Technol 2009; 46(3): 181-9.

[2]   Babu IS, Rao GH. Optimization for the production of lipase in submerged fermentation by *Yarrowia lipolytica* NCIM 3589. Res J Microbiol 2007; 2(1): 88-93.
        [http://dx.doi.org/10.3923/jm.2007.88.93]

[3]   Santis-Navarro A, Gea T, Barrena R, Sánchez A. Production of lipases by solid state fermentation using vegetable oil-refining wastes. Bioresour Technol 2011; 102(21): 10080-4.
        [http://dx.doi.org/10.1016/j.biortech.2011.08.062] [PMID: 21903382]

[4]   Silva JN, Godoy MG, Gutarra ML, Freire DM. Impact of extraction parameters on the recovery of lipolytic activity from fermented babassu cake. PLoS One 2014; 9(8): e103176.
        [http://dx.doi.org/10.1371/journal.pone.0103176] [PMID: 25090644]

[5]   Diaz JC, Rodrıguez JA, Roussos S, *et al.* Lipase from the thermotolerant fungus *Rhizopus homothallicus* is more thermostable when produced using solid state fermentation than liquid fermentation procedures. Enzyme Microb Technol 2006; 39: 1042-50.
        [http://dx.doi.org/10.1016/j.enzmictec.2006.02.005]

[6]   Awan U, Shafiq K, Mirza S, Ali S, Rehman A, Haq I. Mineral constituent of culture medium for lipase production by *Rhizopus oligosporous* fermentation. Asian J Plant Sci 2003; 2(12): 913-5.
        [http://dx.doi.org/10.3923/ajps.2003.913.915]

[7]   Falony G, Armas JC, Mendoza JC, Hernández JL. Production of extracellular lipase from *Aspergillus niger* by solid-state fermentation. Food Technol Biotechnol 2006; 44(2): 235-40.

[8]   Godoya MG, Gutarra ML, Maciel FM, Felix SP, Bevilaqua JV, Machado OL, *et al.* Use of a low-cost methodology for biodetoxification of castor bean waste and lipase production. Enzyme Microb Technol 2009; 44: 317-22.
        [http://dx.doi.org/10.1016/j.enzmictec.2009.01.002]

[9]   Vuchkov I, Stoyanov S. Mathematical modelling and optimization of technological objects. Sofia: Technics 1980; pp. 135-51.

[10]   Dobrev GT, Pishtiyski IG, Stanchev VS, Mircheva R. Optimization of nutrient medium containing agricultural wastes for xylanase production by *Aspergillus niger* B03 using optimal composite experimental design. Bioresour Technol 2007; 98(14): 2671-8.
        [http://dx.doi.org/10.1016/j.biortech.2006.09.022] [PMID: 17092711]

[11]   Saifuddin N, Raziah A. Enhancement of lipase enzyme activity in non-aqueous media through a rapid three phase partitioning and microwave irradiation. E-J Chem 2008; 5(4): 864-71.
        [http://dx.doi.org/10.1155/2008/920408]

[12]   Dobrev G, Zhekova B, Dobreva V, Strinska H, Doykina P, Krastanov A. Lipase biosynthesis by *Aspergillus carbonarius* in a nutrient medium containing products and byproducts from the oleochemical industry. Biocatal Agric Biotechnol 2015; 4(1): 77-82.
        [http://dx.doi.org/10.1016/j.bcab.2014.09.011]

[13]   ICC Standard Methods. Available from: https://www.icc.or.at/standard_methods

[14]   Dubois M, Gilles KA, Hamilton JK, Rebers PA, Smith F. Colorimetric method for determination of sugars and related substances. Anal Chem 1956; 28: 350-6.
        [http://dx.doi.org/10.1021/ac60111a017]

[15]   Pitol LO, Finkler AT, Dias GS, *et al.* Optimization studies to develop a low-cost medium for production of the lipases of *Rhizopus microsporus* by solid-state fermentation and scale-up of the process to a pilot packed-bed bioreactor. Process Biochem 2017; 62: 37-47.
        [http://dx.doi.org/10.1016/j.procbio.2017.07.019]

[16]   Onipe OO, Jideani AIO, Beswa D. Composition and functionality of wheat bran and its application in some cereal food products. Int J Food Sci Technol 2015; 50(12): 2509-18.
        [http://dx.doi.org/10.1111/ijfs.12935]

[17]   Sun SY, Xu Y. Solid-state fermentation for 'whole-cell synthetic lipase' production from *Rhizopus chinensis* and identification of the functional enzyme. Process Biochem 2008; 43: 219-24.
        [http://dx.doi.org/10.1016/j.procbio.2007.11.010]

[18]   Ferraz LR, Oliveira DS, Silva MF, *et al.* Production and partial characterization of multifunctional lipases by *Sporobolomyces ruberrimus* using soybean meal, rice meal and sugarcane bagasse as substrates. Biocatal Agric Biotechnol 2012; 1: 243-52.
[http://dx.doi.org/10.1016/j.bcab.2012.03.008]

[19]   Colla LM, Rizzardi J, Pinto MH, Reinehr CO, Bertolin TE, Costa JA. Simultaneous production of lipases and biosurfactants by submerged and solid-state bioprocesses. Bioresour Technol 2010; 101(21): 8308-14.
[http://dx.doi.org/10.1016/j.biortech.2010.05.086] [PMID: 20580228]

[20]   Gutarra ML, Godoy MG, Maugeri F, Rodrigues MI, Freire DM, Castilho LR. Production of an acidic and thermostable lipase of the mesophilic fungus *Penicillium simplicissimum* by solid-state fermentation. Bioresour Technol 2009; 100(21): 5249-54.
[http://dx.doi.org/10.1016/j.biortech.2008.08.050] [PMID: 19560339]

[21]   Rigo E, Ninow JL, Di Luccio M, *et al.* Lipase production by solid fermentation of soybean meal with different supplements. Lebensm Wiss Technol 2010; 43: 1132-7.
[http://dx.doi.org/10.1016/j.lwt.2010.03.002]

[22]   Rodriguez JA, Mateos JC, Nungaray J, *et al.* Improving lipase production by nutrient source modification using *Rhizopus homothallicus* cultured in solid state fermentation. Process Biochem 2006; 41: 2264-9.
[http://dx.doi.org/10.1016/j.procbio.2006.05.017]

[23]   Veerabhadrappa MB, Shivakumar SB, Devappa S. Solid-state fermentation of *Jatropha* seed cake for optimization of lipase, protease and detoxification of anti-nutrients in *Jatropha* seed cake using *Aspergillus versicolor* CJS-98. J Biosci Bioeng 2014; 117(2): 208-14.
[http://dx.doi.org/10.1016/j.jbiosc.2013.07.003] [PMID: 23958640]

3

Enhancement of Recombinant Antibody Expression Level by Growth Controlled Medium

Le Thi Minh Phuc[1,2,3], Tetsuji Sasaki[4], Hisayo Shimizu[4], Nguyen Thi Minh Huyen[3], Nguyen Thi Thu Thuy[3], Le Quang Huan[3] and Akiyoshi Taniguchi[1,2,*]

[1]Cellular Functional Nanomaterials Group, Research Center for Functional Materials, National Institute for Materials Science, 1-1 Namiki, Tsukuba, Ibaraki 305-0044, Japan

[2]Graduate School of Advanced Science and Engineering, Waseda University, 3-4-1 Okubo, Shinjuku-ku, Tokyo 169-8555, Japan

[3]Institute of Biotechnology, Vietnam Academy of Science and Technology, 18 Hoang Quoc Viet, Cau Giay, Hanoi, Vietnam

[4]Kyokuto Pharmaceutical Industrial Co., Ltd. 3333-26 Aza-Asayama, Kamitezuna, Takahagi-shi, Ibaraki 318-0004, Japan

Abstract:

Background:

Transient expression system is very widely used for protein expression including recombinant protein expression. Normally, in the commercially available expression system, cells were grown rapidly in protein expression. However, over cell growth was not only unnecessary for high level protein expression, but also induced cell death and led to protein degradation.

Aims and Objectives:

To overcome these limitations a new adapted culture method needs to be developed.

Methods:

In this work, we developed growth control medium for transient high protein expression system using NPLAd (Non-Protein and Lipid Medium Adopted) cells. With this system, cell numbers were not increased until 10 days.

Results:

Our results indicated that expression level in NPLAd system was 1.8 times higher than that in Free style system, which was one of the commercially available high expression systems.

Conclusion:

The antibody performance expressed by NPLAd system was almost the same to that of expressed by Free style system. The results suggested that NPLAd system could be useful for protein expression, such as recombinant antibody.

Keywords: Recombinant antibody, Protein expression, CHO cells, Culture medium, NPLAd system, Antibody performance.

* Address correspondence to this author at the Cellular Functional Nanomaterials Group, Research Center for Functional Materials, National Institute for Materials Science, 1-1 Namiki, Tsukuba, Ibaraki 305-0044, Japan; E-mail: taniguchi.akiyoshi@nims.go.jp

1. INTRODUCTION

Monoclonal antibodies are most selling biopharmaceutics drug in drug market [1 - 4]. Many of the recombinant production systems have been developed, such as gram-negative and positive bacteria, yeasts and filamentous fungi, insect cell lines, mammalian cells to transgenic plants and animals [5]. The advanced mammalian folding, secretion and post-translational apparatus are capable of producing antibodies indistinguishable from those in the human body with least concerns for immunogenic modifications. Moreover, it is also highly efficient for secretion of large and complex IgGs [5]. On the other hand, expression system using mammalian cells has some risk such as viral infections [3 - 7].

Chinese Hamster Ovary (CHO) cells are the prominent choice of mammalian expression system for antibodies [8 - 10]. Due to safety concerns in clinical use, the expression system of CHO cells should not have biological ingredients, such as serum [11]. The adapted culture method is one of the good strategies to simplify CHO cell culture medium without any proteins and lipids. The strategy of adapted culture method includes both medium and cell modification. However, in biological ingredient-free culture systems, the rate of cell proliferation is decreased. Proliferation in simplified cell culture medium often requires autocrine factors, such as EGF [12]. To increase proliferation, the signaling efficiency of these autocrine factors should be increased. We have developed a Non-Protein and Lipid Medium Adopted (NPLAd) cell line for biopharmaceutical recombinant protein expression [13]. The proliferation rate of CHO cells in simplified culture medium was improved by insulin and GM3 addition. Our results suggest that this cell line could be useful for biopharmaceutical recombinant protein expression. However, NPLAd system has some problems, such as low protein expression level, because NPLAd medium was good for cell growth, but not good for protein expression.

In this paper, we developed growth control medium for transient high protein expression system using NPLAd cells with our new concept. Our concept was that cell growth was not necessary for protein expression in transient expression system because most of the cells were not transfected by expression vector. Keeping this in view, we developed low growth rate medium for antibody expression. The results indicated that expression level in this system was higher than that in conventional system, FreeStyle System, which is one of commercially high expression system. The results suggested that this system could be useful for recombinant antibody expression.

2. MATERIALS AND METHODS

2.1. Cell Culture and Medium

Free Style Chinese Hamster Ovary (CHO) cell line (FS CHO–S Cells, ThermoFisher Scientific) manipulated to express was used in the study. FreeStyle CHO Expression Medium (ThermoSicher Scientific, USA) was used as the FreeStyle CHO cell growth medium. In addition, an in-house NPLAd cell line, has been previously described [13]. NPL medium, previously described [13] was modified and used as the NPLAd cell line growth medium. The improved NPL medium was named NPLTT medium. Both FreeStyle and NPLAd Systems were inoculated in shaking flask and incubated at 37°C under 5% CO_2, 100% humidity and 120 rpm shaking.

The improved NPL medium was named NPLTT medium (NPL Medium for Transient Transfection). The concentration of L-Isoleucine, L-leucine, L-ornithine and L-threonine was increased as a protein expression resource in NPLTT medium. The NPLTT medium increased glucose and mannose concentrations as energy sources for cells. Ascorbic acid and glutathione were added to suppress cell damages by their antioxidant effects.

2.2. Transfection Efficiency

Transfection efficiency of transient gene expression was measured by GFP expression assay. Green Fluorescent Protein vector, phMGFP (Promega) was amplified in DH5α and purified with QIAGEN Plasmid Midi Kit (QIAGEN, USA). Freestyle CHO-S cells (Thermo fisher scientific, USA) were cultured in Freestyle CHO medium (Thermo fisher scientific) supplemented with 2 mM L-glutamine. NPLAd and FreeStyle CHO-S cells were cultured for semi-confluent condition under standard humidified conditions (37°C and 5% CO_2) and harvested. Cells were resuspended in fresh medium and seeded in 24-well plates by 20 x 10^5 cells in 0.5 mL medium. 3 wells were tested for a cell condition. Culture plates were incubated at 37°C and 5% CO_2 in shaking at 120 rpm. Transfection reagent Lipofectamine LTX & PLUS Reagent (Thermo fisher scientific) and Freestyle MAX were used for NPLAd cells and Freestyle CHO-S cells, respectively, under standard protocol. 1µg of phMGFP plasmid was used for a well and Opti-MEM (Thermo fisher scientific) was used to prepare transfection complex. GFP expression was measured after 6 days from transfection day with LUNA-FL Dual Fluorescence Cell Counter (Logos Biosystems, USA).

2.3. Plasmid Construction

Anti-HER2 antibody sequence was inserted into pCDNA3.1 vector between Nhe I and Xho I restriction site. The order of the sequence was as follows: signal peptide, heavy chain (VH), GS repeat region, light chain (VL) and constant regions (CH2-CH3). The signal peptide was a 19-amino-acid sequence (MKHLWFFLLLVAAPRWVLS) the original of which comes from signal peptide of V-region precursor (Genebank: AAA58803). This sequence was designed to be auto-cleavaged after leading the anti-Her2 antibody throughout cell into the culture medium. The auto-cleavage ability of this sequence was tested using SignalP 4.1 Server (http://www.cbs.dtu.dk/services/SignalP/). VH and VL DNA region were constructed following the synthetic sequence of the author's group (Genbank: AM402973.1). Between VH and VL, a 35-GS repeat region was inserted to create a hinge region which makes VH and VL more flexible. The CH2-CH3 region was directly cloned from Vietnamese blood sample.

2.4. Cell Growth Assay and Recombinant Antibody Expression

In the NPLAd System, NPLAd cells were seeded at 1×10^6 cell / mL in a Shaking flask. Shaking culture was carried out for 24 hours as preculture. After preculture, 2 μL / mL of FuGENE HD transfection reagent (Promega) and 0.4 μg / mL of anti-HER 2 antibody vector were added. Cultured for 4 hours after transfection, 1 mM of valproic acid was added. Shaking culture was continued for 2 weeks at 37°C.

In the FreeStyle System, CHO-s cells were seeded at a cell density of 2×10^5 cells / mL. Transfection was performed by adding 1 μg / mL of Lipofectamine LTX Reagent with PLUS Reagent (Thermo Fisher Scientific) and 0.4 μg / mL of anti-HER 2 antibody vector. After transfection, shaking culture was carried out for 12 days at 37°C.

Samples were taken at regular intervals on both systems and the number of cells and the expression levels of antibody were measured. The cell number was measured by a dye exclusion method with Hemocytometer. Antibody concentration was measured by ELISA using human IgG ELISA quantitation set (Bethyl laboratories inc, USA).

2.5. Purification of Recombinant Antibody

Antibodies expressed by NPLAd System and FreeStyle System were purified by the same method. Culture supernatant was desalted with a Vivaflow 50 ultrafiltration membrane (Sartorius AG, MWCO 30,000) and concentrated to 250 times.

Desalted and concentrated culture supernatant was purified using a commercially available Antibody Purification Kit Protein A (Bio-Rad) according to kits protocol. After affinity purification, the purity and recovered amount of the antibody were measured. The purity of the antibody after purification was calculated by measuring the amount of antibody by ELISA, measuring the protein concentration with a protein assay kit (Bio-Rad), and dividing the amount of antibody by the amount of protein. The recovery rate of the antibody after purification was calculated by dividing the total amount of antibody in the culture supernatant by the total amount of purified antibody.

Purified antibodies were electrophoresed on SDS-PAGE under non-reducing conditions (without DTT). Approximately, 100 ng of purified antibody per lane is supplied to an electrophoresis gel. After electrophoresis, western blotting was performed. Anti-human IgG-HRP was bound to the blotted nitrocellulose membrane, followed by antibody staining with EzWest Blue (ATTO Corporation). Molecular weight was estimated using Marker as Precision Plus Protein Dual Color Standards (MW range from 10 to 150 kD, Bio-Rad).

2.6. ELISA for Binding Activity of Anti-HER2 Antibody

Binding activity of anti-HER2 antibody was determined using human IgG ELISA quantitation set (Bethyl laboratories inc) according to the manufacturer's instructions with some modified. The A431 cells were seeded at density 10^4 cells/well in a 96-well plate at 37°C, 24 h. Next, the cells were washed once with wash solution and fixed with 100 μl 4% formaldehyde at room temperature (rt), 15 min. Then the cells were washed twice with wash solution and blocked with 200 μl blocking solution within 30 min at rt. After removing the blocking solution, cells were washed 5 times and incubated to 100 μl of samples. Cells with no incubation to anti-Her2 antibody were also tested as controls. 100 μl of diluted HRP labeled antibody was added to each well. After 1 hour incubation, cells were washed 5 times. Following 100 μl of TMB substrate solution was added to each well and the plate was developed in the dark for 15 min. The reaction was stopped by adding 100 μl 0.18M H_2SO_4 to each well, absorbance at 450 nm was measured by a plate reader.

3. RESULTS AND DISCUSSION

3.1. The Transfection Efficiency of FS CHO-S and NPLAD Cells

At first, we checked the transfection efficiency of FS CHO-S and NPLAd cells using GFP expression vector. For FS CHO-S cells, Freestyle MAX were used under standard protocol. The transfection efficiency of FS CHO-S was approximately 95% (Fig. **1A**). We also used Freestyle MAX for NPLAd cells, however, transfection efficiency was quite low (data not shown). For NPLAd cells, we checked several kinds of transfection reagents, Lipofectamine LTX & PLUS reagent showed the highest transfection efficiency among these transfection reagents. The transfection efficiency of NPLAd cells using Lipofectamine LTX & PLUS was approximately 21% (Fig. **1B**). The results indicated that the transfection efficiency of NPLAd cells was approximately 4.5-fold lower than that of FS CHO-S. Low transfection efficiency could be due to cell membrane lipid component of NPLAd cells [14].

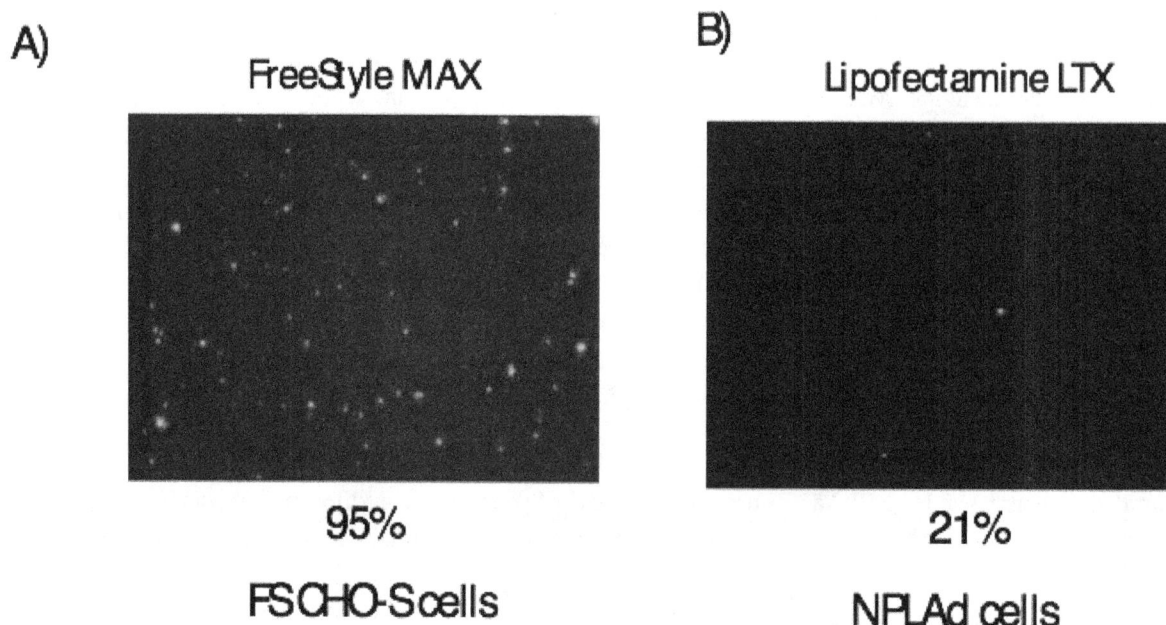

Fig. (1). The transfection efficiency of FS CHO-S (**A**) and NPLAd cells (**B**).

3.2. Preparation of Growth Control Medium

NPLAd medium was found to be good for cell growth, but with low protein expression [13]. We modified the previously reported NPL medium [13] with the aim of maintaining the survival of cells at high density. The improved NPL medium was named NPLTT medium (NPL Medium for Transient Transfection). In NPLTT medium, the concentrations of amino acids were increased as a protein expression resource. The concentrations of sugars were increased as an energy source for cells. Antioxidant reagents, which were ascorbic acid and glutathione were added to suppress cell damage. In addition, 1% Pluronic F-68 Non-ionic Surfactant (Thermo Fisher Scientific Inc.) was added to suppress shock absorption of cells during shaking culture and form cell clumps. In NPLAd system, valproic acid (VPA) was added to medium after transfection to inhibit histone deacetylase activity [15], suppresses gene uptake and cell proliferation [16].

3.3. Comparison of Cell Growth and Antibody Expression Levels in NPLAd and Free Style Systems.

In transient expression system, cell growth was not necessary for protein expression because most of cells were not transfected by expression vector. Keeping this in view, we developed low growth rate medium for antibody expression. To confirm the characterization of our developed expression system, we compared of cell growth and antibody expression levels in NPLAd system to Free style system, which was one of the commercially available useful and high expression systems. The time course of cell numbers is shown in Fig. (**2A**). In the case of Free style system, cell numbers were increased time dependently until 7 days. After that, cell numbers were dramatically decreased. After 12 days, cell numbers were decreased to almost seeding numbers. This cell number decrease could be over cell growth. On the other hand, in case of NPLAd system, cell numbers were not increased until 10 days, after that they were slightly decreased. We seeded cells almost confluent, because NPLTT medium did not support cell growth.

The time course of antibody expression levels is shown in Fig. (**2B**). In the case of Free style system, antibody expression level was increased time dependently from 4 to 7 days, and reached maximum to approximately 2500 ng/ml, after that the expression level was decreased. In NPLAd system, expression level of anti-HER2 antibody was increased until 14 days, and reached maximum to approximately 4500 ng/ml. The results indicated that expression level in NPLAd system was 1.8 times higher than that in Free style system. After 7 days, the expression level in NPLAd system was 1.4 times higher than that in Free style system. In NPLAd system, high numbers of transfected cells could be kept without cell growth within 10 days. These high numbers of cells could express high level of anti-HER2 antibody.

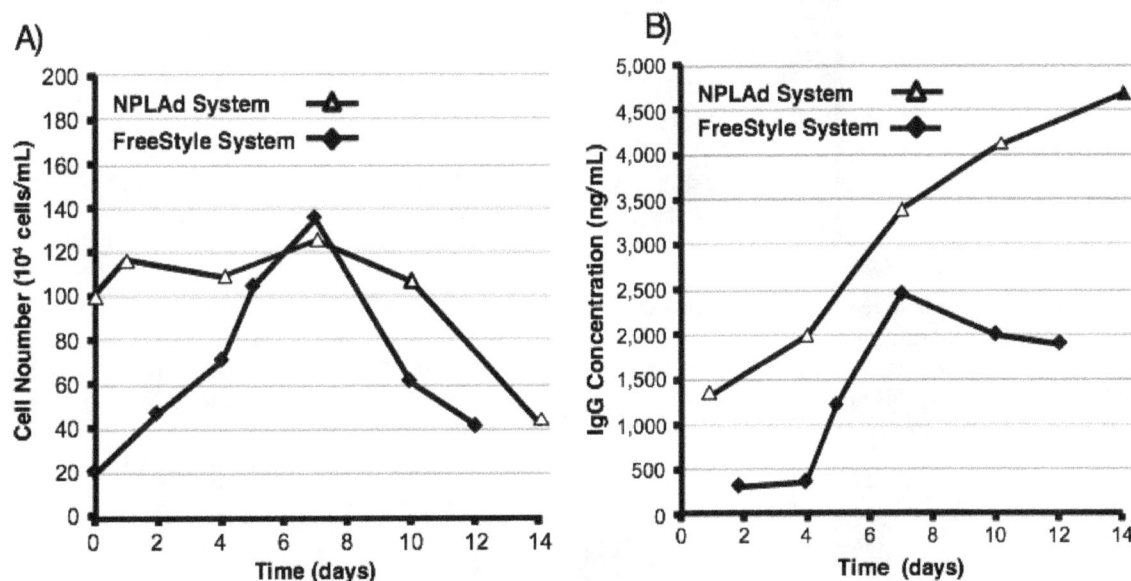

Fig. (2). Comparison of cell growth (**A**) and antibody expression levels (**B**) in NPLAd and Free style systems.

3.4. Purification of Recombinant Antibodies

Antibody recovery rate by Protein A affinity chromatography was 70% or more (NPLAd System 73.3%, FreeStyle System 84.8%), antibody purity was 90% or more (NPLAd System 92%, FreeStyle System 90%). In NPLAd System, NPLTT medium contains almost no protein so antibody protein can be purified to high purity only by concentration and affinity chromatography.

As a result of Western blot after purification, three bands reacting with anti-human IgG were confirmed between approximately 100 and 150 KDa in molecular weight. We estimated that the top band showed full-IgG (147 KDa), the middle band showed IgG without one right chain (122 KDa) and the bottom band showed IgG without two right chains (97 KDa). The migration distance of each bands of antibodies expressed by NPLAd System and FreeStyle System were almost the same (Fig. **3**).

3.5. Comparison of Binding Activities of Recombinant Antibodies

To confirm the antibody performance expressed by NPLAd system, we compared the binding activities of recombinant antibodies with Free style system. The A431 cells, HER2-expressed cell line [17], were seeded in a 96-well plate. And then the anti-HER2 antibodies, which were expressed by NPLAd and Free style system, were added. The binding levels of recombinant antibodies were detected by human IgG ELISA. As shown in Fig. (**4**), binding activities of recombinant antibodies expressed by NPLAd and Free style system were almost the same. The results indicated that the antibody performance expressed by NPLAd system was almost the same to the one expressed by Free style system.

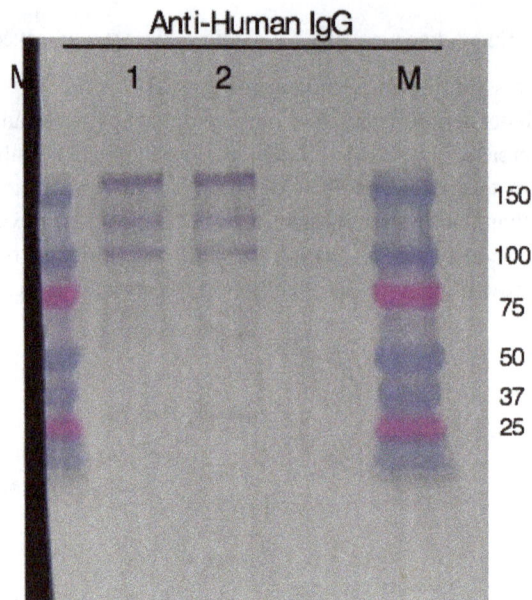

Fig. (3). Western blot analysis of recombinant antibody expressed by NPLAd and Free style systems. In SDS-PAGE, the sample was electrophoresed without reduction with DTT. After electrophoresis, Western blotting and immunostaining with anti-human IgG-HRP were performed. For electrophoresis, molecular weight is estimated using MW 10-150 kDa marker. Lane 1 is purified antibody of NPLAd System (97 ng), lane 2 is purified antibody of Free style system (81 ng). Lane M is the Precision Plus Protein Dual Color Standards Marker, which shows a molecular weight range of 10-150 kDa.

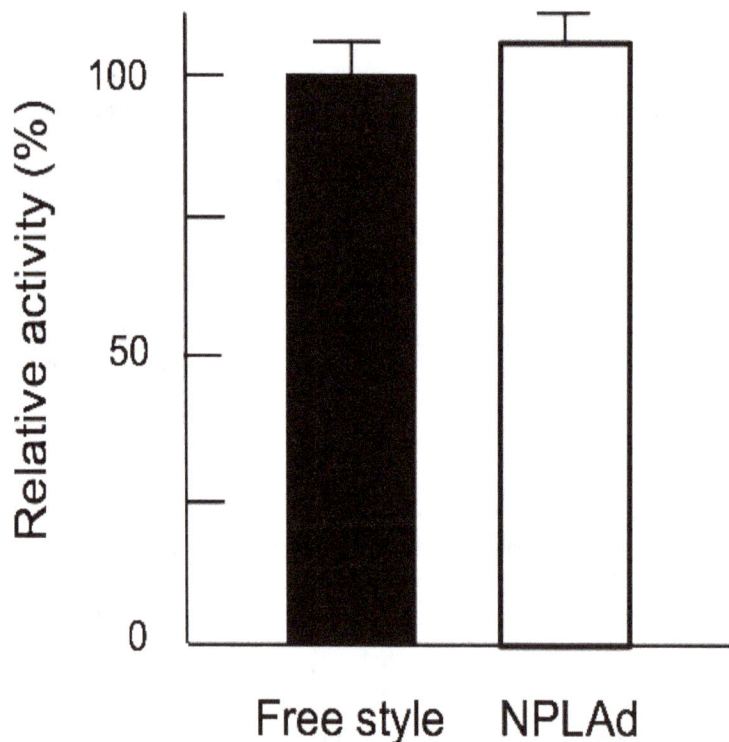

Fig. (4). Relative binding activities of antibodies expressed by NPLAd and Free style systems. 50 ng/ml of antibodies were used for assay. N=3.

CONCLUSION

We developed high antibody expression system namely NPLAd system, which used low growth rate medium. Even at low transfection efficiency, antibody expression level of NPLAd system was 1.4 times higher than that in Free style system. We concluded that our developed growth controlled medium enhanced recombinant antibody expression level. This expression system would be useful for recombinant protein expression.

HUMAN AND ANIMAL RIGHTS

No Animals/Humans were used for studies of this research.

ACKNOWLEDGEMENTS

Declared none.

REFERENCES

[1]   Obradovic M, Mrhar A, Kos M. Market uptake of biologic and small-molecule--targeted oncology drugs in Europe. Clin Ther 2009; 31(12): 2940-52.
        [http://dx.doi.org/10.1016/j.clinthera.2009.12.019] [PMID: 20110034]

[2]   Maggon K. Monoclonal antibody "gold rush". Curr Med Chem 2007; 14(18): 1978-87.
        [http://dx.doi.org/10.2174/092986707781368504] [PMID: 17691940]

[3]   Aggarwal RS. What's fueling the biotech engine-2012 to 2013. Nat Biotechnol 2014; 32(1): 32-9.
        [http://dx.doi.org/10.1038/nbt.2794] [PMID: 24406926]

[4]   Ecker DM, Jones SD, Levine HL. The therapeutic monoclonal antibody market. MAbs 2015; 7(1): 9-14.
        [http://dx.doi.org/10.4161/19420862.2015.989042] [PMID: 25529996]

[5]   Frenzel A, Hust M, Schirrmann T. Expression of recombinant antibodies. Front Immunol 2013; 4: 217.
        [http://dx.doi.org/10.3389/fimmu.2013.00217] [PMID: 23908655]

[6]   Aghajani J, Mirtajani SB, Kojuri SA, Zaheire R, Ayoubi S. Assessment of HIV infection in cells of infected individuals. Banat's J Biotechnol 2017; 8: 24-38.
        [http://dx.doi.org/10.7904/2068-4738-VIII(16)-24]

[7]   Azar OL, Moradi Kor N, Ehsani M, Aiubi S, Rahmani FA. Cytochemical staining for the detection of acute and chronic blood leukemia. Banat's J Biotechnol 2016; 7: 46-52.
        [http://dx.doi.org/10.7904/2068-4738-VII(14)-46]

[8]   Kelley B. Industrialization of mAb production technology: The bioprocessing industry at a crossroads. MAbs 2009; 1(5): 443-52.
        [http://dx.doi.org/10.4161/mabs.1.5.9448] [PMID: 20065641]

[9]   Kunert R, Reinhart D. Advances in recombinant antibody manufacturing. Appl Microbiol Biotechnol 2016; 100(8): 3451-61.
        [http://dx.doi.org/10.1007/s00253-016-7388-9] [PMID: 26936774]

[10]   Ahmadi S, Davami F, Davoudi N, et al. Monoclonal antibodies expression improvement in CHO cells by PiggyBac transposition regarding vectors ratios and design. PLoS One 2017; 12(6): e0179902.
        [http://dx.doi.org/10.1371/journal.pone.0179902] [PMID: 28662065]

[11]   Sunstrom NA, Gay RD, Wong DC, Kitchen NA, DeBoer L, Gray PP. Insulin-like growth factor-I and transferrin mediate growth and survival of Chinese hamster ovary cells. Biotechnol Prog 2000; 16(5): 698-702.
        [http://dx.doi.org/10.1021/bp000102t] [PMID: 11027159]

[12]   Fisher DA. Epidermal growth factor in the developing mammal. Mead Johnson Symp Perinat Dev Med 1988; 33-40.
        [PMID: 2485437]

[13]   Sasaki T, Taniguchi A. Development of a non-protein and lipid medium adopted cell line for biopharmaceutical recombinant protein expression. Open Biotechnol J 2013; 7: 1-6.
        [http://dx.doi.org/10.2174/1874070701307010001]

[14]   Sandbichler AM, Aschberger T, Pelster B. A method to evaluate the efficiency of transfection reagents in an adherent zebrafish cell line. Biores Open Access 2013; 2(1): 20-7.
        [http://dx.doi.org/10.1089/biores.2012.0287] [PMID: 23515475]

[15]   Phiel CJ, Zhang F, Huang EY, Guenther MG, Lazar MA, Klein PS. Histone deacetylase is a direct target of valproic acid, a potent anticonvulsant, mood stabilizer, and teratogen. J Biol Chem 2001; 276(39): 36734-41.
        [http://dx.doi.org/10.1074/jbc.M101287200] [PMID: 11473107]

[16]   Wulhfard S, Baldi L, et al. Valproic acid enhances recombinant mRNA and protein levels in transiently transfected Chinese hamster ovary cells. J Biotechnol 2010; 20; 148(2-3): 128-32.

[17]   McCluskey AJ, Olive AJ, Starnbach MN, Collier RJ. Targeting HER2-positive cancer cells with receptor-redirected anthrax protective antigen. Mol Oncol 2013; 7(3): 440-51.
        [http://dx.doi.org/10.1016/j.molonc.2012.12.003] [PMID: 23290417]

4

Experimental Study of Thermal Restraint in Bio-Protectant Disaccharides by FTIR Spectroscopy

S. Magazù[1,2,*], E. Calabrò[1] and M.T. Caccamo[1,2]

[1]*Department of Mathematical and Informatics Sciences, Physical Sciences and Earth Sciences of Messina University, Viale Ferdinando Stagno D' Alcontres 31, 98166 Messina, Italy*
[2]*Istituto Nazionale di Alta Matematica "F. Severi" – INDAM - Gruppo Nazionale per la Fisica Matematica – GNFM, Messina, Italy*

Abstract:

Background:

In the present paper, InfraRed (IR) spectra on water mixtures of two homologous disaccharides, *i.e.* sucrose and trehalose, as a function of temperature have been collected.

Methods:

In particular, IR spectra were registered, in the spectral range from 4000 cm^{-1} to 400 cm^{-1}, to investigate the thermal response of the water mixtures of two homologous disaccharides, through positive thermal scans, *i.e.* by increasing the temperature from the value of 25°C to the value of 50°C. The OH-stretching region has been analyzed by means of two simple and straightforward procedures, *i.e.* by evaluating the shift of the intramolecular OH stretching center frequency and the Spectral Distance (SD).

Result and Conclusion:

Both the analyses indicate that trehalose water mixture have a higher thermal response than that of the sucrose-water mixture.

Keywords: Sucrose, Trehalose, Bio-protection, Temperature, FTIR spectroscopy, Infrared.

1. INTRODUCTION

It is well known that homologous disaccharides i. e. sucrose and trehalose although have the same chemical formula ($C_{12}H_{22}O_{11}$), they present different bio-protective properties. In particular, trehalose shows a higher bio-protective effectiveness in comparison with sucrose, playing a key role in cryptobiotic-activating substances [1 - 11]. Sucrose is constituted by a glucose ring in the α configuration and a fructose ring in the β configuration; the α and β structures of the same monosaccharide differ only in the orientation of the OH groups at some carbon atom in the ring itself [12 - 18]. Trehalose is a disaccharide of glucose constituted by two pyranose rings in the same configuration, linked by a glycosidic bond between the chiral carbon atoms C1 of the two rings [19 - 28].

In Fig. (**1**), the chemical structures of sucrose and trehalose are reported.

More precisely, trehalose, due to its bioprotective properties allows to many organisms, which synthesize it to undergo in a state of "suspended life" and to re-activate the vital functions when the external conditions come back favourable [29 - 42].

---

[*] Address correspondence to this author at the Department of Mathematical and Informatics Sciences, Physical Sciences and Earth Sciences of Messina University, Viale Ferdinando Stagno D' Alcontres 31, 98166 Messina, Italy; E-mail: smagazu@unime.it

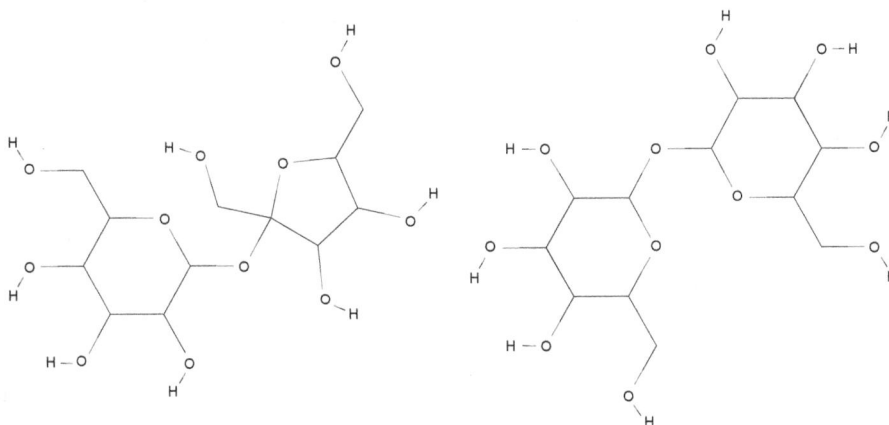

Fig. (1). Chemical structure of sucrose (left) and chemical structure of trehalose (right).

Thus, such organisms can resist dehydration by means of the formation of a glass of intracellular carbohydrates, that, with its high viscosity and hydrogen bonding, is able to stabilize and protect the integrity of molecules. For these reasons, the trehalose can be employed to preserve red blood cells [43 - 55].

It is should be stressed that trehalose possesses an unusual ability to stabilize biomolecules since it is able to stabilize the structure without interfering with the functionality of the protein [56 - 64].

Furthermore, the thermal properties of bioprotectant systems have been the subject of attention not only for the physical, biological and chemical research, but also for the food products and for the cultural heritage field [65 - 74]. Although a lot of studies have been addressed the anhydrobiosis and biopreservation phenomena; the molecular mechanisms are not yet clear [75 - 87]. As far as the employed spectroscopic technique is concerned, in our case, IR allows to characterize the system vibrational motions and hence the system structural arrangements and their changes with temperature. In this study, the Mid IR (MIR) region spanning from 4000 cm^{-1} to 400 cm^{-1} has been analyzed [88 - 92]. In particular, to highlight the thermal properties of these disaccharides mixtures, two straightforward approaches have been performed: the shift of the OH- stretching region (3700 - 3000cm^{-1}) as a function of temperature and the analysis of the temperature dependence of the Spectral Distance (SD). From the first approach, it is possible to see changes in the spectral band position while the SD reflects the structural changes in disaccharides mixtures. Both the approaches suggest that trehalose water mixture has a higher thermal restraint than sucrose-water mixture [93 - 97].

2. EXPERIMENTAL SETUP AND SAMPLE PREPARATION

Sucrose and trehalose were purchased from Aldrich-Chemie. Infrared spectra were collected in the 25÷50°C temperature range, by means of the FTIR Vertex 70 V spectrometer (Bruker Optics) using Platinum diamond ATR. The investigated concentration value, expressed as weight fraction, *i.e.* (grams of disaccharide)/(grams of disaccharide+ grams of water), was: 0.30. For each spectrum, 32 scans were repeated at a resolution of 4 cm^{-1} for the spectral range between 4000 and 400 cm^{-1}. Such experimental conditions were kept constant for all the measurements. First to proceed with the spectra analysis, data pre-processing has been performed:

1. baseline treatment
2. smoothing treatment
3. first and second derivatives
4. normalization of spectra

Then, spectral analysis has been performed by means of the software MATLAB 2016a (Mathworks, Natick, USA) and the software Origin 9 (OriginLab Co., Northampton, USA).

3. EXPERIMENTAL DATA RESULTS AND DISCUSSION

Fig. (2a) reports, as an example, the IR spectra in the OH-stretching region (3700 $<\Delta\omega<$ 3000 cm^{-1} spectral range) for different values of temperature, *i.e.* T=25°C, 30°C, 35°C, 40°C, 45°C and 50°C for sucrose water mixture while Fig. (2b) shows, as an example, the IR spectra in the OH-stretching region (3700 $<\Delta\omega<$ 3000 cm^{-1} spectral range) for different values of temperature, *i.e.* T=25°C, 30°C, 35°C, 40°C, 45°C and 50°C for trehalose water mixture.

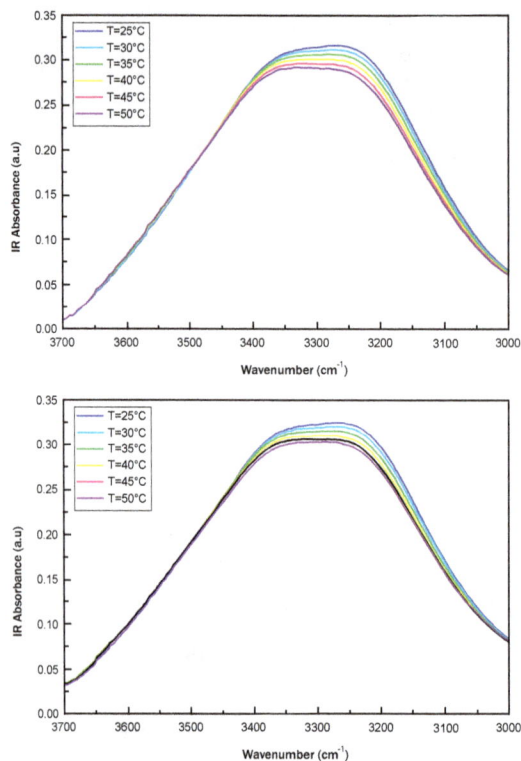

Fig. (2). a) OH-stretching region (3700 <Δω< 3000 cm-1 spectral range) for different values of temperature, *i.e.* T=25°C, 30°C, 35°C, 40°C, 45°C and 50°C for sucrose water mixture and **b**) OH-stretching region (3700 <Δω< 3000 cm-1 spectral range) for different values of temperature, *i.e.* T=25°C, 30°C, 35°C, 40°C, 45°C and 50°C for trehalose water mixture.

To quantitatively characterizethe thermal response of the IR spectra, we have first evaluated the OH stretching center frequency shift and then the spectral distance.

In Fig. (**3**) the trend of the OH stretching center frequency shift for sucrose water mixture and for trehalose water mixture is reported. As it can be seen the sucrose in the temperature range 25,0°C<ΔT<50,0°C changes the intramolecular band center frequency from 3271 cm^{-1} to 3310 cm^{-1}, while trehalose changes the intramolecular band center frequency from 3264 cm^{-1} to 3285 cm^{-1}.

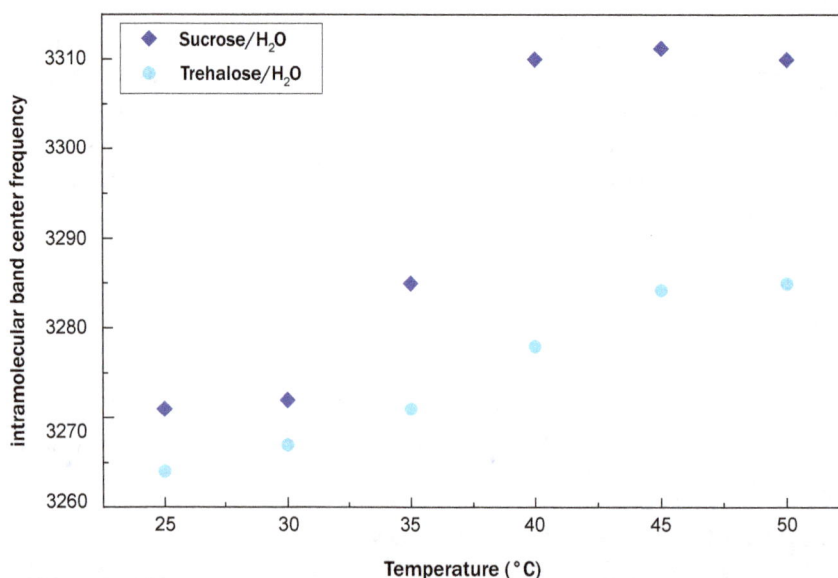

Fig. (3). Trend of the OH stretching center frequency shift for sucrose water mixture and for trehalose water mixture.

Such a result suggests a higher thermal restraint for trehalose water mixture than sucrose water mixture. In this framework, in order to extract a quantitative value and to perform a comparison between the two homologous disaccharides, an interpretative model making reference to the well known logistic function has been applied; such a procedure allows to characterize the thermal restraint of the system:

$$S(T) = A \left(1 - \frac{1}{1+e^{-B(T-T_0)}} \right) + (C - DT)$$

(1)

where A is the relaxation amplitude, B represents the relaxation stepness, T is the temperature value of the sigmoid inflection point and the contribution C-DT takes into account the low-temperature contribution. Furthermore, it is important to remember that the inverse of the relaxation amplitude is connected with the thermal restraint of the system. In Fig. (**4**), the fit, performed by means of equation (1), of OH-stretching band center frequency as a function of temperature for sucrose water mixture is shown.

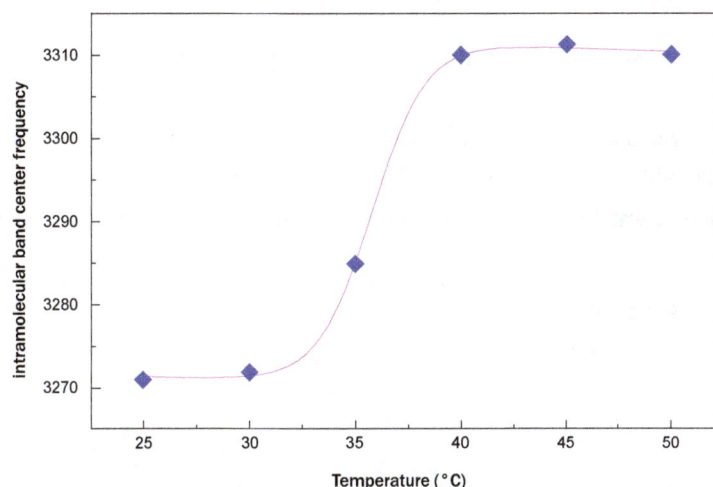

Fig. (4). Intramolecular band center frequency versus temperature for sucrose water mixture. Blue diamonds: Experimental data; magenta continuous line: Fitting curve obtained by equation (1).

Fig. (**5**) reports the fit, performed by means of equation (1), of the OH-stretching band center frequency as a function of temperature for trehalose water mixture.

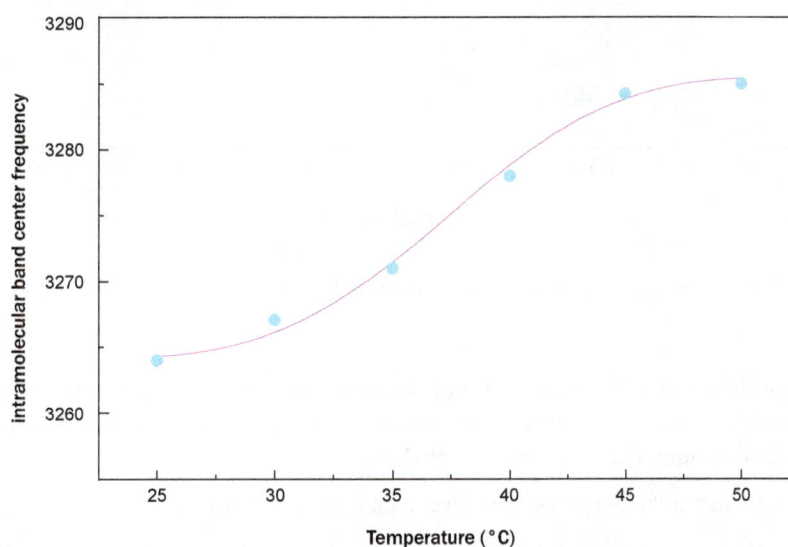

Fig. (5). Intramolecular band center frequency versus temperature for trehalose water mixture. Light blue circles: Experimental data; magenta continuous line: Fitting curve by equation (1).

What emerges from this analysis is that the data arrange themselves along an increasing sigmoidal curve providing a higher thermal restraint, together with a higher T_0 value for trehalose water mixture in respect to sucrose water mixture. More precisely, the value of thermal restraint for the sucrose-water mixture is equal to 0.024 that is lower than that of trehalose water solution whose thermal restraint value is 0.031. As far as the crossover temperature value is concerned, the value of trehalose water mixture, 37.32°C, is higher than that of sucrose water mixture, which is equal to 35.77°C. These results show that dynamics of trehalose water mixture is slower than sucrose water mixture; the trehalose has a low sensitivity to temperature changes.

In order to catch small changes in intramolecular band center frequency, the SD has been calculated by means of the following expression:

$$SD = \left(\sum_v [I(\omega T) - I(\omega, T = 25°C]^2 \Delta\omega\right)^{1/2}$$
(2)

where I(ω,t) represents the absorbance as a function of frequency ω and temperature T, while Δω is the frequency resolution instrument.

Such a method is able to evaluate the difference in the spectra collected at different temperatures from the spectrum measured at the lowest temperature, that in our case, equal to 25 °C.

In Fig. (6), the SD of sucrose water mixture (blue diamond) and trehalose water mixture (light blue circle) is shown.

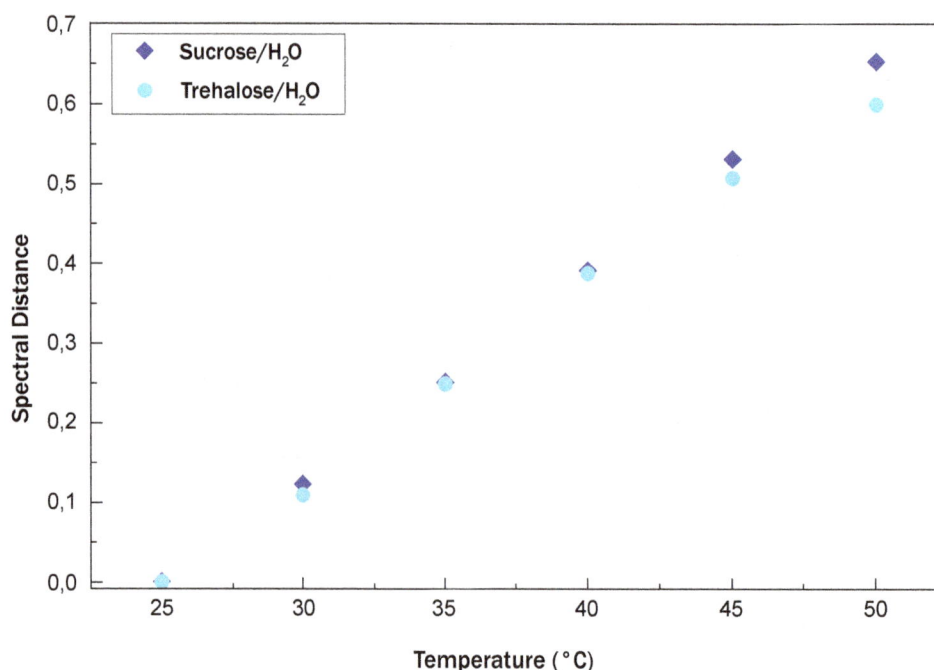

Fig. (6). SD evaluated in the center frequency of sucrose water mixture (blue diamond) and of trehalose water mixture (light blue circle).

As it can be seen that the trend is linear; in order to obtain a quantitative value for the thermal response of the system, we have calculated the straight line slope. For the sucrose-water mixture, the straight line slope value is 0.02648 while for the trehalose water mixture, the slope value is equal to 0.02244.

This latter value suggests that trehalose is less sensitive to changes in temperature.

CONCLUSION

The aim of this work was to characterize the thermal response for two homologous disaccharides, *i.e.* sucrose and trehalose in water mixtures by FTIR spectroscopy. The spectral range investigated spanned from 4000 cm^{-1} to 400 cm^{-1} with the intramolecular OH vibrational contribution, covering the spectral range from 3000 to 3800 cm^{-1}, in the temperature range between 25°C and 50°C.

The registered spectra have been analyzed by means of two straightforward approaches, *i.e.* by evaluating both the shift of the intramolecular OH stretching center frequency and the Spectral Distance (SD). Although the two different approaches are simple, both the analyses furnish the same results confirming that trehalose water mixture has a higher thermal restraint with respect to the sucrose-water mixture.

The agreement between the results obtained by the two different approaches provides coherent information on the thermal stability of the disaccharide water mixtures investigated.

HUMAN AND ANIMAL RIGHTS

No animals/humans were used for studies that are the basis of this research.

ACKNOWLEDGEMENTS

Declared none.

REFERENCES

[1]   Crowe JH, Carpenter JF, Crowe LM. The role of vitrification in anhydrobiosis. Annu Rev Physiol 1998; 60: 73-103.
        [http://dx.doi.org/10.1146/annurev.physiol.60.1.73] [PMID: 9558455]

[2]   Crowe JH, Crowe LM, Oliver AE, Tsvetkova N, Wolkers W, Tablin F. The trehalose myth revisited: Introduction to a symposium on stabilization of cells in the dry state. Cryobiology 2001; 43(2): 89-105.
        [http://dx.doi.org/10.1006/cryo.2001.2353] [PMID: 11846464]

[3]   Green JL, Angell CA. Phase relations and vitrification in saccharide-water solutions and the trehalose anomaly. J Phys Chem 1989; 93: 2880-2.
        [http://dx.doi.org/10.1021/j100345a006]

[4]   Magazù S, Migliardo F, Benedetto A, Calabrò E, La Torre R, Caccamo MT. Bioprotective effects of sucrose and trehalose on proteins in Sucrose: Properties. Biosynthesis and Health Implications 2013; pp. 35-46.

[5]   Chen T, Fowler A, Toner M. Literature review: Supplemented phase diagram of the trehalose-water binary mixture. Cryobiology 2000; 40(3): 277-82.
        [http://dx.doi.org/10.1006/cryo.2000.2244] [PMID: 10860627]

[6]   Kandror O, DeLeon A, Goldberg AL. Trehalose synthesis is induced upon exposure of *Escherichia coli* to cold and is essential for viability at low temperatures. Proc Natl Acad Sci USA 2002; 99(15): 9727-32.
        [http://dx.doi.org/10.1073/pnas.142314099] [PMID: 12105274]

[7]   Baptista RP, Pedersen S, Cabrita GJ, Otzen DE, Cabral JM, Melo EP. Thermodynamics and mechanism of cutinase stabilization by trehalose. Biopolymers 2008; 89(6): 538-47.
        [http://dx.doi.org/10.1002/bip.20926] [PMID: 18213692]

[8]   Sola-Penna M, Meyer-Fernandes JR. Stabilization against thermal inactivation promoted by sugars on enzyme structure and function: Why is trehalose more effective than other sugars? Arch Biochem Biophys 1998; 360(1): 10-4.
        [http://dx.doi.org/10.1006/abbi.1998.0906] [PMID: 9826423]

[9]   Magazù S, Migliardo F, Parker SF. Vibrational properties of bioprotectant mixtures of trehalose and glycerol. J Phys Chem B 2011; 115(37): 11004-9.
        [http://dx.doi.org/10.1021/jp205599a] [PMID: 21830802]

[10]   Iannelli E, Tettamanti E, Galantini L, Magazù S. An integrated quasi-elastic light-scattering, pulse-gradient-spin-echo study on the transport properties of α,α-trehalose, sucrose, and maltose deuterium oxide solutions. J Phys Chem B 2001; 105(48): 12143-9.
        [http://dx.doi.org/10.1021/jp011275j]

[11]   Tanaka M, Machida Y, Niu S, *et al.* Trehalose alleviates polyglutamine-mediated pathology in a mouse model of Huntington disease. Nat Med 2004; 10(2): 148-54.
        [http://dx.doi.org/10.1038/nm985] [PMID: 14730359]

[12]   Magazù S, Migliardo F, Mondelli C, Vadalà M. Correlation between bioprotective effectiveness and dynamic properties of trehalose-water, maltose-water and sucrose-water mixtures. Carbohydr Res 2005; 340(18): 2796-801.
        [http://dx.doi.org/10.1016/j.carres.2005.09.026] [PMID: 16246316]

[13]   Lerbret A, Bordat P, Affouard F, Descamps M, Migliardo F. How homogeneous are the trehalose, maltose, and sucrose water solutions? An insight from molecular dynamics simulations. The Journal of Physical Chemistry B 2005; 109(21): 11046-57.

[14]   Magazù S, Migliardo F, Mondelli C, Romeo G. Temperature dependence of mean square displacement by IN13: A comparison between trehalose and sucrose water mixtures. Chem Phys 2003; 292: 247-51.
        [http://dx.doi.org/10.1016/S0301-0104(03)00101-0]

[15]   O'Brien J. Stability of trehalose, sucrose and glucose to nonenzymatic browning in model systems. J Food Sci 1996; 61: 679-82.
        [http://dx.doi.org/10.1111/j.1365-2621.1996.tb12180.x]

[16]   Liu Q, Schmidt RK, Teo B, Karplus PA, Bray JW. Molecular dynamics studies of the hydration of α,α-trehalose. J Am Chem Soc 1997; 119: 7851-62.
        [http://dx.doi.org/10.1021/ja970798v]

[17]   Affouard F, Bordat P, Descamps M, et al. A combined neutron scattering and simulation study on bioprotectant systems. Chem Phys 2005; 317: 258-66.
        [http://dx.doi.org/10.1016/j.chemphys.2005.05.033]

[18]   Arai C, Kohguchi M, Akamatsu S, et al. Trehalose suppresses lipopolysaccharide-induced osteoclastogenesis bone marrow in mice. Nutr Res 2001; 21(7): 993-9.
        [http://dx.doi.org/10.1016/S0271-5317(01)00315-3] [PMID: 11446983]

[19]   Minutoli L, Altavilla D, Bitto A, et al. Trehalose: A biophysics approach to modulate the inflammatory response during endotoxic shock. Eur J Pharmacol 2008; 589(1-3): 272-80.
        [http://dx.doi.org/10.1016/j.ejphar.2008.04.005] [PMID: 18555988]

[20]   Pagnotta SE, Ricci MA, Bruni F, McLain S, Magazù S. Water structure around trehalose. Chem Phys 2008; 345(2-3): 159-63.
        [http://dx.doi.org/10.1016/j.chemphys.2007.07.052]

[21]   Magazù S, Migliardo F, Telling MTF. Structural and dynamical properties of water in sugar mixtures. Food Chem 2008; 106(4): 1460-6.
        [http://dx.doi.org/10.1016/j.foodchem.2007.05.097]

[22]   Magazù S, Migliardo F, Barreca D, Bellocco E, Laganà G. Aggregation processes of biomolecules in presence of trehalose. J Mol Struct 2007; 840: 114-8.
        [http://dx.doi.org/10.1016/j.molstruc.2006.11.028]

[23]   Calabrò E, Magazù S. On the bioprotective effectiveness of trehalose against man-made electromagnetic fields. Adv Chem Res 2015; 24: 25-50.

[24]   Magazù S, Migliardo F, Ramirez-Cuesta AJ. Concentration dependence of vibrational properties of bioprotectant/water mixtures by inelastic neutron scattering. J R Soc Interface 2007; 4(12): 167-73.
        [http://dx.doi.org/10.1098/rsif.2006.0162] [PMID: 17018423]

[25]   Magazù S, Migliardo F, Telling MTF. Study of the dynamical properties of water in disaccharide solutions. Eur Biophys J 2007; 36(2): 163-71.
        [http://dx.doi.org/10.1007/s00249-006-0108-0] [PMID: 17109123]

[26]   Magazù S, Migliardo F, Ramirez-Cuesta AJ. Changes in vibrational modes of water and bioprotectants in solution. Biophys Chem 2007; 125(1): 138-42.
        [http://dx.doi.org/10.1016/j.bpc.2006.07.003] [PMID: 16887256]

[27]   Minutoli L, Altavilla D, Bitto A, et al. The disaccharide trehalose inhibits proinflammatory phenotype activation in macrophages and prevents mortality in experimental septic shock. Shock 2007; 27(1): 91-6.
        [http://dx.doi.org/10.1097/01.shk.0000235092.76292.bc] [PMID: 17172986]

[28]   Magazù S, Migliardo F, Telling MTF. α,α-trehalose-water solutions. VIII. Study of the diffusive dynamics of water by high-resolution quasi elastic neutron scattering. J Phys Chem B 2006; 110(2): 1020-5.
        [http://dx.doi.org/10.1021/jp0536450] [PMID: 16471637]

[29]   Branca C, Maccarrone S, Magazù S, Maisano G, Bennington SM, Taylor J. Tetrahedral order in homologous disaccharide-water mixtures. J Chem Phys 2005; 122(17): 174513.
        [http://dx.doi.org/10.1063/1.1887167] [PMID: 15910051]

[30]   Magazù S, Migliardo F, Ramirez-Cuesta AJ. Inelastic neutron scattering study on bioprotectant systems. J R Soc Interface 2005; 2(5): 527-32.
        [http://dx.doi.org/10.1098/rsif.2005.0059] [PMID: 16849211]

[31]   Crowe LM. Lessons from nature: The role of sugars in anhydrobiosis. Comp Biochem Physiol A Mol Integr Physiol 2002; 131(3): 505-13.
        [http://dx.doi.org/10.1016/S1095-6433(01)00503-7] [PMID: 11867276]

[32]   Magazù S, Maisano G, Migliardo F, Mondelli C. α,α-trehalose/water solutions. VII: An elastic incoherent neutron scattering study on fragility. J Phys Chem B 2004; 108(36): 13580-5.
        [http://dx.doi.org/10.1021/jp035973a]

[33]   Magazù S, Maisano G, Migliardo F, Mondelli C. Mean-square displacement relationship in bioprotectant systems by elastic neutron scattering. Biophys J 2004; 86(5): 3241-9.
        [http://dx.doi.org/10.1016/S0006-3495(04)74372-6] [PMID: 15111437]

[34]   Magazu S, Migliardo F, Mondelli C. Elastic incoherent neutron scattering from homologous disaccharides/H2O mixtures. J Chem Phys 2004; 119(24): 13033-8.
[http://dx.doi.org/10.1063/1.1625369]

[35]   Abate L, Blanco I, Branca C, et al. Homologous disaccharide properties at low temperatures. J Mol Liq 2003; 103-104(SPEC.): 177-80.
[http://dx.doi.org/10.1016/S0167-7322(02)00137-X]

[36]   Magazù S, Migliardo F, Vertessy BG, Caccamo MT. Investigations of homologous disaccharides by elastic incoherent neutron scattering and wavelet multiresolution analysis. Chem Phys 2013; 424: 56-61.
[http://dx.doi.org/10.1016/j.chemphys.2013.05.004]

[37]   Migliardo F, Caccamo MT, Magazù S. Elastic incoherent neutron scatterings wavevector and thermal analysis on glass-forming homologous disaccharides. J Non-Cryst Solids 2013; 378: 144-51.
[http://dx.doi.org/10.1016/j.jnoncrysol.2013.06.030]

[38]   Marchese N, Cannuli A, Caccamo MT, Pace C. 2016 New generation non-stationary portable neutron generators for biophysical applications of neutron activation analysis Biochimica et Biophysica Acta (BBA) BBAGEN-28495 1861 Issue 1 Part B 3661-3670.
[http://dx.doi.org/10.1016/j.bbagen.2016.05.023.]

[39]   Magazù S, Villari V, Faraone A, Maisano G, Heenan RK, King S. α,α-trehalose-water solutions VI. A view of the structural and dynamical properties of OβG micelles in the presence of trehalose. J Phys Chem B 2002; 106(27): 6954-60.
[http://dx.doi.org/10.1021/jp020698y]

[40]   Branca C, Magazù S, Maisano G, Migliardo F. Vibrational and relaxational contributions in disaccharide/H2O glass formers. Phys Rev B 2001; 64(22): 224204.
[http://dx.doi.org/10.1103/PhysRevB.64.224204]

[41]   Elbein AD, Pan YT, Pastuszak I, Carroll D. New insights on trehalose: A multifunctional molecule. Glycobiology 2003; 13(4): 17R-27R.
[http://dx.doi.org/10.1093/glycob/cwg047] [PMID: 12626396]

[42]   Garg AK, Kim J-K, Owens TG, et al. Trehalose accumulation in rice plants confers high tolerance levels to different abiotic stresses. Proc Natl Acad Sci USA 2002; 99(25): 15898-903.
[http://dx.doi.org/10.1073/pnas.252637799] [PMID: 12456878]

[43]   Higashiyama T. Novel functions and applications of trehalose. Pure Appl Chem 2002; 74: 1263-9.
[http://dx.doi.org/10.1351/pac200274071263]

[44]   Kilburn D, Townrow S, Meunier V, Richardson R, Alam A, Ubbink J. Organization and mobility of water in amorphous and crystalline trehalose. Nat Mater 2006; 5(8): 632-5.
[http://dx.doi.org/10.1038/nmat1681] [PMID: 16845422]

[45]   Chen Q, Haddad GG. Role of trehalose phosphate synthase and trehalose during hypoxia: From flies to mammals. J Exp Biol 2004; 207(Pt 18): 3125-9.
[http://dx.doi.org/10.1242/jeb.01133] [PMID: 15299033]

[46]   Magazù S, Migliardo F, Benedetto A, Mondelli C, Gonzalez MA. Thermal behaviour of hydrated lysozyme in the presence of sucrose and trehalose by EINS. J Non-Cryst Solids 2011; 357(2): 664-70.
[http://dx.doi.org/10.1016/j.jnoncrysol.2010.06.075]

[47]   Migliardo F, Caccamo MT, Magazù S. Thermal analysis on bioprotectant disaccharides by elastic incoherent neutron scattering. Food Biophys 2014; 9(2): 99-104.
[http://dx.doi.org/10.1007/s11483-013-9322-3]

[48]   Caccamo MT, Cannuli A, Calabrò E, Magazù S. Acoustic levitator power device: Study of ethylene-glycol water mixtures IOP conference series: Materials science and engineering. 2017; 199.(1) Article number: 012119
[http://dx.doi.org/10.1088/1757-899X/199/1/012119]

[49]   Magazu S, Villari V, Migliardo P, Maisano G, Telling MTF. Diffusive dynamics of water in the presence of homologous disaccharides: A comparative study by quasi elastic neutron scattering. IV J Phys Chem B 2001; 105(9): 1851-5.
[http://dx.doi.org/10.1021/jp002155z]

[50]   Branca C, Magazù S, Maisano G, Migliardo P. Experimental study of the hydration properties of homologous disaccharides. J Biol Phys 2000; 26(4): 295-306.
[http://dx.doi.org/10.1023/A:1010344122760] [PMID: 23345728]

[51]   Ballone P, Marchi M, Branca C, Magazú S. Structural and vibrational properties of trehalose: A density functional study. J Phys Chem B 2000; 104(26): 6313-7.
[http://dx.doi.org/10.1021/jp994346b]

[52]   Lokotosh TV, Magazù S, Maisano G, Malomuzh NP. Nature of self-diffusion and viscosity in supercooled liquid water. Phys Rev E Stat Phys Plasmas Fluids Relat Interdiscip Topics 2000; 62(3 Pt A): 3572-80.
[http://dx.doi.org/10.1103/PhysRevE.62.3572] [PMID: 11088858]

[53]   Varga B, Migliardo F, Takacs E, Vertessy B, Magazù S, Mondelli C. Neutron scattering studies on dUTPase complex in the presence of bioprotectant systems. Chem Phys 2008; 345(2-3): 250-8.
[http://dx.doi.org/10.1016/j.chemphys.2007.07.004]

[54]   Hennet L, Cristiglio V, Kozaily J, *et al.* Aerodynamic levitation and laser heating: Applications at synchrotron and neutron sources. Eur Phys J Spec Top 2011; 196(1): 151-65.
[http://dx.doi.org/10.1140/epjst/e2011-01425-0]

[55]   Branca C, Magazù S, Migliardo F, Migliardo P. Destructuring effect of trehalose on the tetrahedral network of water: A Raman and neutron diffraction comparison. Physica A 2002; 304(1-2): 314-8.
[http://dx.doi.org/10.1016/S0378-4371(01)00517-9]

[56]   Magazù S, Migliardo F, Benedetto A. Mean square displacements from elastic incoherent neutron scattering evaluated by spectrometers working with different energy resolution on dry and hydrated (H2O and D2O) lysozyme. J Phys Chem B 2010; 114(28): 9268-74.
[http://dx.doi.org/10.1021/jp102436y] [PMID: 20575549]

[57]   Magazù S, Calabrò E, Caccamo MT, Cannuli A. The shielding action of disaccharides for typical proteins in aqueous solution against static, 50 Hz and 1800 MHz frequencies electromagnetic fields. Curr Chem Biol 2016; 10(1)
[http://dx.doi.org/10.2174/2212796810666160419153722]

[58]   Jovanović N, Bouchard A, Hofland GW, Witkamp GJ, Crommelin DJ, Jiskoot W. Distinct effects of sucrose and trehalose on protein stability during supercritical fluid drying and freeze-drying. Eur J Pharm Sci 2006; 27(4): 336-45.
[http://dx.doi.org/10.1016/j.ejps.2005.11.003] [PMID: 16338123]

[59]   Calabrò E, Condello S, Currò M, *et al.* Effects of low intensity static magnetic field on FTIR spectra and ROS production in SH-SY5Y neuronal-like cells. Bioelectromagnetics 2013; 34(8): 618-29.
[http://dx.doi.org/10.1002/bem.21815] [PMID: 24217848]

[60]   Barreca D, Laganà G, Ficarra S, *et al.* Spectroscopic determination of lysozyme conformational changes in the presence of trehalose and guanidine. Cell Biochem Biophys 2013; 66(2): 297-307.
[http://dx.doi.org/10.1007/s12013-012-9485-4] [PMID: 23184705]

[61]   Barreca D, Laganà G, Ficarra S, *et al.* Anti-aggregation properties of trehalose on heat-induced secondary structure and conformation changes of bovine serum albumin. Biophys Chem 2010; 147(3): 146-52.
[http://dx.doi.org/10.1016/j.bpc.2010.01.010] [PMID: 20171005]

[62]   Barreca D, Laganà G, Bruno G, Magazù S, Bellocco E. Diosmin binding to human serum albumin and its preventive action against degradation due to oxidative injuries. Biochimie 2013; 95(11): 2042-9.
[http://dx.doi.org/10.1016/j.biochi.2013.07.014] [PMID: 23886889]

[63]   Magazù S, Calabrò E, Campo S, Interdonato S. New insights into bioprotective effectiveness of disaccharides: an FTIR study of human haemoglobin aqueous solutions exposed to static magnetic fields. J Biol Phys 2012; 38(1): 61-74.
[http://dx.doi.org/10.1007/s10867-010-9209-1] [PMID: 23277670]

[64]   Branca C, Magazù S, Maisano G, Migliardo P, Villari V, Sokolov AP. The fragile character and structure-breaker role of α, α-trehalose: Viscosity and Raman scattering findings. J Phys Condens Matter 1999; 11(19): 3823-32.
[http://dx.doi.org/10.1088/0953-8984/11/19/305]

[65]   Branca C, Magazù S, Maisano G, Migliardo P, Tettamanti E. On the bioprotective effectiveness of trehalose: Ultrasonic technique, Raman scattering and NMR investigations. J Mol Struct 1999; 480-481: 133-40.
[http://dx.doi.org/10.1016/S0022-2860(98)00626-7]

[66]   Branca C, Magazù S, Maisano G, Migliardo P. α,α-trehalose-water solutions. 3. Vibrational dynamics studies by inelastic light scattering. J Phys Chem B 1999; 103(8): 1347-53.
[http://dx.doi.org/10.1021/jp983470c]

[67]   Magazù S, Maisano G, Migliardo P, Middendorf HD, Villari V. Hydration and transport properties of aqueous solutions of α-α-trehalose. J Chem Phys 1998; 109(3): 1170-4.
[http://dx.doi.org/10.1063/1.476662]

[68]   Magazu S, Maisano G, Middendorf HD, Migliardo P, Musolino M, Villari V. α,α--trehalose-water solutions. II. Influence of hydrogen bond connectivity on transport properties. J Phys Chem B 1998; 102(11): 2060-3.
[http://dx.doi.org/10.1021/jp980235l]

[69]   Duong T, Barrangou R, Russell WM, Klaenhammer TR. Characterization of the tre locus and analysis of trehalose cryoprotection in Lactobacillus acidophilus NCFM. Appl Environ Microbiol 2006; 72(2): 1218-25.
[http://dx.doi.org/10.1128/AEM.72.2.1218-1225.2006] [PMID: 16461669]

[70]   Magazù S, Maisano G, Migliardo F, *et al.* Characterization of molecular motions in biomolecular systems by elastic incoherent neutron scattering. J Chem Phys 2008; 129(15): 155103.
[http://dx.doi.org/10.1063/1.2989804] [PMID: 19045233]

[71]   Blazhnov IV, Magazù S, Maisano G, Malomuzh NP, Migliardo F. Macro- and microdefinitions of fragility of hydrogen-bonded glass-forming liquids. Phys Rev E Stat Nonlin Soft Matter Phys 2006; 73(3 Pt 1): 031201.
[http://dx.doi.org/10.1103/PhysRevE.73.031201] [PMID: 16605509]

[72]   Magazù S, Migliardo F, Malomuzh NP, Blazhnov IV. Theoretical and experimental models on viscosity: I. Glycerol. J Phys Chem B 2007; 111(32): 9563-70.
[http://dx.doi.org/10.1021/jp071949b] [PMID: 17655214]

[73]   Leszek B. Dimensional changes of waterlogged archaeological hardwoods pre-treated with aqueous mixtures of lactitol/trehalose and mannitol/trehalose before freeze-drying. J Cult Herit 2015; 16(6): 876-82.
[http://dx.doi.org/10.1016/j.culher.2015.03.010]

[74]   Magazù S, Maisano G, Migliardo F, Benedetto A. Elastic incoherent neutron scattering on systems of biophysical interest: mean square displacement evaluation from self-distribution function. J Phys Chem B 2008; 112(30): 8936-42.
[http://dx.doi.org/10.1021/jp711930b] [PMID: 18610954]

[75]   Lombardo D, Kiselev MA, Magazù S, Calandra P. Amphiphiles self-assembly: Basic concepts and future perspectives of supramolecular approaches. Adv Condens Matter Phys 2015; 2015: 151683.
[http://dx.doi.org/10.1155/2015/151683]

[76]   Nagase H, Endo T, Ueda H, Nakagaki M. An anhydrous polymorphic form of trehalose. Carbohydr Res 2002; 337(2): 167-73.
[http://dx.doi.org/10.1016/S0008-6215(01)00294-4] [PMID: 11814449]

[77]   Magazù S, Calabrò E, Campo S, Interdonato S. New insights into bioprotective effectiveness of disaccharides: an FTIR study of human haemoglobin aqueous solutions exposed to static magnetic fields. J Biol Phys 2012; 38(1): 61-74.
[http://dx.doi.org/10.1007/s10867-010-9209-1] [PMID: 23277670]

[78]   Crowe JH, Crowe LM, Chapman D. Preservation of membranes in anhydrobiotic organisms: The role of trehalose. Science 1984; 223(4637): 701-3.
[http://dx.doi.org/10.1126/science.223.4637.701] [PMID: 17841031]

[79]   Magazù S, Migliardo F, Affouard F, Descamps M, Telling MTF. Study of the relaxational and vibrational dynamics of bioprotectant glass-forming mixtures by neutron scattering and molecular dynamics simulation. J Chem Phys 2010; 132(18): 184512.
[http://dx.doi.org/10.1063/1.3407428]

[80]   Magazù S, Maisano G, Migliardo F, Benedetto A. Mean square displacement evaluation by elastic neutron scattering self-distribution function. Phys Rev E Stat Nonlin Soft Matter Phys 2008; 77(6 Pt 1): 061802.
[http://dx.doi.org/10.1103/PhysRevE.77.061802] [PMID: 18643290]

[81]   Magazù S, Maisano G, Migliardo F, et al. Characterization of molecular motions in biomolecular systems by elastic incoherent neutron scattering. J Chem Phys 2008; 129(15): 155103.
[http://dx.doi.org/10.1063/1.2989804] [PMID: 19045233]

[82]   Magazù S. IQENS - Dynamic light scattering complementarity on hydrogenous systems. Phys B 1996; 226(1-3): 92-106.
[http://dx.doi.org/10.1016/0921-4526(96)00255-4]

[83]   Lupi L, Comez L, Paolantoni M, Fioretto D, Ladanyi BM. Dynamics of biological water: insights from molecular modeling of light scattering in aqueous trehalose solutions. J Phys Chem B 2012; 116(25): 7499-508.
[http://dx.doi.org/10.1021/jp301988f] [PMID: 22651571]

[84]   Branca C, Magazù S, Maisano G, Migliardo P, Villari V. Conformational distribution of poly(ethylene oxide) in molten phase and in aqueous solution by quasi-elastic and inelastic light scattering. J Phys Condens Matter 1998; 10(45): 10141-57.
[http://dx.doi.org/10.1088/0953-8984/10/45/004]

[85]   Faraone A, Magazù S, Maisano G, Ponterio R, Villari V. Experimental evidence of slow dynamics in semidilute polymer solutions. Macromol 1999; 32(4): 1128-33.
[http://dx.doi.org/10.1021/ma9809684]

[86]   Magazu' S, Maisano G, Mallamace F, Micali N. Growth of fractal aggregates in water solutions of macromolecules by light scattering. Phys Rev A Gen Phys 1989; 39(8): 4195-200.
[http://dx.doi.org/10.1103/PhysRevA.39.4195] [PMID: 9901749]

[87]   Jannelli MP, Magazù S, Migliardo P, Aliotta F, Tettamanti E. Transport properties of liquid alcohols investigated by IQENS, NMR and DLS studies. J Phys Cond Matter. J Phys Condens Matter 1996; 8(43): 8157-71.
[http://dx.doi.org/10.1088/0953-8984/8/43/012]

[88]   Caccamo MT, Magazù S. Tagging the oligomer-to-polymer crossover on EG and PEGs by infrared and Raman spectroscopies and by wavelet cross-correlation spectral analysis. Vib Spectrosc 2016; 2016(85): 222-7.
[http://dx.doi.org/10.1016/j.vibspec.2016.04.017]

[89]   Magazù S, Migliardo F, Caccamo MT. Innovative wavelet protocols in analyzing elastic incoherent neutron scattering. J Phys Chem B 2012; 116(31): 9417-23.
[http://dx.doi.org/10.1021/jp3060087] [PMID: 22793379]

[90]   Cannuli A, Caccamo M T, Castorina G, Colombo F, Magazù S. Laser Techniques on Acoustically Levitated Droplets EPJ Web of Conferences 167, article number 05010, 2018.
[http://dx.doi.org/10.1051/epjconf/201816705010]

[91]   Caccamo MT, Magazù S. Ethylene Glycol - Polyethylene Glycol (EG-PEG) mixtures: Infrared spectra wavelet cross-correlation analysis. Appl Spectrosc 2017; 71(3): 401-9.
[http://dx.doi.org/10.1177/0003702816662882] [PMID: 27558367]

[92]   Caccamo MT, Magazù S. Multiscaling wavelet analysis of infrared and raman data on polyethylene glycol 1000 aqueous solutions. Spectrosc Lett 2017; 50(3): 130-6.
[http://dx.doi.org/10.1080/00387010.2017.1291524]

[93]   Migliardo F, Magazù S, Caccamo MT. Infrared, raman and INS studies of poly-ethylene oxide oligomers. J Mol Struct 2013; 1048: 261-6.
         [http://dx.doi.org/10.1016/j.molstruc.2013.05.060]

[94]   Magazù S, Calabrò E, Campo S. FTIR spectroscopy studies on the bioprotective effectiveness of trehalose on human hemoglobin aqueous
         solutions under 50 Hz electromagnetic field exposure. J Phys Chem B 2010; 114(37): 12144-9.

[95]   Branca C, Magazù S, Maisano G. S. M. Bennington A, Fåk B. Vibrational studies on disaccharide/H2O systems by inelastic neutron
         scattering, raman, and IR spectroscopy. J Phys Chem B 2003; 107(6): 1444-51.
         [http://dx.doi.org/10.1021/jp026255b]

[96]   Caccamo MT, Zammuto V, Gugliandolo C, Madeleine-Perdrillat C, Spanò A, Magazù S. Thermal restraint of a bacterial exopolysaccharide of
         shallow vent origin. Int J Biol Macromol 2018; 114: 649-55.
         [http://dx.doi.org/10.1016/j.ijbiomac.2018.03.160] [PMID: 29601879]

[97]   Caccamo MT, Magazù S. Thermal restraint on PEG-EG mixtures by FTIR investigations and wavelet cross-correlation analysis. Polym Test
         2017; 62: 311-8.
         [http://dx.doi.org/10.1016/j.polymertesting.2017.07.008]

5

Nanomaterials as Protein, Peptide and Gene Delivery Agents

Anika Guliani[1,2] and Amitabha Acharya[1,2,*]

[1]*Biotechnology Division, CSIR-Institute of Himalayan Bioresource Technology, Palampur (H.P.) 176061, India*
[2]*Academy of Scientific & Innovative Research (AcSIR), CSIR-Institute of Himalayan Bioresource Technology, Palampur (H.P.) 176061, India*

Abstract:

Background:

Nanomaterials offer significant advantages in delivery of different biomolecules which suffer from drawbacks like poor bioavailability, low stability and retention time, degradation in biological systems *etc*. Nanotechnological approach has shown promising results for the sustained release of these biomolecules with minimal toxicity concerns. The present review describes a comprehensive outlook of the different nanomaterials used for the delivery of these biomolecules.

Methods:

Current literature reports related to protein, peptide and gene delivery agents have been reviewed and classified according to their applications.

Results:

Studies suggested that the nanomaterial based delivery agents can be broadly classified in to five categories which include metallic NPs, polymeric NPs, magnetic NPs, liposomes and micelles. All these materials provided significant improvement in the targeted delivery of biomolecules.

Conclusion:

Concerns regarding the bioavailability, stability and delivery of proteins, peptides, genes need to be investigated to improve their therapeutic potential in the biological milieu. The use of nanoparticles as drug delivery vehicles may avoid undesirable hazards and may increase their pharmaceutical efficacy.

Keywords: Nanoparticles, Bioactive peptide, SiRNA, Gene delivery, Transfection, Toxicity.

1. INTRODUCTION

Proteins and peptides are long being used to treat a wide variety of human diseases. Proteins play an irreplaceable role in life, taking part in all vital life activities at the molecular level. There are different varieties of proteinaceous molecules which act as antibodies, growth factors, antigens and possess excellent antioxidant, antibacterial, antiviral activities, provide support for cholesterol synthesis and boost immune system [1]. But factors like aggregation, unfolding/misfolding, small size *etc.* restrict the therapeutic potential of these biomolecules [2, 3]. Likewise, the delivery of genes to the organs is also considered as one of the major issues since this can lead to enhanced expression of the already present genes. Gene delivery has proven to be effective in the repair of many tissues by gene knockdown method. It has been also proved to be beneficial in causing apoptosis of the damaged cells by silencing the mRNA

---

* Address correspondence to this author at the Academy of Scientific & Innovative Research (AcSIR), CSIR-Institute of Himalayan Bioresource Technology, Palampur (H.P.) 176061, India; E-mails: amitabhachem@gmail.com; amitabha@ihbt.res.in

produced and inhibiting the production of the functional protein. The conventional methods for the transfection are associated with the use of viral vectors which have several disadvantages *viz.,* high manufacturing cost, mammalian cells compatibility issues, opsonization by body cells, increased immunogenic responses, *etc.* [4 - 7]. Nanomaterials can be used for the oral delivery of the proteins, peptides and genes, since these offer advantages *viz.,* increase the solubility of poorly water-soluble molecules, help in targeting gastrointestinal tract, allow transcytosis across the mucosal layer, protect encapsulated biomolecules from adverse environmental conditions, *etc.* [8]. Nanoparticles (NPs) being small in size and highly biocompatible in nature, allow slow and sustained release of biomolecules which can cross the epithelial barrier and provide effective therapeutic outcomes with minimal dosage. The delivery of the genes with the help of non-viral vectors is now easy since these are biocompatible, non-immunogenic in nature. There are a number of gene carriers which have been designed on the basis of nanotechnology and have proven to be capable of delivering the cargo at the targeted sites with minimal toxicity concerns [9]. This review will cover a description of different protein, peptide and gene delivery agents along with their potential therapeutic applications.

2. NANOMATERIALS FOR PROTEIN AND PEPTIDE DELIVERY

Here in this section, different classes of nanomaterials used for protein and peptide delivery have been discussed.

2.1. Metallic Nanoparticles

Metallic NPs like gold (Au), silver (Ag), copper (Cu), *etc.* have been used for the delivery of various proteins and peptides. In the literature, gold nanoparticles (AuNPs) have been used for nucleus targeting by functionalizing their surface with transactivator of transcription (TAT) peptides [10, 11]. In general, peptide conjugated NPs were synthesized by immobilization of thiol groups of the peptide on the AuNPs surface. Bovine Serum Albumin (BSA) and streptavidin have been covalently conjugated on the AuNPs surface [10, 12]. AuNPs passivated with poly-ethylene-glycol (PEG) were functionalized with peptide-toxin conantokin-G. These functionalized NPs can selectively bind to *N*-methyl-D-aspartate (NMDA) receptors present on the neuron surface and have been found to block the cell death pathways [13]. AuNPs attached with amyloid growth inhibitor peptide (-CLPFFD-NH$_2$-) showed an easy recognition of murine bone marrow macrophages in comparison to the bare AuNPs. These peptide conjugated NPs showed enhanced macrophage response by secreting more inflammatory cytokines like TNF-α, IL-6 [14]. Cecropin-melittin, an Antimicrobial Peptide (AMP) (KWKLFKKIGAVLKVLC) containing cysteine amino acid at its terminal end, was conjugated with AuNPs. The conjugated AuNPs exhibited low polydispersity and narrow size distribution of 14 nm and was tested against *S.aureus* and *E.coli* for their antibacterial activity and *in vivo* studies. Results suggested better antibacterial activity for peptide conjugated AuNPs [15]. Another AMP, esculentin-1a has been conjugated to AuNPs with the help of PEG and has been found to be effective against *P. aeruginosa* at a low concentration. The conjugated NPs were non toxic to human keratinocytes and could break the bacterial membrane easily [16]. An AMP odorranain-A-OA1 (OA1) (VVKCSYRLGSPDSQCN) was conjugated to citrate-capped silver nanoparticles (AgNPs) (size of ~12-14 nm). The peptides attached to AgNPs were found to be more stable and displayed better antibacterial activity as compared to the native peptide or the bare AgNPs. The conjugates were also less toxic to the mammalian cells [17]. Breast cancer cells (MCF-7) have been treated with AgNPs attached with Cell Penetrating Peptides (CPP). The CPP aided in enhanced intake of AgNPs across the cellular membranes, thereby causing the cancer cell death [18]. The effect of AgNPs along with MBP-1 peptide was studied on the disease-causing bacteria *S.aureus* which is commonly resistant to many antibiotics. The wound caused by the bacteria has been found to heal and the bacterial colonization also decreased at the wound site due to mutual effect of both the AgNPs and MBP-1 peptide [19]. The specific peptide sequence of arginine, glycine, aspartic acid (RGD) interacts with integrin receptors in various adhesion processes for cell targeting. Fluorescent peptide RGD was immobilized on AgNPs to enhance the NP entry into the leukemia ($\alpha_5\beta_1$) and neuroblastoma ($\alpha_v\beta_3$) cells overexpressing their specific integrins [20]. Silicon NPs (SiNPs) conjugated with RGD have been effectively used as a tracking agent as well as for the killing of the specific malignant cancer cell U87MG [21]. SiNPs coated with bovine serum albumin/ human serum albumin (BSA/HSA) proteins have been studied for cholesterol effluxing as well as for fluorescence imaging in HUVEC and HCAEC cell lines. The HSA coated SiNPs have been reported to show enhanced permeation as well as cholesterol effluxing in both the cells [22]. A report on the functionalization of the SiNPs has described conjugation of SiNPs with chitosan undecylenic acid which was further modified with L-cysteine and CPP. These modified L-cysteine NPs have been found to be effective for the oral delivery of the insulin and have also been found to increase the bioavailability of insulin by ~ 1.86 times as compared to bare NPs [23]. Extracellularly synthesized copper nanoparticles (CuNPs) by stem latex of *Euphorbia nivulia* have been exploited for *in vitro* administration of NPs in A549 cancer cell lines. CuNPs were found to be toxic depending upon the dose while the latex coated CuNPs have been found to be non toxic and have been used as a delivery agent in the cancer cell lines [24]. Table 1 summarizes reports of peptide conjugated or encapsulated metallic NPs along with their specific applications.

Table 1. Metallic NPs conjugated/encapsulated to different peptide sequences.

Metallic NPs	Peptide Sequence	Functions	References
AgNPs	Odorranain-A-OA1 (OA1) (VVKCSYRLGSPDSQCN)	Antimicrobial peptide	[17]
AgNPs	CRGDKGPDC	Increased penetration in the glioma cells	[78]
AgNPs	CSG-LL37	Antimicrobial activity, wound healing, neovascularization and angiogenesis	[79]
AgNPs	GGGRRRRRRYGRKKRRQRR (G_3R_6TAT)	Cell penetrating peptide, reducing agent and surfactant	[80]
AgNPs	Ac-LIVAGK-NH$_2$	Antibacterial	[81]
AuNPs	CALNN–AALRRASLG	cAMP-dependent protein kinase A	[82]
AuNPs	Ac–Cys[Ac]–Gly–Dphe–Pro–Arg–Gly–Cys[Ac] –OH	Substrate for thrombin	[83]
AuNPs	KWKLFKKIGAVLKVL	Antibacterial	[84]
AuNPs	Pro-His-Cys-Lys-Arg-Met; Pep-A	Antioxidant	[85]
AuNPs	LA-WKRAKLAK	Anticancer	[86]

2.2. Polymeric Nanoparticles

The use of polymeric NPs has also been extensively studied for encapsulation, attachment of various proteins/peptides for targeted delivery, increase the bioavailability, retention time, stability, *etc.* of the attached biomolecules. Polymers like poly (lactic-*co*-glycolic acid) (PLGA), chitosan, poly (lactic acid) (PLA), polycaprolactone (PCL) have been used for the attachment of different proteins and peptides. Lysozyme and a transforming growth factor (TGF-β) have been encapsulated into PLGA polymer. The encapsulated proteins were found to be more stable; their structural integrity as well as the biological activity were also found to retain in the encapsulated form [25]. Many enzymes degrade when administered into the body and thus are unstable in nature. The enzyme laccase was attached on the surface of chitosan NPs and found that the enzyme sensitivity at different pH and temperature was similar to that of the free enzyme. The enzyme was found to be more stable and showed enhanced activity towards degradation of microbes than the free enzyme [26]. The surface of PLA NPs was modified with Polyethylene Glycol (PEG) and to this an anticancer peptide NuBCP-9 was encapsulated. The spacer length of PEG and the molecular weight of PLA varied which aided in more effective delivery of NuBCP-9 and caused apoptosis of the cancer cells [27]. Another polymer (poly(d,l lactic-cohydroxymethyl glycolic acid) (pLHMGA) has been used for the preparation of vaccines against the cancer cells. A long synthetic peptide isolated from HPV16 E7 oncoprotein together with toll like receptor 3 has been encapsulated into the NPs and were found to be effective in decreasing the tumor and suggested no adverse effects on the body as compared to the free adjuvants administered with vaccines [28]. Similarly, BSA was encapsulated into a PLGA with an encapsulation efficiency of 65%. It was observed that the concentration of the polymer used was directly proportional to the BSA encapsulation efficiency [29].

2.3. Liposomes

Liposomes act as good carriers for both water-soluble as well as water-insoluble molecules. A liposome-based drug delivery system which can cross the Blood Brain Barrier (BBB) has been developed. The surface of the PEG-modified liposomes was further conjugated with Angiopep-2 peptide. This peptide acted as a targeting ligand and could bind with bEnd.3 cells and can cross the BBB when injected intravascularly into the body. These liposomes have been further checked for their ability to entrap ultrasound contrast gases for bioimaging applications [30]. Mannosylated liposomes have been used as an encapsulating agent for MBP peptides. These peptides could retard the growth of encephalomyelitis. These liposomes have been administered into the body of multiple sclerosis patients where they have been found to be useful in curing the disease by decreasing MCP-1/CCL2, MIP-1β/CCL4, IL-7, and IL-2 factors in the serum of the patients [31]. Again, DAMGO(H-Tyr-D-Ala- Gly-MePhe-Gly-ol), an opioid peptide with poor brain penetrating properties have been successfully encapsulated in glutathione (GSH)-coated liposomes. The biological activity was found to enhance in comparison to the empty GSH-PEG liposomes and the corresponding results suggested that the PEG liposomal DAMGO has been efficient in targeting brain without the use of any specific ligand [32]. Hydrophobic polyampholyte-modified liposomes, synthesized using a combination ofε - poly-L-lysine- dodecylsuccinic anhydride- succinic anhydride were compared with unmodified liposomes for the delivery of lysozyme. The polyampholyte- modified liposomes exhibited high encapsulation efficiency, stability and low cytotoxicity. These liposomes have been found to bypass the endocytic pathway and thus serve as a very efficient tool for delivery of the protein [33]. An antidiabetic peptide from Atlantic salmon (Salmo salar) protein hydrolysates (SPH) has been loaded into liposomes. The surface of the liposomes was coated with chitosan which enhanced the stability and could preserve the SPH loaded liposomes even at 4°C [34]. The induction of adaptive immune response by a subunit vaccine, encapsulated in liposomes, with the iNKT adjuvant α-GalCer, has been found to be responsible for the enlarged production of IFN-γ and cytotoxic T-cell responses. The vaccines administered intravenously induced the anti-tumor responses and caused immune stimulation [35]. BSA has been encapsulated in to pH responsive polymer based liposomes. These liposomes when administered in the rat bladder at pH 6.5, showed better delivery of the loaded protein at the bladder site of MB49 cells and macrophages as compared to the physiological pH (7.4) [36]. The cationic AMPs play vital roles to fight against several pathogenic bacteria as well as these are helpful in innate immunity response. But these lack stability, infer cytotoxicity and are rarely available to the body. To increase the bioavailability and retention time, two cationic AMPs *viz.*, LL-37 and indolicidin, have been encapsulated into liposomes and tested against herpes simplex 1 virus. The LL-37 formulation formed ~110 nm sized liposomes which were found to be more stable.These nanoformulations were efficiently taken up by HaCaT cells and showed decreased cytotoxicity level compared to the free peptide (Fig. **1**) [37].

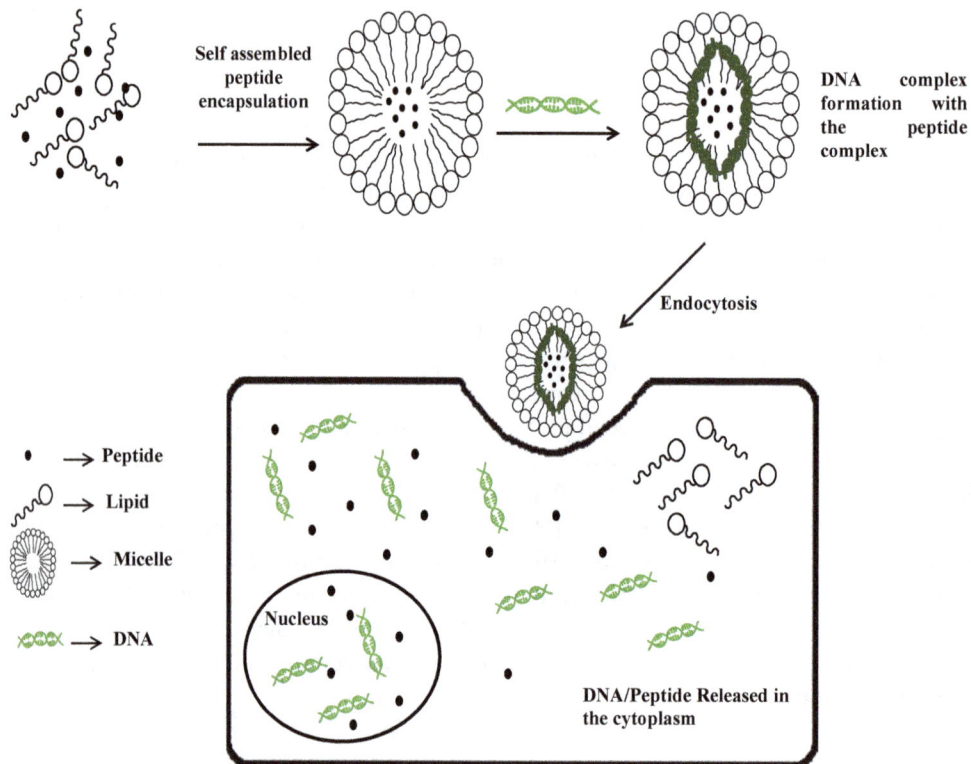

Fig. (1). Nanoencapsulation of peptides and DNA into liposomes.

2.4. Micelles

Albumin, a negatively charged protein found in blood, has been used as biocompatible material for the encapsulation of proteins like lysozyme, Spry 1, *etc.* The surface of the albumin protein was modified with poly(oligo (ethylene glycol) methyl ether methacrylate) which enhanced the stability of the NPs formed. This PEGylated albumin polyion complex micelle has been found to bind with cationic polymers and also retained the functional aspect of the encapsulated molecule. The Spyr 1 showed better anticancer activity in MDA-MB-231 and MCF-7 cell lines, by hampering the growth of three dimensional multicellular tumor spheroids [38]. In order to achieve higher drug payload at the tumor sites, the surface of the doxorubicin (Dox)-loaded DSPE–PEG micelles have been modified with GE11 peptide which helped in the targeted delivery of Dox at the tumor site [39]. Cyclic peptide RGDfK can effectively bind with $α_v β_3$ which are overexpressed on the cancer cells. The strong affinity resulted from the presence of carboxylate, guanidinium, and hydrophobic groups of c(RGDfK). In this respect, c(RGDfK)-PEG-PLA/PEG-PLA/DTX has been prepared which showed strong potential to act as anti-cancer agents to the tumor cells,over expressing $\alpha_v\beta_3$ on the surface [40]. The fatty acid was covalently attached to the surface of proteins which further served as an anchoring platform for plasma membrane binding. Reverse micelle approach has been used to encapsulate two myristoylated proteins namely recoverin and HIV-1 and was found to retain their functional aspects [41]. The reverse micelles of polyurathene have been loaded with BSA, with and without heptakis(2,6-di-O-methyl)-b-cyclodextrin (DM-b-CD) conjugation. The BSA encapsulation efficiency, release rate was found to be significantly higher for the DM-b-CD-containing micelles [42]. Reverse micelles have also been reported to improve the stability and half-life of the lipase from *R. delemar* and *C. rugose* encapsulated in to it [43]. PLA micelles loaded with Dox and surface coated with PEG have been evaluated against Caco-2 cell. The micelle were attached with a twelve amino acid peptide, TWYKIAFQRNRK, which had a strong binding affinity for $\alpha_6\beta_1$ receptor, abundantly expressed on Caco-2 cells [44].

2.5. Magnetic Nanoparticles

The superparamagnetic iron oxide NPs (SPIONPs) have been synthesized and their derivatization was done using a TAT-derived peptide Gly-Arg-Lys-Lys-Arg-Arg-Gln-Arg-Arg-Arg-Gly-Tyr-Lys(FITC)-Cys-NH$_2$.These NPs were found to enter the hematopoietic and neural progenitor cells at a concentration of 10-30 pg of iron per cell. These NPs were also found to be present in the cytoplasm of CD34$^+$ cells and showed no cytotoxic effects [45]. SPIONPs have been surface modified with PEG and was used as MRI contrast agent. These were further attached with TAT for quick cellular internalization [46]. The surface of IONPs has been coated with PEG,followed by further functionalization using heparin at a concentration of 35.4 µg of heparin/mg of Fe. These NPs have been found to bind with protamine with loading capacity of 22.9 ± 4.7 µg/mg of Fe. Again, these NPs showed better retention ability in the blood and accumulated in a smaller amount in the liver, suggesting their easy recognition by the immune system [47]. The magnetic NPs coated with 2,3 –dimercaptosuccinic acid have been covalently attached with a GEBP-11 peptide (GEBP11-DMSA-MNPs)via amide bond coupling reaction. Further, these were attached to Cy5.5 via ester bond formation. Cell lines *viz.*, HUVEC and SGC7901 were used to evaluate the magnetic as well as fluorescence imaging potential of the prepared NPs. The GEBP-11 peptides aided in higher uptake of the NPs and no cytotoxic effects were found. These GEBP11-DMSA-MNPs have been used as an imaging probe for evaluating the angiogenesis in gastric cancer cells with high specificity and sensitivity [48]. The monocrystalline IONPs have been attached with TAT peptide (GRKKRRQRRRGYK) which helped in the internalization of the IONPs and allowed magnetic resonance imaging of the targeted site. The biodistribution studies have been done using BalB C mice by injecting these NPs and major accumulation was found in lymph, spleen and liver [49]. To increase the intracellular effectiveness of IONPs, TAT peptide (YGRKKRRQRRR) has been attached to IONPs. The effect of these NPs has been studied in lung cancer cell lines *viz.*, A549 and H358 and was found that the conjugation of CPP enhanced the intracellular uptake. These TAT conjugated to dextran coated IONPs, caused reactive oxygen species production when exposed to alternating magnetic field and resulted cell death. The treatment also caused the destabilization of the lysosomal membrane and eventually apoptosis of the cells [50]. Gold shell has been created at the surface of magnetic NPs, derivatized with dithiocarbamate. The C-terminus of the AMP (KWKLFKKIGAVLKVLC) was used for the immobilization of the peptide on gold surface (AMP-NPs).These NPs have been tested against *E. coli* and *S. aureus*. Results suggested that these NPs were internalized by phagocytes as well as by endothelial cells, and caused no inflammatory reaction at a concentration upto 200 µg/mL [51].

3. NANOMATERIALS FOR GENE DELIVERY

This section deals with different nanomaterials used for the delivery of genetic materials.

3.1. Metallic Nanoparticles

AuNPs functionalized with PEP peptide were studied for their transfection activity in the mesenchymal stem cells. Both the transfection activity for Vascular Endothelial Growth Factor (VEGF) and antibacterial activity were found to be greatly enhanced with the increase in the concentration of the peptide [52]. Unimer polyion complex-assembled (uPIC) AuNPs attached to cRGD peptide (cRGD-AuNPs-uPIC) were used for the delivery of siRNA. These peptides helped in effective targeted delivery of siRNA in HeLa cell lines [53]. CD4$^+$ and CD25$^+$regulatory T-cells (Tregs) play an important role in maintaining the immunogenic responses in the body. AuNPs conjugated with eGFP-siRNA penetrated inside Tregs and caused a reduction in eGFP expression and modulated the genetic expression [54]. Quercetin attached to AgNPs was tested for antimicrobial activity against various pathogenic microbes (AgNPs-Qe). These AgNPs-Qe were stabilized with siRNA. The siRNA/AgNPs-Qe were found effective in silencing the targeted gene and caused reduction in bacterial propagation. The intravenous injection of siRNA/ AgNPs-Qe to nude mice showed a decrease in bacterial content in blood [55]. One-step synthesis was employed for the preparation of AgNPs using chitosan-g-polyacrylamide. These NPs were stabilized using PEG and functionalized with RGDS peptide. The DNA transfection ability of NPs was found to be greatly enhanced in HeLa and A549 cells with minimal cellular toxicity [56]. The Se@MIL-101 and Ru@MIL-101NPs were formed by linking Se and Ru to Fe attached with siRNA. These NPs were tested against multi-drug resistant bacteria. In MCF-7/T,these NPs caused silencing of MDR genes, which led to enhanced apoptosis and instability of microtubules in cellular cytoskeleton [57].

3.2. Polymeric Nanoparticles

Cationic polymer polyethylenimine (PEI) and glycol chitosan have been reported to encapsulate siRNA. These formed complexes were found to be stable and inhibited Red Fluorescent Protein (RFP) gene expressions in cells and mice bearing tumors [58]. The synergistic effect for gene expression and cancer treatment has been assessed by treating the cells with nano-carriers of PEI and PCL. Both PEI and PCL were encapsulated with Dox and siRNA, which were surface modified with poly(ethylene glycol)-block-poly(glutamic acid) for targeted delivery to cancer cells. The cancer cells showed enhanced delivery of the BCL-2 siRNA showing RNA interference with increased Dox activity [59]. The surface of the dendrimer was modified with a peptide RGD and AuNPs were entrapped in it.The naïve, the modified dendrimers as well as the dendrimers encapsulating AuNPs were used as a carrier to transfer pDNA in human mesenchymal cells. This pDNA encoded human bone morphogenetic protein-2 (hBMP-2) gene, green fluorescent protein and luciferase. The PEG-RGD entrapped AuNPs dendrimer was found to be more efficient in transfecting hBMP-2 gene [60]. AgCHS (Chitin synthase gene), a double-stranded RNA was delivered using chitosan NPs in order to increase its retention stability as well as efficacy in the gut epithelial cells of larva. The larva of Anopheles gambiae when fed with chitosan/AgCHS NPs suppressed AgCHS1 and AgCHS2 genes, responsible for synthesis of chitin. The reduction in chitin level was found to be 38% and larvae were more susceptible to different pesticides compared to control [61]. The rat skeletal muscle was injected with positively charged lipids microbubbles with DNA attached to them. On exposing to ultrasound waves, the DNA released from bubbles and deposited in the skeletal muscle tissues. An enhanced gene expression has been found for echo contrast bubbles as compared to the naked DNA [62]. A polycationic NP composed of 2-(diethylamino)ethyl methacrylate were prepared by loading anionic siRNA in it. The *in vitro* efficacy was evaluated in embryonic kidney cell line HEK293T and the murine macrophage cell line RAW264.7. These cell lines showed an increase internalization of siRNA [63]. For transdermal delivery of DNA, NPs of chitosan and poly-g-glutamic acid were prepared. Compared to the control; the chitosan NPs with DNA showed enhanced gene expression in the mouse skin due to increased penetration [64]. The encapsulation of luciferase targeted siRNA was done in PLGA polymer. The *in vitro* efficacy was higher for silencing of the model gene, luciferase in MDA-kb2 cell lines [65].

3.3. Liposomes

Liposomes conjugated with bi-ligands have been exploited for the delivery of genes to cross the BBB. The β-gal plasmid was loaded into bi-ligand (transferrin-poly-L-arginine) liposomes and were labeled with 1,1'-dioctadecyl-3,3,3',3'-tetramethyl-indocarbocyanine iodide(DiR). The bi-ligand liposomes crossed the brain and showed enhanced expression of β-gal in the brain tissue of rat as compared to transferrin ligated liposome [66].The fibrosis can be decreased by knocking out the Myocardin-Related Transcription Factors (MRTF). A receptor-targeted liposome-peptide-siRNA NPs has been designed for silencing MRT and to treat conjunctival fibrosis. Two different liposomes LYR (non-PEGylated liposome-peptide Y-siRNA) and LER (non-PEGylated liposome-peptide ME27-siRNA) were found to be effective in causing approximately 76% silencing of the MRTF-B. This effect was found to be increased when compared with siRNA alone or the non-targeting peptides [67]. The liposomes loaded with VEGF siRNA and doclitaxel were surface modified with Angiopep-2 and neuropilin-1 receptor (tLyP-1) peptides. The effect of these liposomes were observed in human glioblastoma cells (U87 MG) and murine Brain Microvascular Endothelial Cells (BMVEC). These modified liposomes showed augmented penetration, gene silencing and antiproliferative activity [68].A Multi-Functional Nano Device (MEND) was synthesized for the *in vivo* delivery of siRNA to the Tumor Endothelial Cells (TEC). The MEND made up of a cationic lipid YSK05 was loaded with TEC siRNA which was further attached with cyclic RGD peptide. The cRGD aided in targeted delivery of the MEND towards TEC cells and caused gene silencing [69].The liposomal-siRNA, surface modified with PEG linked to hyaluronic acid (PEG-HA-NP)were reported for enhanced delivery of siRNA. The increase in siRNA delivery was due to fast degradation of hyaluronic acid in endosomes, causing an endosome escape and finally siRNA release. The p-glycoprotein expression on the surface of MCF-7 cells has been downregulated by the transfection of anti-P glycoprotein siRNA and showed better silencing [70]. The tumor model generated in mice was treated with pDNA encoding firefly luciferase. The tumors treated with lipid-pDNA along with ultra-sonication showed maximum levels of transfection as compared to conventional transfection method. Same results have been reported when the tumors were treated with pDNA IL-12 gene using lipids. There was suppression in the growth of the tumors [71].

3.4. Magnetic Nanoparticles

The MNPs have been well explored as transfecting agents. A technique termed as magnetofection has been introduced which works on the principle of combining the genetic material with MNPs. For the *in vitro* transfection, the MNPs along with the genetic material are added to the cell culture where the effect of magnetic field generated by the metals causes an increment in the transfection rate. In case of *in vivo* studies, the magnetic field generated causes both the targeted delivery as well as the speed of transfection. By altering the magnetic fields, the release of the genes at the targeted site has been achieved [72]. HeLa cell line induced with reporter gene luciferase was evaluated for the knockdown activity by the magnetofection of siRNA [73]. An antisense oligodeoxynucleotides have been delivered in HUVEC as well as in femoral arteries of mice. The transfection of oligodeoxynucleotides was enhanced by 84%. The effect of oligodeoxynucleotides and siRNA has been shown in the signaling pathway where NADPH oxidase activity was reduced by Sh-2 domain of phosphatase 1 in HUVEC cells [74]. For targeted delivery of DNA, heparin sulfate linker was used to attach Adeno-Associated Virus (AAV) encoding Green Fluorescent Protein (GFP) at the surface of MNPs [75]. The IONPs were coated with lipids in order to increase their stability by adding a solvent N-methyl-2-pyrrolidone (NMP) to stimulate adhesion. The cationic lipid coating on the surface of NPs was exploited to adsorb DNA and siRNA on its surface by electrostatic interaction. The *in vitro* effectiveness evaluated in HeLa cell lines and the transfection for DNA delivery was found to be maximum for 50-100 nm sized NPs whereas 40 nm sized NPs were appropriate for delivery of siRNA [76].The IONPs containing plasmid DNA were surface coated with PEI and further attached with bis(cysteinyl) histidine-rich TAT peptide. The synergistic effect of magnetofection and TAT mediated-delivery was enhanced by 4 times both *in vitro* as well under *in vivo* conditions [77].

CONCLUSION

The hunt to design advanced nanomaterials for targeted delivery of biologically important molecules has been gaining tremendous interest amongst researchers. The delivery of proteins, peptides and genes is very promising in performing various functions of the body. The impediment in the delivery of these molecules is their low stability, degradation by various enzymes in the biological milieu, less bioavailability, *etc.* Nanotechnology-based systems have been developed to address these problems and to increase the efficacy of these molecules. Further, specific delivery of the biomolecules to the infected sites via targeted approach can be also achieved using nanodelivery agents for the treatment of acute and chronic ailments. The encapsulation/conjugation has helped in maintaining the structural as well as functional identity of the molecules. The nano-encapsulated siRNA has been found to be more stable and their retention time has also been found to increase. But keeping in mind that "Technology-Yes, but Safety-Must", prior to implementing this new class of delivery agents, we also need to evaluate the new kinds of toxicity concerns associated with these delivery agents.

ACKNOWLEDGEMENTS

The authors are grateful to the Director, CSIR-IHBT for his constant support and encouragement. AA acknowledges financial assistance in the form of project grant MLP-0201 from CSIR and GAP-0214 (EMR/2016/003027) from DST, Government of India. AG is thankful to the Academy of Scientific and Innovative Research (AcSIR). The CSIR-IHBT communication number of this manuscript is 4250.

REFERENCES

[1]   Campos-Vega R, Pool H, Vergara-Castañeda H. Micro and nanoencapsulation: A new hope to combat the effects of chronic degenerative diseases In: Foods Bioactives. Processing, Quality and Nutrition 2013.

[2]   Li X, Zhang Y, Yan R, *et al.* Influence of process parameters on the protein stability encapsulated in poly-DL-lactide-poly(ethylene glycol) microspheres. J Control Release 2000; 68(1): 41-52.
[http://dx.doi.org/10.1016/S0168-3659(00)00235-2] [PMID: 10884578]

[3]   Li J, Huang L. Targeted delivery of RNAi therapeutics for cancer therapy. Nanomedicine (Lond) 2010; 5(10): 1483-6.
[http://dx.doi.org/10.2217/nnm.10.124] [PMID: 21143026]

[4]   Singh S. Nanomaterials as non-viral siRNA delivery agents for cancer therapy. Bioimpacts 2013; 3(2): 53-65.
[PMID: 23878788]

[5]   Reis CP, Neufeld RJ, Ribeiro AJ, Veiga F, Nanoencapsulation II. Biomedical applications and current status of peptide and protein nanoparticulate delivery systems. Nanomedicine (Lond) 2006; 2(2): 53-65.
[http://dx.doi.org/10.1016/j.nano.2006.04.009] [PMID: 17292116]

[6]   Mohan A, Rajendran SR, He QS, Bazinet L, Udenigwe CC. Encapsulation of food protein hydrolysates and peptides: A review. RSC Advances 2015; 5: 79270-8.
[http://dx.doi.org/10.1039/C5RA13419F]

[7]   Choi YK, Poudel BK, Marasini N, *et al.* Enhanced solubility and oral bioavailability of itraconazole by combining membrane emulsification and spray drying technique. Int J Pharm 2012; 434(1-2): 264-71.
[http://dx.doi.org/10.1016/j.ijpharm.2012.05.039] [PMID: 22643224]

[8]   Fathi M, Mozafari MR, Mohebbi M. Nanoencapsulation of food ingredients using lipid based delivery systems. Trends Food Sci Technol 2012; 23: 13-27.
[http://dx.doi.org/10.1016/j.tifs.2011.08.003]

[9]   Shen H, Sun T, Ferrari M. Nanovector delivery of siRNA for cancer therapy. Cancer Gene Ther 2012; 19(6): 367-73.
[http://dx.doi.org/10.1038/cgt.2012.22] [PMID: 22555511]

[10]   Tkachenko AG, Xie H, Liu Y, *et al.* Cellular trajectories of peptide-modified gold particle complexes: Comparison of nuclear localization signals and peptide transduction domains. Bioconjug Chem 2004; 15(3): 482-90.
[http://dx.doi.org/10.1021/bc034189q] [PMID: 15149175]

[11]   Sun L, Liu D, Wang Z. Functional gold nanoparticle-peptide complexes as cell-targeting agents. Langmuir 2008; 24(18): 10293-7.
[http://dx.doi.org/10.1021/la8015063] [PMID: 18715022]

[12]   Liu Y, Franzen S. Factors determining the efficacy of nuclear delivery of antisense oligonucleotides by gold nanoparticles. Bioconjug Chem 2008; 19(5): 1009-16.
[http://dx.doi.org/10.1021/bc700421u] [PMID: 18393455]

[13]   Maus L, Dick O, Bading H, Spatz JP, Fiammengo R. Conjugation of peptides to the passivation shell of gold nanoparticles for targeting of cell-surface receptors. ACS Nano 2010; 4(11): 6617-28.
[http://dx.doi.org/10.1021/nn101867w] [PMID: 20939520]

[14]   Bastús NG, Sánchez-Tilló E, Pujals S, *et al.* Homogeneous conjugation of peptides onto gold nanoparticles enhances macrophage response. ACS Nano 2009; 3(6): 1335-44.
[http://dx.doi.org/10.1021/nn8008273] [PMID: 19489561]

[15]   Rai A, Pinto S, Evangelista MB, *et al.* High-density antimicrobial peptide coating with broad activity and low cytotoxicity against human cells. Acta Biomater 2016; 33: 64-77.
[http://dx.doi.org/10.1016/j.actbio.2016.01.035] [PMID: 26821340]

[16]   Casciaro B, Moros M, Rivera-Fernández S, Bellelli A, de la Fuente JM, Mangoni ML. Gold-nanoparticles coated with the antimicrobial peptide esculentin-1a(1-21)NH$_2$ as a reliable strategy for antipseudomonal drugs. Acta Biomater 2017; 47: 170-81.
[http://dx.doi.org/10.1016/j.actbio.2016.09.041] [PMID: 27693686]

[17]   Pal I, Brahmkhatri VP, Bera S, *et al.* Enhanced stability and activity of an antimicrobial peptide in conjugation with silver nanoparticle. J Colloid Interface Sci 2016; 483: 385-93.
[http://dx.doi.org/10.1016/j.jcis.2016.08.043] [PMID: 27585423]

[18]   Mussa Farkhani S, Asoudeh Fard A, Zakeri-Milani P, Shahbazi Mojarrad J, Valizadeh H. Enhancing antitumor activity of silver nanoparticles by modification with cell-penetrating peptides. Artif Cells Nanomed Biotechnol 2017; 45(5): 1029-35.
[http://dx.doi.org/10.1080/21691401.2016.1200059] [PMID: 27357085]

[19]   Salouti M, Mirzaei F, Shapouri R, Ahangari A. Synergistic antibacterial activity of plant peptide MBP-1 and silver nanoparticles combination on healing of infected wound due to *Staphylococcus aureus*. Jundishapur J Microbiol 2016; 9(1): e27997.
[http://dx.doi.org/10.5812/jjm.27997] [PMID: 27099683]

[20]   Di Pietro P, Zaccaro L, Comegna D, *et al.* Silver nanoparticles functionalized with a fluorescent cyclic RGD peptide: A versatile integrin targeting platform for cells and bacteria. RSC Adv 2016; 6: 112381-92.
[http://dx.doi.org/10.1039/C6RA21568H]

[21]   Song C, Zhong Y, Jiang X, *et al.* Peptide-conjugated fluorescent silicon nanoparticles enabling simultaneous tracking and specific destruction of cancer cells. Anal Chem 2015; 87(13): 6718-23.
         [http://dx.doi.org/10.1021/acs.analchem.5b00853] [PMID: 26021403]

[22]   Walia S, Guliani A, Acharya A. A theragnosis probe based on BSA/HSA-conjugated biocompatible fluorescent silicon nanomaterials for simultaneous *in vitro* cholesterol effluxing and cellular imaging of macrophage cells. ACS Sustain Chem& Eng 2017; 5: 1425-35.
         [http://dx.doi.org/10.1021/acssuschemeng.6b01998]

[23]   Shrestha N, Araújo F, Shahbazi MA, *et al.* Drug delivery: Thiolation and cell penetrating peptide surface functionalization of porous silicon nanoparticles for oral delivery of insulin. Adv Funct Mater 2016; 26: 3374.
         [http://dx.doi.org/10.1002/adfm.201670124]

[24]   Valodkar M, Jadeja RN, Thounaojam MC, Devkar RV, Thakore S. Biocompatible synthesis of peptide capped copper nanoparticles and their biological effect on tumor cells. Mater Chem Phys 2011; 128: 83-9.
         [http://dx.doi.org/10.1016/j.matchemphys.2011.02.039]

[25]   Swed A, Cordonnier T, Fleury F, Boury F. Protein encapsulation into PLGA nanoparticles by a novel phase separation method using non-toxic solvents. J Nanomed Nanotechnol 2014; 5: 241.

[26]   Koyani RD, Vazquez-Duhalt R. Laccase encapsulation in chitosan nanoparticles enhances the protein stability against microbial degradation. Environ Sci Pollut Res Int 2016; 23(18): 18850-7.
         [http://dx.doi.org/10.1007/s11356-016-7072-8] [PMID: 27318485]

[27]   Kumar M, Gupta D, Singh G, *et al.* Novel polymeric nanoparticles for intracellular delivery of peptide Cargos: Antitumor efficacy of the BCL-2 conversion peptide NuBCP-9. Cancer Res 2014; 74(12): 3271-81.
         [http://dx.doi.org/10.1158/0008-5472.CAN-13-2015] [PMID: 24741005]

[28]   Rahimian S, Fransen MF, Kleinovink JW, *et al.* Polymeric nanoparticles for co-delivery of synthetic long peptide antigen and poly IC as therapeutic cancer vaccine formulation. J Control Release 2015; 203: 16-22.
         [http://dx.doi.org/10.1016/j.jconrel.2015.02.006] [PMID: 25660830]

[29]   Gaudana R, Khurana V, Parenky A, Mitra AK. Encapsulation of protein-polysaccharide HIP complex in polymeric nanoparticles. J Drug Deliv 2011.
         [http://dx.doi.org/10.1155/2011/458128] [PMID: 21603214]

[30]   Endo-Takahashi Y, Ooaku K, Ishida K, Suzuki R, Maruyama K, Negishi Y. Preparation of angiopep-2 peptide-modified bubble liposomes for delivery to the brain. Biol Pharm Bull 2016; 39(6): 977-83.
         [http://dx.doi.org/10.1248/bpb.b15-00994] [PMID: 27251499]

[31]   Lomakin Y, Belogurov A, Glagoleva I, *et al.* Administration of myelin basic protein peptides encapsulated in mannosylated liposomes normalizes level of serum TNF-α and IL-2 and chemoattractants ccl2 and ccl4 in multiple sclerosis patients. Mediators Inflamm 2016; 2016: 2847232.
         [http://dx.doi.org/10.1155/2016/2847232] [PMID: 27239100]

[32]   Lindqvist A, Rip J, van Kregten J, Gaillard PJ, Hammarlund-Udenaes M. *In vivo* functional evaluation of increased brain delivery of the opioid peptide DAMGO by Glutathione-PEGylated liposomes. Pharm Res 2016; 33(1): 177-85.
         [http://dx.doi.org/10.1007/s11095-015-1774-3] [PMID: 26275529]

[33]   Ahmed S, Fujita S, Matsumura K. Enhanced protein internalization and efficient endosomal escape using polyampholyte-modified liposomes and freeze concentration. Nanoscale 2016; 8(35): 15888-901.
         [http://dx.doi.org/10.1039/C6NR03940E] [PMID: 27439774]

[34]   Li Z, Paulson AT, Gill TA. Encapsulation of bioactive salmon protein hydrolysates with chitosan-coated liposomes. J Funct Foods 2015; 19: 733-43.
         [http://dx.doi.org/10.1016/j.jff.2015.09.058]

[35]   Neumann S, Young K, Compton B, Anderson R, Painter G, Hook S. Synthetic TRP2 long-peptide and α-galactosylceramide formulated into cationic liposomes elicit CD8+ T-cell responses and prevent tumour progression. Vaccine 2015; 33(43): 5838-44.
         [http://dx.doi.org/10.1016/j.vaccine.2015.08.083] [PMID: 26363382]

[36]   Vila-Caballer M, Codolo G, Munari F, *et al.* A pH-sensitive stearoyl-PEG-poly(methacryloyl sulfadimethoxine)-decorated liposome system for protein delivery: An application for bladder cancer treatment. J Control Release 2016; 238: 31-42.
         [http://dx.doi.org/10.1016/j.jconrel.2016.07.024] [PMID: 27444816]

[37]   Ron-Doitch S, Sawodny B, Kühbacher A, *et al.* Reduced cytotoxicity and enhanced bioactivity of cationic antimicrobial peptides liposomes in cell cultures and 3D epidermis model against HSV. J Control Release 2016; 229: 163-71.
         [http://dx.doi.org/10.1016/j.jconrel.2016.03.025] [PMID: 27012977]

[38]   Jiang Y, Lu H, Chen F, *et al.* PEGylated albumin-based polyion complex micelles for protein delivery. Biomacromolecules 2016; 17(8): 8-17.
         [http://dx.doi.org/10.1021/acs.biomac.5b01537] [PMID: 26809948]

[39]   Fan M, Liang X, Yang D, *et al.* Epidermal growth factor receptor-targeted peptide conjugated phospholipid micelles for doxorubicin delivery. J Drug Target 2016; 24(2): 111-9.
         [http://dx.doi.org/10.3109/1061186X.2015.1058800] [PMID: 26176268]

[40]   Li C, Wang W, Xi Y, *et al.* Design, preparation and characterization of cyclic RGDfK peptide modified poly(ethylene glycol)-block-

poly(lactic acid) micelle for targeted delivery. Mater Sci Eng C 2016; 64: 303-9.
[http://dx.doi.org/10.1016/j.msec.2016.03.062] [PMID: 27127057]

[41]   Valentine KG, Peterson RW, Saad JS, *et al.* Reverse micelle encapsulation of membrane-anchored proteins for solution NMR studies. Structure 2010; 18(1): 9-16.
[http://dx.doi.org/10.1016/j.str.2009.11.010] [PMID: 20152148]

[42]   Du X, Song N, Yang YW, Wu G, Ma J, Gao H. Reverse micelles based on β-cyclodextrin-incorporated amphiphilic polyurethane copolymers for protein delivery. Polym Chem 2014; 5: 5300-9.
[http://dx.doi.org/10.1039/C4PY00278D]

[43]   Hayes DG, Gulari E. Improvement of enzyme activity and stability for reverse micellar-encapsulated lipases in the presence of short-chain and polar alcohols. Biocatalysis 1994; 11: 223-31.
[http://dx.doi.org/10.3109/10242429408998142]

[44]   Ren Y, Mu Y, Song Y, *et al.* A new peptide ligand for colon cancer targeted delivery of micelles. Drug Deliv 2016; 23(5): 1763-72.
[http://dx.doi.org/10.3109/10717544.2015.1077293] [PMID: 26289214]

[45]   Lewin M, Carlesso N, Tung CH, *et al.* Tat peptide-derivatized magnetic nanoparticles allow *in vivo* tracking and recovery of progenitor cells. Nat Biotechnol 2000; 18(4): 410-4.
[http://dx.doi.org/10.1038/74464] [PMID: 10748521]

[46]   Nitin N, LaConte LE, Zurkiya O, Hu X, Bao G. Functionalization and peptide-based delivery of magnetic nanoparticles as an intracellular MRI contrast agent. J Biol Inorg Chem 2004; 9(6): 706-12.
[http://dx.doi.org/10.1007/s00775-004-0560-1] [PMID: 15232722]

[47]   Zhang J, Shin MC, David AE, *et al.* Long-circulating heparin-functionalized magnetic nanoparticles for potential application as a protein drug delivery platform. Mol Pharm 2013; 10(10): 3892-902.
[http://dx.doi.org/10.1021/mp400360q] [PMID: 24024964]

[48]   Su T, Wang Y, Wang J, *et al.* *In vivo* magnetic resonance and fluorescence dual-modality imaging of tumor angiogenesis in rats using GEBP11 peptide targeted magnetic nanoparticles. J Biomed Nanotechnol 2016; 12(5): 1011-22.
[http://dx.doi.org/10.1166/jbn.2016.2233] [PMID: 27305822]

[49]   Wunderbaldinger P, Josephson L, Weissleder R. Tat peptide directs enhanced clearance and hepatic permeability of magnetic nanoparticles. Bioconjug Chem 2002; 13(2): 264-8.
[http://dx.doi.org/10.1021/bc015563u] [PMID: 11906263]

[50]   Hauser AK. Peptide-functionalized magnetic nanoparticles for cancer therapy applications. 2016.
[http://dx.doi.org/10.13023/ETD.2016.157]

[51]   Maleki H, Rai A, Pinto S, *et al.* High antimicrobial activity and low human cell cytotoxicity of core–shell magnetic nanoparticles functionalized with an antimicrobial peptide. ACS Appl Mater Interfaces 2016; 8(18): 11366-78.
[http://dx.doi.org/10.1021/acsami.6b03355] [PMID: 27074633]

[52]   Peng LH, Huang YF, Zhang CZ, *et al.* Integration of antimicrobial peptides with gold nanoparticles as unique non-viral vectors for gene delivery to mesenchymal stem cells with antibacterial activity. Biomaterials 2016; 103: 137-49.
[http://dx.doi.org/10.1016/j.biomaterials.2016.06.057] [PMID: 27376562]

[53]   Yi Y, Kim HJ, Mi P, *et al.* Targeted systemic delivery of siRNA to cervical cancer model using cyclic RGD-installed unimer polyion complex-assembled gold nanoparticles. J Control Release 2016; 244(Pt B): 247-56.
[http://dx.doi.org/10.1016/j.jconrel.2016.08.041] [PMID: 27590214]

[54]   Gamrad L, Rehbock C, Westendorf AM, Buer J, Barcikowski S, Hansen W. Efficient nucleic acid delivery to murine regulatory T cells by gold nanoparticle conjugates. Sci Rep 2016; 6: 28709.
[http://dx.doi.org/10.1038/srep28709] [PMID: 27381215]

[55]   Sun D, Zhang W, Li N, *et al.* Silver nanoparticles-quercetin conjugation to siRNA against drug-resistant *Bacillus subtilis* for effective gene silencing: *In vitro* and *in vivo*. Mater Sci Eng C 2016; 63: 522-34.
[http://dx.doi.org/10.1016/j.msec.2016.03.024] [PMID: 27040247]

[56]   Sarkar K, Banerjee SL, Kundu PP, Madras G, Chatterjee K. Biofunctionalized surface-modified silver nanoparticles for gene delivery. J Mater Chem B Mater Biol Med 2015; 3: 5266-76.
[http://dx.doi.org/10.1039/C5TB00614G]

[57]   Chen Q, Xu M, Zheng W, Xu T, Deng H, Liu J. Se/Ru-Decorated porous metal-organic framework nanoparticles for the delivery of pooled siRNAs to reversing multidrug resistance in taxol-resistant breast cancer cells. ACS Appl Mater Interfaces 2017; 9(8): 6712-24.
[http://dx.doi.org/10.1021/acsami.6b12792] [PMID: 28191840]

[58]   Huh MS, Lee SY, Park S, *et al.* Tumor-homing glycol chitosan/polyethylenimine nanoparticles for the systemic delivery of siRNA in tumor-bearing mice. J Control Release 2010; 144(2): 134-43.
[http://dx.doi.org/10.1016/j.jconrel.2010.02.023] [PMID: 20184928]

[59]   Cao N, Cheng D, Zou S, Ai H, Gao J, Shuai X. The synergistic effect of hierarchical assemblies of siRNA and chemotherapeutic drugs co-delivered into hepatic cancer cells. Biomaterials 2011; 32(8): 2222-32.
[http://dx.doi.org/10.1016/j.biomaterials.2010.11.061] [PMID: 21186059]

[60] Kong L, Alves CS, Hou W, *et al.* RGD peptide-modified dendrimer-entrapped gold nanoparticles enable highly efficient and specific gene delivery to stem cells. ACS Appl Mater Interfaces 2015; 7(8): 4833-43.
[http://dx.doi.org/10.1021/am508760w] [PMID: 25658033]

[61] Zhang X, Zhang J, Zhu KY. Chitosan/double-stranded RNA nanoparticle-mediated RNA interference to silence chitin synthase genes through larval feeding in the African malaria mosquito (*Anopheles gambiae*). Insect Mol Biol 2010; 19(5): 683-93.
[http://dx.doi.org/10.1111/j.1365-2583.2010.01029.x] [PMID: 20629775]

[62] Christiansen JP, French BA, Klibanov AL, Kaul S, Lindner JR. Targeted tissue transfection with ultrasound destruction of plasmid-bearing cationic microbubbles. Ultrasound Med Biol 2003; 29(12): 1759-67.
[http://dx.doi.org/10.1016/S0301-5629(03)00976-1] [PMID: 14698343]

[63] Forbes DC, Peppas NA. Polycationic nanoparticles for siRNA delivery: Comparing ARGET ATRP and UV-initiated formulations. ACS Nano 2014; 8(3): 2908-17.
[http://dx.doi.org/10.1021/nn500101c] [PMID: 24548237]

[64] Lee PW, Peng SF, Su CJ, *et al.* The use of biodegradable polymeric nanoparticles in combination with a low-pressure gene gun for transdermal DNA delivery. Biomaterials 2008; 29(6): 742-51.
[http://dx.doi.org/10.1016/j.biomaterials.2007.10.034] [PMID: 18001831]

[65] Patil Y, Panyam J. Polymeric nanoparticles for siRNA delivery and gene silencing. Int J Pharm 2009; 367(1-2): 195-203.
[http://dx.doi.org/10.1016/j.ijpharm.2008.09.039] [PMID: 18940242]

[66] Sharma G, Modgil A, Layek B, *et al.* Cell penetrating peptide tethered bi-ligand liposomes for delivery to brain *in vivo*: Biodistribution and transfection. J Control Release 2013; 167(1): 1-10.
[http://dx.doi.org/10.1016/j.jconrel.2013.01.016] [PMID: 23352910]

[67] Yu-Wai-Man C, Tagalakis AD, Manunta MD, Hart SL, Khaw PT. Receptor-targeted liposome-peptide-siRNA nanoparticles represent an efficient delivery system for MRTF silencing in conjunctival fibrosis. Sci Rep 2016; 6: 21881.
[http://dx.doi.org/10.1038/srep21881] [PMID: 26905457]

[68] Yang ZZ, Li JQ, Wang ZZ, Dong DW, Qi XR. Tumor-targeting dual peptides-modified cationic liposomes for delivery of siRNA and docetaxel to gliomas. Biomaterials 2014; 35(19): 5226-39.
[http://dx.doi.org/10.1016/j.biomaterials.2014.03.017] [PMID: 24695093]

[69] Sakurai Y, Hatakeyama H, Sato Y, *et al.* RNAi-mediated gene knockdown and anti-angiogenic therapy of RCCs using a cyclic RGD-modified liposomal-siRNA system. J Control Release 2014; 173: 110-8.
[http://dx.doi.org/10.1016/j.jconrel.2013.10.003] [PMID: 24120854]

[70] Ran R, Liu Y, Gao H, *et al.* Enhanced gene delivery efficiency of cationic liposomes coated with PEGylated hyaluronic acid for anti P-glycoprotein siRNA: A potential candidate for overcoming multi-drug resistance. Int J Pharm 2014; 477(1-2): 590-600.
[http://dx.doi.org/10.1016/j.ijpharm.2014.11.012] [PMID: 25448564]

[71] Negishi Y, Endo-Takahashi Y, Maruyama K. Gene delivery systems by the combination of lipid bubbles and ultrasound. Drug Discov Ther 2016; 10(5): 248-55.
[http://dx.doi.org/10.5582/ddt.2016.01063] [PMID: 27795481]

[72] Dobson J. Gene therapy progress and prospects: Magnetic nanoparticle-based gene delivery. Gene Ther 2006; 13(4): 283-7.
[http://dx.doi.org/10.1038/sj.gt.3302720] [PMID: 16462855]

[73] Schillinger U, Brill T, Rudolph C, *et al.* Advances in magnetofection-magnetically guided nucleic acid delivery. J Magn Magn Mater 2005; 293: 501-8.
[http://dx.doi.org/10.1016/j.jmmm.2005.01.032]

[74] Krötz F, de Wit C, Sohn HY, *et al.* Magnetofection-a highly efficient tool for antisense oligonucleotide delivery *in vitro* and *in vivo*. Mol Ther 2003; 7(5 Pt 1): 700-10.
[http://dx.doi.org/10.1016/S1525-0016(03)00065-0] [PMID: 12718913]

[75] McBain SC, Yiu HH, Dobson J. Magnetic nanoparticles for gene and drug delivery. Int J Nanomedicine 2008; 3(2): 169-80.
[PMID: 18686777]

[76] Jiang S, Eltoukhy AA, Love KT, Langer R, Anderson DG. Lipidoid-coated iron oxide nanoparticles for efficient DNA and siRNA delivery. Nano Lett 2013; 13(3): 1059-64.
[http://dx.doi.org/10.1021/nl304287a] [PMID: 23394319]

[77] Song HP, Yang JY, Lo SL, *et al.* Gene transfer using self-assembled ternary complexes of cationic magnetic nanoparticles, plasmid DNA and cell-penetrating Tat peptide. Biomaterials 2010; 31(4): 769-78.
[http://dx.doi.org/10.1016/j.biomaterials.2009.09.085] [PMID: 19819012]

[78] Toome K, Willmore AM, Säälik P, *et al.* Ddel-19 penetration of homing peptide-functionalized nanoparticles to glioma spheroids *in vitro*. Neuro Onco 2015; 17(suppl-5): v77-7.

[79] Steinstraesser L, Koehler T, Jacobsen F, *et al.* Host defense peptides in wound healing. Mol Med 2008; 14(7-8): 528-37.
[http://dx.doi.org/10.2119/2008-00002.Steinstraesser] [PMID: 18385817]

[80] Liu L, Yang J, Xie J, *et al.* The potent antimicrobial properties of cell penetrating peptide-conjugated silver nanoparticles with excellent selectivity for gram-positive bacteria over erythrocytes. Nanoscale 2013; 5(9): 3834-40.

[http://dx.doi.org/10.1039/c3nr34254a] [PMID: 23525222]

[81]   Reithofer MR, Lakshmanan A, Ping AT, Chin JM, Hauser CA. *In situ* synthesis of size-controlled, stable silver nanoparticles within ultrashort peptide hydrogels and their anti-bacterial properties. Biomaterials 2014; 35(26): 7535-42.
[http://dx.doi.org/10.1016/j.biomaterials.2014.04.102] [PMID: 24933510]

[82]   Wang Z, Lévy R, Fernig DG, Brust M. Kinase-catalyzed modification of gold nanoparticles: A new approach to colorimetric kinase activity screening. J Am Chem Soc 2006; 128(7): 2214-5.
[http://dx.doi.org/10.1021/ja058135y] [PMID: 16478166]

[83]   Guarise C, Pasquato L, De Filippis V, Scrimin P. Gold nanoparticles-based protease assay. Proc Natl Acad Sci USA 2006; 103(11): 3978-82.
[http://dx.doi.org/10.1073/pnas.0509372103] [PMID: 16537471]

[84]   Rai A, Pinto S, Velho TR, *et al.* One-step synthesis of high-density peptide-conjugated gold nanoparticles with antimicrobial efficacy in a systemic infection model. Biomaterials 2016; 85: 99-110.
[http://dx.doi.org/10.1016/j.biomaterials.2016.01.051] [PMID: 26866877]

[85]   Kalmodia S, Vandhana S, Tejaswini Rama BR, *et al.* Bio-conjugation of antioxidant peptide on surface-modified gold nanoparticles: A novel approach to enhance the radical scavenging property in cancer cell. Cancer Nanotechnol 2016; 7: 1.
[http://dx.doi.org/10.1186/s12645-016-0013-x] [PMID: 26900409]

[86]   Akrami M, Balalaie S, Hosseinkhani S, *et al.* Tuning the anticancer activity of a novel pro-apoptotic peptide using gold nanoparticle platforms. Sci Rep 2016; 6: 31030.
[http://dx.doi.org/10.1038/srep31030] [PMID: 27491007]

6

Protease-, Pectinase- and Amylase- Producing Bacteria from a Kenyan Soda Lake

Kevin Raymond Oluoch[1,2,*], Patrick Wafula Okanya[2], Rajni Hatti-Kaul[1], Bo Mattiasson[1] and Francis Jakim Mulaa[2]

[1]Department of Biotechnology, Center for Chemistry and Chemical Engineering, Lund University, Lund, Sweden
[2]Department of Biochemistry, University of Nairobi, Nairobi, Kenya

Abstract:

Background:

Alkaline enzymes are stable biocatalysts with potential applications in industrial technologies that offer high quality products.

Objective:

The growing demand for alkaline enzymes in industry has enhanced the search for microorganisms that produce these enzymes.

Methods:

Eighteen bacterial isolates from Lake Bogoria, Kenya, were screened for alkaline proteases, pectinases and amylases; characterized and subjected to quantitative analysis of the enzymes they produced.

Results:

The screening analysis ranked 14, 16 and 18 of the bacterial isolates as potent producers of alkaline proteases, pectinases and amylases, respectively. The isolates were classified into two groups: Group 1 (16 isolates) were facultatively alkaliphilic *B. halodurans* while group 2 (2 isolates) were obligately alkaliphilic *B. pseudofirmus*. Further analysis revealed that group 1 isolates were divided into two sub-groups, with sub-group I (4 isolates) being a phenotypic variant sub-population of sub-group II (12 isolates). Variation between the two populations was also observed in their enzymatic production profiles *e.g.* sub-group I isolates did not produce alkaline proteolytic enzymes while those in sub-group II did so (0.01-0.36 U/ml). Furthermore, they produced higher levels of the alkaline pectinolytic enzyme polygalacturonase (0.12-0.46 U/ml) compared to sub-group II isolates (0.05-0.10 U/ml), which also produced another pectinolytic enzyme - pectate lyase (0.01 U/ml). No clear distinction was however, observed in the production profiles of alkaline amylolytic enzymes by the isolates in the two sub-populations [0.20-0.40 U/ml (amylases), 0.24-0.68 U/ml (pullulanases) and 0.01-0.03 U/ml (cyclodextrin glycosyl transferases)]. On the other hand, group 2 isolates were phenotypically identical to one another and also produced similar amounts of proteolytic (0.38, 0.40 U/ml) and amylolytic [amylases (0.06, 0.1 U/ml), pullulanases (0.06, 0.09 U/ml) and cyclodextrin glycosyl transferases (0.01, 0.02 U/ml)] enzymes.

Conclusion:

The facultatively alkaliphilic *B. halodurans* and obligately alkaliphilic *B. pseudofirmus* isolates are attractive biotechnological sources of industrially important alkaline enzymes.

Keywords: Soda lake, Alkaliphiles, Proteases, Pectinases, Amylases, *Bacillus halodurans* , *Bacillus pseudofirmus*.

* Address correspondence to this author at the Department of Biochemistry, University of Nairobi, Nairobi, Kenya; E-mail: kevin.oluoch@uonbi.ac.ke

1. INTRODUCTION

Soda lakes are some of the main types of alkaline environments that occur naturally on earth. They are characterized by high alkalinity (pH 8-12), large amounts of Na_2CO_3 and NaCl [5% (w/v) NaCl to saturation], and high and low concentrations of Na^+ and Mg^{2+}/Ca^{2+} ions, respectively, due to evaporative concentration [1, 2]. Although soda lakes have a wide geographical distribution on earth, their hostile nature and inaccessibility have contributed to only a few of them being explored from the limnological and microbiological point of view. Among the most studied soda lakes are those fed by carbonated hot springs in the Kenyan Rift valley area. These alkaline water bodies also support a considerably diverse population of alkaliphiles i.e. microorganisms that are able to grow at optimal pH values > 9.0, but fail to grow/grow slowly at around pH 7 [3]. They include archea, Gram negative protobacteria (*e.g. Pseudomonas, Halomonas* and *Deleya*) and Gram positive bacteria/eubacteria (*e.g. Bacillus* and *Clostridium*) [4].

Alkaliphilic bacteria, especially those belonging to the genus *Bacillus* have generated a lot of interest because of their capability to thrive under harsh alkaline conditions and to produce alkali- stable enzymes [5]. The unique characteristics of these biocatalysts, therefore, present an opening for their utilization in industrial technologies that offer high quality products. For example, alkaline enzymes have been shown to have a great impact in laundry and dishwashing detergents [6]. Other areas of application of alkaline enzymes are in the leather, silver recovery, pharmaceutical, chemical, food and feed, pulp and paper, and textile industries as well as in wastewater treatment and cyclodextrin production [6 - 10].

A considerable number of bioprospection studies for bacteria have been carried out world-wide for production of alkaline enzymes, but only a few such studies have been forthcoming from Kenya. In this article, we report on the quantitative screening of bacteria isolated from different samples collected in Lake Bogoria (a Kenyan soda lake) hot spring wells for production of alkaline proteases, pectinases and amylases; characterization of the bacterial isolates and qualitative determination of the enzymes they produce.

2. MATERIALS AND METHODS

2.1. Source of Microorganisms

Soil, water, polymer and maize-cob samples were collected in sterile bottles from different sites within alkaline hot spring wells and their drainage channels leading to Lake Bogoria (00° 15'N & 36°06' E), Kenya. Samples were transported to the Department of Biochemistry, University of Nairobi, Kenya, where they were stored at -20 °C before use. Reference strains *Bacillus halodurans* DSM 497 and *Bacillus pseudofirmus* DSM 8715 were bought from Deutsche Sammlung von Mikroorganismen und Zellkulturen (DSMZ), Braunschweig, Germany. All chemicals were purchased from standard sources.

2.2. Isolation

One hundred and ninety soda lake bacterial isolates, designated as LBW, LBS or LBK, were isolated from the Lake Bogoria samples by serial dilution on modified Horikoshi I solid medium (pH 10) [11]. Eighteen isolates were chosen for this study based on their dominance in growth during the isolation process. Table 1 shows the origins of the chosen bacterial isolates and the physical characteristics recorded at the sampling sites.

Table 1. The origins of the soda lake bacterial isolates used in this study.

Isolate	Sampling Site	Sample Type	pH	Temp (°C)
LBW 2719	Hot spring well no. 2	Soil	9	65
LBW 226	Hot spring well no. 2	Polymer	9	65
LBW 318	Hot spring well no. 3	Maize-cob	9	63
LBS 77	Upstream drainage channel of hot spring well no. 7	Water	9	60
LBK 261	Periphery of hot spring well no. 2	Water	9	54
LBW 4512	Periphery of hot spring well no. 4	Polymer	9	50
LBW 327	Periphery of hot spring well no. 3	Polymer	9	57
LBW 434	Periphery of hot spring well no. 4	Polymer	9	52
LBW 7526a	Periphery of hot spring well no. 7	Polymer	9.5	55
LBW 39	Periphery of hot spring well no. 3	Polymer	9	58
LBW 7526b	Periphery of hot spring well no. 7	Polymer	9.5	56

(Table 1) contd.....

Isolate	Sampling Site	Sample Type	pH	Temp ($^\circ$C)
LBW 328	Periphery of hot spring well no. 3	Polymer	9	57
LBS 16	Downstream drainage channel of hot spring well no. 1	Soil	9	45
LBW 625	Periphery of hot spring well no. 6	Soil	9	55
LBW 446	Periphery of hot spring well no. 4	Polymer	9	50
LBW 5117	Periphery of hot spring well no. 5	Soil	8	51
LBW 313	Periphery of hot spring well no. 3	Maize-cob	9	50
LBW 317	Periphery of hot spring well no. 3	Maize-cob	9	48

2.3. Screening

The soda lake bacterial isolates were subjected to plate-test screening for production of alkaline proteases, pectinases and amylases, respectively. Screening for protease production was performed as described by Juwon and Emanuel [12]. The solid gelatin-nutrient agar medium [adjusted to pH 10.5 with 20% (w/v) Na_2CO_3] was spot-inoculated with 1 µl of each isolate and cultures incubated at 37 $^\circ$C for 72 h. Complete degradation of gelatin by the isolates is manifested by the formation of halos around positive protease producing colonies after flooding the plate with an aqueous solution of saturated ammonium sulfate (769 g^{-1}).

Screening for pectinase production was performed as described by Soares *et al*. [13]. The solid pectin medium was prepared according to Kelly and Fogarty [14] and its pH adjusted to 10.5 with 20% (w/v) Na_2CO_3. The medium was spot-inoculated with 1 µl of each isolate and the plate incubated at 37° C for 72 h. The assay plate was developed by staining with Lugol's iodine (5.0 g potassium iodide, 1.0 g iodine and 330 ml H_2O) dye solution and the presence of halos around the colonies used as an indication of pectinase producing isolates.

Amylase production by the isolates was evaluated by inoculating 1 µl of each isolate on Horikoshi II solid medium [15] (pH 10.5) and incubating at 37 $^\circ$C for 72 h. The assay plate was flooded with Gram's iodine (1.27g iodine in 10 ml distilled water containing 2 g potassium iodide, and diluted to 300ml with distilled water) dye solution and the presence of halos around the colonies used as an indication of amylase producing isolates [16].

The efficiency of the proteases, pectinases and amylases to solubilize gelatin, pectin and starch respectively, and thus form halos around the colonies during the screening processes above, was determined qualitatively in terms of solubilization index (SI) value for each isolate calculated using the formula: SI = mean Diameter of halo (Dh) (mm)/mean Diameter of colony (Dc) (mm) [17]. The obtained mean SI values were used to rank the isolates as excellent producers of the enzymes when the colonies presented halo sizes > 3.0; very good producers when the halos were 2.0 < SI ≤ 3.0; good producers when halos were 1.0 < SI ≤ 2.0; weak producers when halos were 0 < SI ≤ 1.0 (or not clear); producers of minor and unquantifiable enzymes when halo sizes cannot be determined and poor producers when no halos were observed.

2.4. Phenotypic and Molecular Characterization

Identification of the soda lake bacterial isolates was performed on the basis of their phenotypic and molecular characterization after growing them on solid nutrient media (adjusted to pH 10.5 with 20% (w/v) Na_2CO_3) at 37 $^\circ$C for different periods of time as described below:

2.4.1. Morpholgical, Biochemical and Physiological Characterization

For cell shape determination and classification as Gram positive/Gram negative, 12 or 48 h old colonies were subjected to Gram's staining by the method described by Gerhardt *et al*. [18]. The KOH test was also done to corroborate the Gram stain results [19]. For the detection of spores, 72 h old colonies were treated with saline [0.9% (w/v)] solution on microscope slides and observed under a phase contrast microscope (Leica EZA D, Cambridge, UK). To study the colony morphology of the isolates, 48 h old cultures were evaluated with the naked eye.

Catalase test was performed by the method described by Gerhardt *et al*. [18] while oxidase test was carried out using oxidase reagent according to Gordon-McLeod (Sigma-Aldrich) on 12 or 48 h old colonies, as recommended by the manufacturer. Production of effervescence due to catalase-catalyzed breakdown of H_2O_2 to molecular O_2 indicates a positive reaction while change in the color of bacterial colony to purple is a positive oxidase reaction.

The effects of pH (7.0 – 11.0) (alkaliphily test), temperature (25 - 65 $^\circ$C), and salinity [0 – 18% (w/v) NaCl] on growth of the isolates, as well as oxygen requirement- and cell motility- tests, were carried out according to Martins *et al.* [20] on 48 h old cultures.

2.4.2. 16S rDNA Amplification, DNA Sequencing, Phylogenetic Analysis and Accession Numbers

Genomic DNA was extracted from 24 h old soda lake bacterial colonies using PureLink® Genomic DNA extraction kit (Invitrogen Life Technologies™, CA, USA) as per the manufacturer's instructions and their A260/A280 and A260/A230 ratios checked to assess their purity [21]. 16S rDNAs were PCR amplified using universal forward 8-27F (5'-AGAGTTTGATCCTGGCTCAG-3') and reverse 1492R (5'-CTACGGCTACCTTGTTACGA-3') primers in a Gene Amp PCR system 9700 (Applied Biosystems, CA, USA [22]. Amplified 16S rDNA fragments were purified from a 1% (w/v) agarose gel using a Qiagen DNA purification kit (Limburg, Netherlands) as recommended by the manufacturer and sequenced on both strands at Macrogen Europe Laboratory (Amsterdam, Netherlands). DNA Baser Sequence Assembler (version 4.20) was used to edit and assemble contigs from chromatograms. Assembled nucleotide sequences were aligned with those obtained in the GeneBank of NCBI in MUSCLE [23]. A Phylogenetic tree was then constructed from the aligned nucleotide sequences using the Bayesian phylogenetic method executed in MrBayes bioinformatics software version 3.1.2 [24] and visualised using Fig Tree software version 1.2.2 (http://tree.bio.ed.ac.uk/). The 16S rDNA sequences determined in this study have been deposited in the GenBank at NCBI database under accession numbers KU321024 to KU321040. The sequence of isolate LBK 261 whose accession number is AY423275 was also deposited in the GenBank.

2.5. Production Media and Assay Procedure for Enzyme Activity Determination

The soda lake bacterial isolates were inoculated separately in 50 ml alkaline broth medium (in 250 ml conical flasks) prepared as described in the screening section for amylases and pectinases above, but without agar. Each isolate was similarly inoculated in separate 50 ml alkaline broth media (pH adjusted to 10.5 using 20% Na_2CO_3) containing 0.5 g chicken quill feathers and 300 μl L^{-1} trace elements [(g/L^{-1}): $MnCl_2$ $4H_2O$ (15.1), $CuSO_4$ $5H_2O$ (0.125), $Co(NO_3)_2$ $6H_2O$ (0.23), $CaCl_2$ $2H_2O$ (1.7), Na_2MoO_4 $2H_2O$ (0.125), H_3BO_3 (2.5), $FeSO_4$ $7H_2O$ (1.3), $ZnSO_4$ $7H_2O$ 2.5 ml of 18.3 M H_2SO_4]. All cultures were cultivated at 37 °C and 200 rpm for 48 h, after which they were centrifuged at 4,000 rpm and 4 °C for 30 min. Cell-free culture supernatants from the chicken quill-feather, pectin and starch media were used as crude sources for quantitative determination of protease, pectinase and amylase activities, respectively, in duplicates, as described below:

2.5.1. Protease Activity Assay

Protease activity determination was performed by the method described by Bakhtiar *et al.* [25], but with a slight modification. 0.5 ml enzyme was added to 0.5 ml 1% (w/v) casein in 50 mM glycine-NaOH buffer pH 10.5 and the mixture incubated at 50 °C for 20 min. The reaction was stopped by adding 0.5 ml 10% (w/v) ice-cold trichloroacetic acid, left to stand for 5 min at 4 °C and centrifuged at 8,000 rpm at the same temperature for 2 min. 0.5 ml supernatant was obtained and 2.5 ml 0.5 M sodium carbonate and 0.5 ml three-fold diluted Folin-Ciocalteau phenol reagent added to it in that order. The reaction mixture was left undisturbed at room temperature for 30 min and the absorbance read at 600 nm. One unit of protease activity was defined as the amount of enzyme that liberated 1 μmol of tyrosine per min under the standard assay conditions.

2.5.2. Pectinase Activity Assay

Polygalacturonase (PGase) activity determination was carried out by adding 0.2 ml enzyme to 0.8 ml 0.5% (w/v) polygalacturonic acid sodium salt in 50 mM glycine-NaOH buffer pH 10.5 and incubating the reaction mixture at 55 °C. After 10 min of incubation, the amount of reducing sugars formed was tracked down by a modified version of the DNS method [26]. One unit of PGase activity was defined as the amount of enzyme that liberated 1 μmol of reducing sugars as monogalacturonic acid per min under the standard assay conditions.

Pectate lyase (PecL) activity was determined by the method described by Soriano *et al.* [27], but with a slight modification. 0.1 ml crude enzyme was added to 1.9 ml 0.2% (w/v) polygalacturonic acid sodium salt in 50 mM glycine NaOH buffer pH 10.5, which also contained 0.5 mM $CaCl_2$. The assay mixture was incubated at 55 °C for 2.5 min and increase in A 232 nm measured at intervals of 0.5 min after terminating the reaction with 2 ml 50 mM HCl and centrifuging the tube at 14,000 rpm (5 min). One unit of PecL activity was defined as the amount of enzyme that liberated 1 μmole unsaturated galacturonates under the standard assay conditions. Molar extinction coefficient of the unsaturated product was assumed to be 4,600 $M^{-1}cm^{-1}$.

2.5.3. Amylase Activity Assay

Amylase and pullulanase activity determination was performed as described by Martins *et al*. [20], but with some modification. 0.1 ml crude enzyme was added to 0.4 ml 0.3% (w/v) starch in 50 mM glycine-NaOH buffer pH 10.5 and 0.4 ml 0.5% (w/v) pullulan in 50 mM glycine-NaOH buffer pH 10.5, respectively. Reaction mixtures were incubated at 55 °C and after 10 min, the amount of reducing sugars formed in both assays were tracked down by a modified version of the DNS method [26]. One unit of amylase (or pullulanase) activity was defined as the amount of enzyme that liberated 1 µmol of reducing sugars per min under the standard assay conditions.

Cyclodextrin glycosyltransferase (CGTase) activity determination was performed as described by Martins *et al*. [20], but with slight modification. 0.1 ml enzyme was mixed with 0.7 ml 5% (w/v) pre-incubated (50 °C, 5 min) maltodextrin in 50 mM glycine-NaOH buffer pH 10.5. The reaction mixture was incubated at 50 °C and after 25 min the reaction was stopped by adding 375 µl 0.15 M NaOH. This was followed by a further addition of 0.1 ml 0.02% (w/v) phenolphthalein in 5 mM Na_2CO_3, after which the reaction mixture was left to stand at room temperature for 15 min and the absorbance read at 550 nm. One unit of CGTase activity was defined as the amount of enzyme that liberated 1 µmol of β-cyclodextrin per min under the standard assay conditions.

3. RESULT AND DISCUSSION

Eighteen soda lake bacteria were isolated from different samples collected in hot spring wells and their drainage channels in L. Bogoria, Kenya, and screened qualitatively on solid gelatin, pectin and starch media (pH 10.5) for production of alkaline proteases, pectinases and amylases, respectively. The screening data analysis showed that 14, 16 and 18 isolates formed clear halos around their respective colonies on solid gelatin, pectin and starch media, respectively (Fig. **1**), thus suggesting that all were potent producers of alkaline amylases, while the majority were producers of proteases and pectinases. The size of halos formed (SI) depended on the efficiencies of the enzymes produced by the isolates to solubilize the starch, gelatin and pectin in their respective media, resulting in the decolorization of the dye around the positive colonies on both the starch and pectin plates, and failure of the $(NH_4)_2SO_4$ solution to precipitate proteins around the positive colonies.

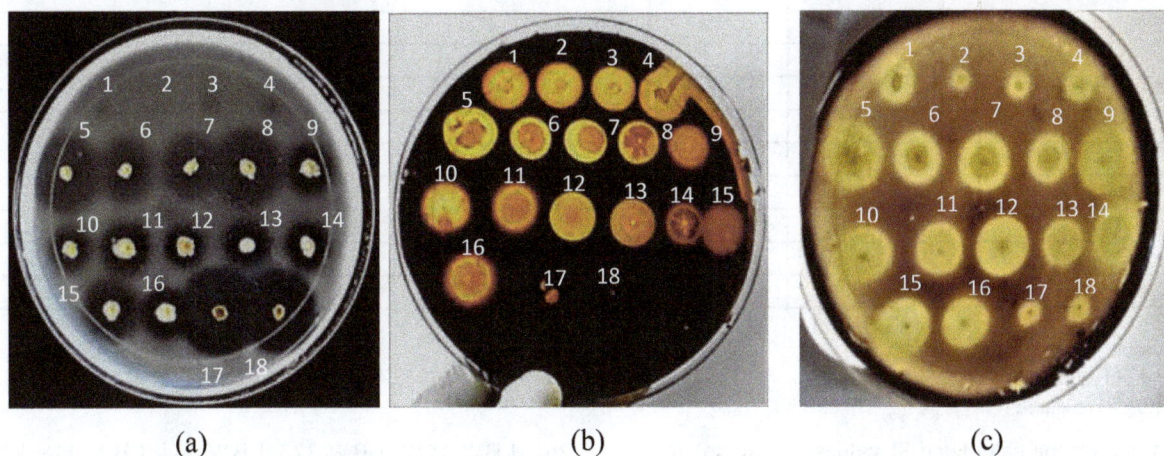

(a)　　　　　　　　(b)　　　　　　　　(c)

Fig. (1). Screening plate assays for alkaline protease, pectinase and amylase production by the soda lake bacterial isolates as seen on solid gelatin (a), pectin (b) and starch (c) media (pH 10.5), respectively. 1 = LBW 2719, 2 = LBW 226, 3 = LBW 318, 4 = LBS 77, 5 = LBK 261, 6 = LBW 4512, 7 = LBW 327, 8 = LBW 434, 9 = LBW 7526a, 10 = LBW 39, 11 = LBW 7526b, 12 = LBW 328, 13 = LBS 16, 14 = LBW 625, 15 = LBW 446, 16 = LBW 5117, 17 = LBW 313 and 18 = LBW 317.

The diameter of the colonies (Dc) and that of their respective halos (Dh) were then measured (Table **2**) and used to calculate the size of the halos or amount of enzymes (SI values) the isolates produced (Table **2** and Fig. **2**).

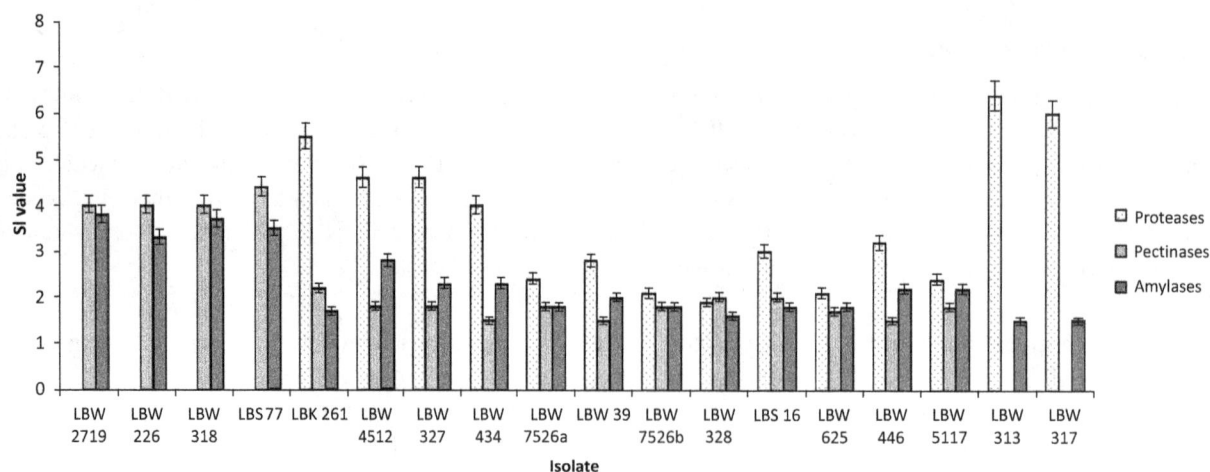

Fig. (2). Production of alkaline proteases, pectinases and amylases by the soda lake bacterial isolates following plate-test screening on gelatin, pectin and starch media (pH 10.5).

Table 2. SI values obtained for alkaline proteases, pectinases and amylases following plate-test screening.

Isolate	Proteases			Pectinases			Amylases		
Isolate	Dc	Dh	SI	Dc	Dh	SI	Dc	Dh	SI
Isolate	(mm)	(mm)	(Dh/Dc)	(mm)	(mm)	(Dh/Dc)	(mm)	(mm)	(Dh/Dc)
LBW 2719	a	2	b	5	20	4	4	15	3.8
LBW 226	a	2	b	5	20	4	3	10	3.3
LBW 318	a	2	b	5	20	4	3	11	3.7
LBS 77	a	2	b	5	22	4.4	4	14	3.5
LBK 261	5	28	5.6	10	22	2.2	13	22	1.7
LBW 4512	5	23	4.6	10	18	1.8	8	22	2.8
LBW 327	5	23	4.6	10	18	1.8	9	21	2.3
LBW 434	7	28	4	12	18	1.5	8	18	2.3
LBW 7526a	8	19	2.4	10	18	1.8	13	23	1.8
LBW 39	6	17	2.8	12	18	1.5	11	22	2
LBW 7526b	8	17	2.1	10	18	1.8	11	20	1.8
LBW 328	8	15	1.9	10	20	2	14	23	1.6
LBS 16	5	15	3	10	20	2	11	20	1.8
LBW 625	7	15	2.1	10	17	1.7	11	20	1.8
LBW 446	6	19	3.2	15	22	1.5	13	28	2.2
LBW 5117	7	17	2.4	10	18	1.8	11	24	2.2
LBW 313	5	32	6.4	2	No halo	c	8	12	1.5
LBW 317	5	30	6	1	No halo	c	8	12	1.5

Key:

[a], No visible growth.

[b], Minor and unquantifiable enzyme.

[c], No enzyme produced.

   Based on the calculated SI values, seven isolates (LBW 261, LBW 4512, LBW 327, LBW 434, LBW 446, LBW 313 and LBW 317) were ranked as excellent producers of proteases and four (LBW 2719, LBW 226, LBW 318 and LBS 77) as excellent producers of both pectinases and amylases ($SI > 3.0$) (Fig. **2**). The rest of the isolates were either very good ($2.0 < SI \leq 3.0$), good ($1.0 < SI \leq 2.0$) or poor producers (no halos formed) of these enzymes. In addition, four isolates (LBW 2719, LBW 226, LBW 318 and LBS 77) were identified as producers of minor and unquantifiable proteases as they presented unclear halos due to poor (invisible) bacterial growths.

   Based on their morphological and biochemical tests, all the soda lake bacterial isolates were motile, rod shaped, Gram positive, catalase- and oxidase- positive endospore-forming bacterial cells, implying that they belonged to the genus *Bacillus* [28]. The colonies of these bacilli isolates were predominantly white, transluscent, glistening, circular, flat, smooth and had entire margins (Table **3**).

Table 3. Colony morphological characteristics of the soda lake bacterial isolates.

Isolate	Pigment	Shape, height, texture and margin
LBW 2719[a]	White, opaque and dull	Circular, raised, rough and dry, entire
LBW 226[a]	White, opaque and dull	Circular, raised, rough and dry, entire
LBW 318[a]	White, opaque and dull	Circular, raised, rough and dry, entire
LBS 77[a]	White, opaque and dull	Circular, raised, rough and dry, entire
LBK 261	White, transluscent and glistening	Circular, flat, smooth, entire
LBW 4512	White, transluscent and glistening	Circular, flat, smooth, entire
LBW 327	White, transluscent and glistening	Circular, flat, smooth, entire
LBW 434	White, transluscent and glistening	Circular, flat, smooth, entire
LBW 7526a	White, transluscent and glistening	Circular, flat, smooth, entire
LBW 39	White, transluscent and glistening	Circular, flat, smooth, entire
LBW 7526b	White, transluscent and glistening	Circular, flat, smooth, entire
LBW 328	White, transluscent and glistening	Circular, flat, smooth, entire
LBS 16	White, transluscent and glistening	Circular, flat, smooth, entire
LBW 625	White, transluscent and glistening	Circular, flat, smooth, entire
LBW 446	White, transluscent and glistening	Circular, flat, smooth, entire
LBW 5117	White, transluscent and glistening	Circular, flat, smooth, entire
LBW 313	Yellow, transluscent and glistening	Circular, flat, smooth, entire
LBW 317	Yellow, transluscent and glistening	Circular, flat, smooth, entire

Key:

[a] Grew into the agar and gradually became irregular and undulate in their later stages of growth.

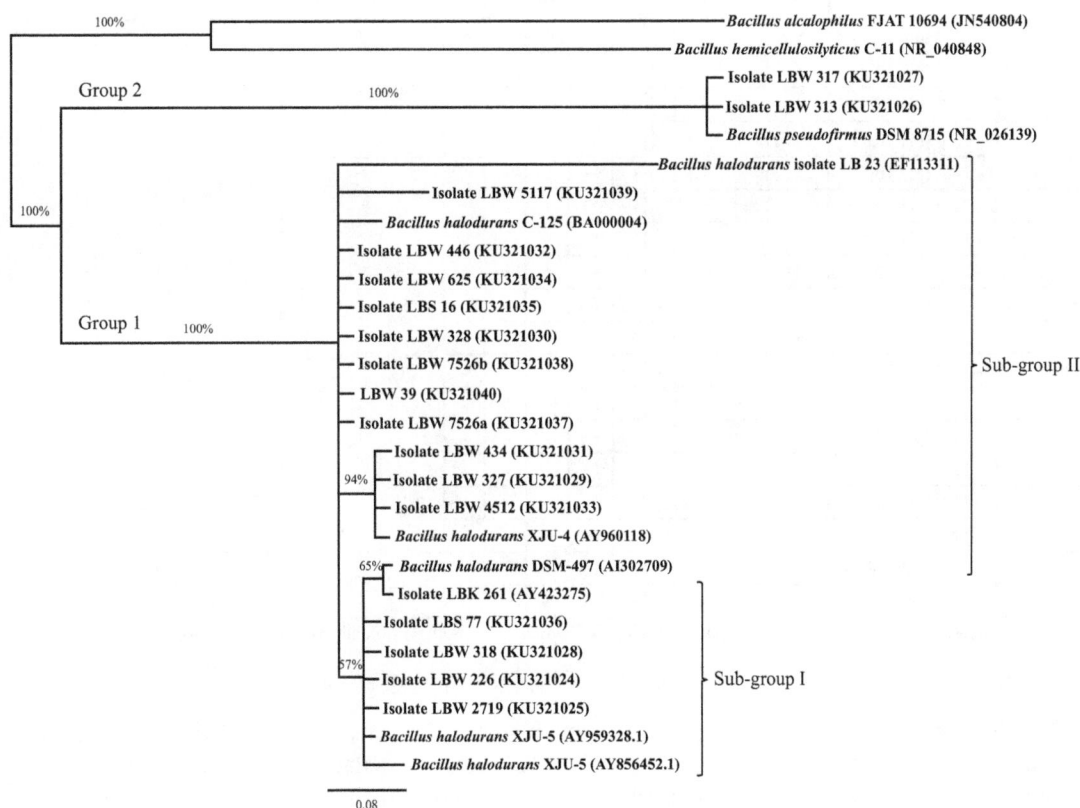

Fig. (3). An unrooted phylogenetic tree constructed from the analysis of the 16S rRNA gene sequences of the alkaliphilic soda lake Bacillus isolates and their closest relatives obtained from the NCBI database. Scale represents the average number of nucleotide substitutions per site. Bootstrap values for every 1000 trees generated are shown at the nodes. The accession numbers of the 16S rRNA sequences for each isolate is shown in brackets.

Tests based on physiological characterisation revealed that the bacilli isolates grew optimally at pH 10.5, thus, implying that they were alkaliphiles (Table **4**). These alkaliphiles were further divided into two groups: Group I (comprising 16 isolates) were able to grow in the pH range 7.0-11.0 while those in group 2 (comprising 2 isolates) did so in the pH range >7.0-11.0, implying that they were facultatively alkaliphilic- and obligately alkaliphilic- bacilli isolates, respectively (Table **4**). Additional physiological tests revealed that group 1 isolates were further divided into two sub-groups: sub-group I (comprising 4 isolates) were able to grow up to 65 °C (optimum 37 °C) and in the presence of up to 5% (w/v) NaCl [optimum 0-4% NaCl (w/v)] while sub-group II (comprising 12 isolates) grew up to 55 °C (optimum 37-45 °C) and in the presence of up to 10% (w/v) NaCl [optimum 0-5% NaCl (w/v)] (Table **4**). On the other hand, group 2 (comprising 2 isolates) were able to grow up to 45 °C (optimum 37-45 °C) and in the presence of up to 15% (w/v) NaCl [optimum 0-5% NaCl (w/v)]. Sub-group I isolates exhibited physiological characteristics similar to those of the facultatively alkaliphilic *Bacillus* sp. C-125 (re-identified as *B. halodurans* C-125), *Bacillus* sp. AH-101 (re-identified as *B. halodurans*) and *B. halodurans* M29 [29 - 31] while those in sub-group II exhibited physiological characteristics similar to those of the facultatively alkaliphilic *B. halodurans* DSM 497 reference/type strain [32] thus, suggesting that sub-group I isolates were stable phenotypic variants of the reference strain. This fact is also supported by the differences observed in the colony morphologies of the isolates in the two sub-groups (Table **3**). On the other hand, group 2 isolates exhibited physiological and colony morphological characteristics similar to those of the obligately alkaliphilic *B. pseudofirmus* DSM 8715 reference strain/type strain [32]. Overall, these results suggest that the facultatively- and obligately alkaliphilic bacilli isolates in this study are tolerant to varying moderately high temperatures- and saline- conditions, which are common features of soda lakes and the hot springs there-in [1].

Table 4. Physiological characteristics of the soda lake bacterial isolates.

Isolate	pH of Growth (Optimum Growth)	Growth at (°C)					Growth in Presence of NaCl (w/v)					Grouping	
		25	37	45	55	65	0	5	10	15	18		
LBW 2719	7.0 - 11.0 (pH 10.5)	+	++	+	+	$+^a$	++	$+^b$	-	-	-	Sub-group I	
LBW 226	7.0 - 11.0 (pH 10.5)	+	++	+	+	$+^a$	++	$+^b$	-	-	-		
LBW 318	7.0 - 11.0 (pH 10.5)	+	++	+	+	$+^a$	++	$+^b$	-	-	-		
LBS 77	7.0 - 11.0 (pH 10.5)	+	++	+	+	$+^a$	++	$+^b$	-	-	-		
LBK 261	7.0 - 11.0 (pH 10.5)	+	++	++	+	-	++	++	+	-	-	Sub-group II	Group 1 Facultatively alkaliphilic bacillus isolates
LBW 4512	7.0 - 11.0 (pH 10.5)	+	++	++	+	-	++	++	+	-	-		
LBW 327	7.0 - 11.0 (pH 10.5)	+	++	++	+	-	++	++	+	-	-		
LBW 434	7.0 - 11.0 (pH 10.5)	+	++	++	+	-	++	++	+	-	-		
LBW 7526a	7.0 - 11.0 (pH 10.5)	+	++	++	+	-	++	++	+	-	-		
LBW 39	7.0 - 11.0 (pH 10.5)	+	++	++	+	-	++	++	+	-	-		
LBW 7526b	7.0 - 11.0 (pH 10.5)	+	++	++	+	-	++	++	+	-	-		
LBW 328	7.0 - 11.0 (pH 10.5)	+	++	++	+	-	++	++	+	-	-		
LBS 16	7.0 - 11.0 (pH 10.5)	+	++	++	+	-	++	++	+	-	-		
LBW 625	7.0 - 11.0 (pH 10.5)	+	++	++	+	-	++	++	+	-	-		
LBW 446	7.0 - 11.0 (pH 10.5)	+	++	++	+	-	++	++	+	-	-		
LBW 5117	7.0 - 11.0 (pH 10.5)	+	++	++	+	-	++	++	+	-	-		
LBW 313	> 7.0 - 11.0 (pH 10.5)	+	++	++	-	-	++	++	+	+	-	Group 1 Obligately alkaliphilic bacillus isolates	
LBW 317	> 7.0 - 11.0 (pH 10.5)	+	++	++	-	-	++	++	+	+	-		

All isolates grew aerobically, except for isolates LBW 313 and LBW 317 which grew both aerobically and anaerobically.
Key:
+, Growth; ++, Optimum growth; +a, Poor growth after 72 h, better growth after 96 h; +b, Weak growth; -, no growth

On the basis of molecular characterization, a phylogenetic tree was constructed using 16S rDNA sequences of the sixteen facultatively alkaliphilic- and two obligately alkaliphilic- bacilli isolates and those of their closest relatives obtained from the NCBI database. The tree divided the isolates into two distinct clusters (Fig. **3**): The largest cluster (designated as group 1) comprising the sixteen facultative alkaliphiles were found to be closely related to *B. halodurans* DSM 497 reference strain, with sub-group I isolates showing a high sequence similarity of 99.9% and those in sub-group II showing high sequence similarities that ranged from 98.7 (isolate LBW 5117) –to- 99.8% (isolates LBW 4512, LBW 327, LBW 434, LBW 7526a, LBW 39, LBW 7526b, LBW 328, LBS 16, LBW 625 and LBW 446) –to- 100% (isolate LBK 261) while the smallest cluster (designated as group 2) comprising the two obligate alkaliphiles were found to be closely related to *B. pseudofirmus* DSM 8715 reference strain with 100% sequence similarity. Both *B* *halodurans* and *B. pseudofirmus* species are among the Gram positive facultative- and obligate- alkaliphiles commonly found in the Kenyan Rift valley soda lakes, respectively [33]. The limited number of alkaliphilic bacterial species in this study is attributed to the sampling methods and limited isolation conditions employed in the field and laboratory, respectively. It is interesting to note that although the facultative alkaliphilic bacterial isolates in sub-groups I and II shared high identities with *B. halodurans* DSM 497 reference strain, they exhibited significantly altered enzymatic production profiles (plate-test screening), and different colony morphological - and physiological- characteristics from one another (Tables **2** , **3** and **4**). These differences can be attributed to phenotypic switching of sub-group II *B. halodurans* isolates to enable them adapt to new local ecological niches in the hot spring wells of Lake Bogoria where temperatures were relatively higher than in their original habitats (*i.e.* peripheries of hot spring wells or downstream their drainage channels) (Table **1**), thus giving rise to a variant sub-population (sub-group I) of this bacterial species with different gene expressions that lead to significantly altered enzymatic production profiles and new phenotypic characteristics, as was also reported by Rodriguez [34] and Sousa *et al.* [35].

Alkaline proteolytic, pectinolytic and amylolytic activities of the isolates were determined quantitatively in their cell-free culture supernatants at pH 10.5 following 48 h of cultivation at 37 °C and 200 rpm using chicken quill feather, pectin and starch as carbon or carbon/nitrogen sources in broth media, respectively. Majority of the isolates produced alkaline proteolytic enzymes when cultured in liquid media containing chicken quill feathers, hence providing an economical and readily available substrate for cultivation of the isolates. Among the positive protease producers were isolates closely associated with *B. halodurans* in sub-group II (0.01-0.36 U/ml) and *B. pseudofirmus* in group 2 (0.38-0.40 U/ml), the majority of which produced titers that were within the range of those produced by their respective reference strains (Fig. **4a**). On the contrary, isolates associated with *B. halodurans* in sub-group I showed no detectable proteolytic activities in their culture supernatants, hence confirming that they were a variant sub-population of sub-group II isolates with an altered proteolytic production profile. The overall pattern of protease production by the isolates appear to correlate with that observed by the same isolates on solid gelatin medium during screening (Table **2**). The best protease producer was isolated LBW 313 with a titer of 0.40 U/ml. This enzyme can be used to catalyze the hydrolysis or synthesis of peptide bonds at alkaline pH and can therefore, find applications in a) detergent industry - as an additive to detergents, for hydrolysis of proten-based stains in laundry or dishes during washing b) leather industry - to dehair hide and produce high quality leather c) waste management - to solubilize proteins in protein-containing food wastes in order to recover liquid concentrates or dry solids of nutritional values for incorporation in fish or livestock feeds and d) chemical industry - to synthesize fine speciality chemicals.

All isolates closely related to *B. halodurans* were able to produce the alkaline pectinolytic enzyme PGase (0.05-0.46 U/ml) while those closely related to *B. pseudofirmus* showed no detectable activity of this enzyme (Fig. **4b**). A similar pattern of PGase production was observed with their respective reference strains. Like on solid pectin medium, these pectinolytic titers were relatively higher for sub-group I isolates (0.12-0.56 U/ml) compared to those of sub-group II (0.05-0.10 U/ml), despite the fact that isolates in both sub-groups were *B. halodurans* (Fig. **4b**). This altered pectinolytic production profile further confirms that sub-group I isolates were indeed a variant sub--population of sub-group II *B. halodurans* isolates. PecL is another pectinolytic enzyme that was only detected in the cell-free culture supernatants of sub-group II isolates (Fig. **4b**). The absence of this enzyme in the cell-free culture supernatants of sub-group I isolates also demonstrates that sub-group I isolates were sub-population variants of sub-group II *B. halodurans* isolates. Isolate LBK 261 was the best producer of both PGase and PecL (0.12 and 0.1U/ml, respectively) while isolate LBW 318 was the best producer of PGase (0.46 U/ml). PGases and PecLs cleave α-(1 → 4) glycosidic linkages in polygalacturonic acid polymers via hydrolytic and trans-elimination reactions, respectively. Thus, the presence of both alkaline enzymes in the cell-free culture supernatants of sub-group II isolates is interesting, since this cocktail of enzymes can be very effective in degrading the polymer, and for this reason they can find applications in the a) pulp and paper industry - to effectively depolymerize polygalacturonic acid polymers in pulps for enhanced bleachability and consequent production of good quality paper with high opacity and good printability b) textile industry - to degum, ret or scour natural plant fibers for textile manufacture.

All isolates closely associated with *B. halodurans* and *B. pseudofirmus* produced a cocktail of alkaline amylolytic enzymes, with the former isolates producing higher titers of amylases and pullulanases compared to those produced by the latter isolates and the reference strains while CGTase titers remained low for all isolates (Fig. **4c**). The low levels of CGTases in the culture supernatants of the isolates can be attributed to unoptimized bioconversion conditions (*e.g.* temperature, pH, cultivation time as well as product inhibition) during cell culture. Interestingly, these amylolytic titers did not correlate with those produced by the same isolates on solid alkaline starch medium (Table **2**). This difference can be attributed to differences in the a) uptake of nutrients by the isolates in the two types of media and/or b) activity detection rule *e.g.* in the solid medium the disappearance of the substrate is measured while in liquid medium the appearance of the product is measured, as was also reported by Castro *et al.* [36]. The best producer of the alkaline amylolytic enzymes was isolated LBW 5117 (0.32, 0.68 and 0.02 U/ml amylases, pullulanases and CGTases, respectively). Amylases hydrolyze internal α-1,4 linkages in starch while pullulanases (debranching enzymes) hydrolyze α-1,6 linkages in starch, amylopectin and pullulan. Thus, the finding of both alkaline amylases and pullulanases in the culture supernatants of the isolates is interesting because this combination of enzymes can be very effective in removing both starch-based sizes (desizing) from cotton yarn during the modern production of textiles and starch-based stains from laundry and dishes. CGTases on the other hand, convert starch into cyclic cyclodextrins whose hydrophobic cavities can form complexes with organic foods and, pharmaceutical- and cosmetic- products to improve their solubility and stability properties.

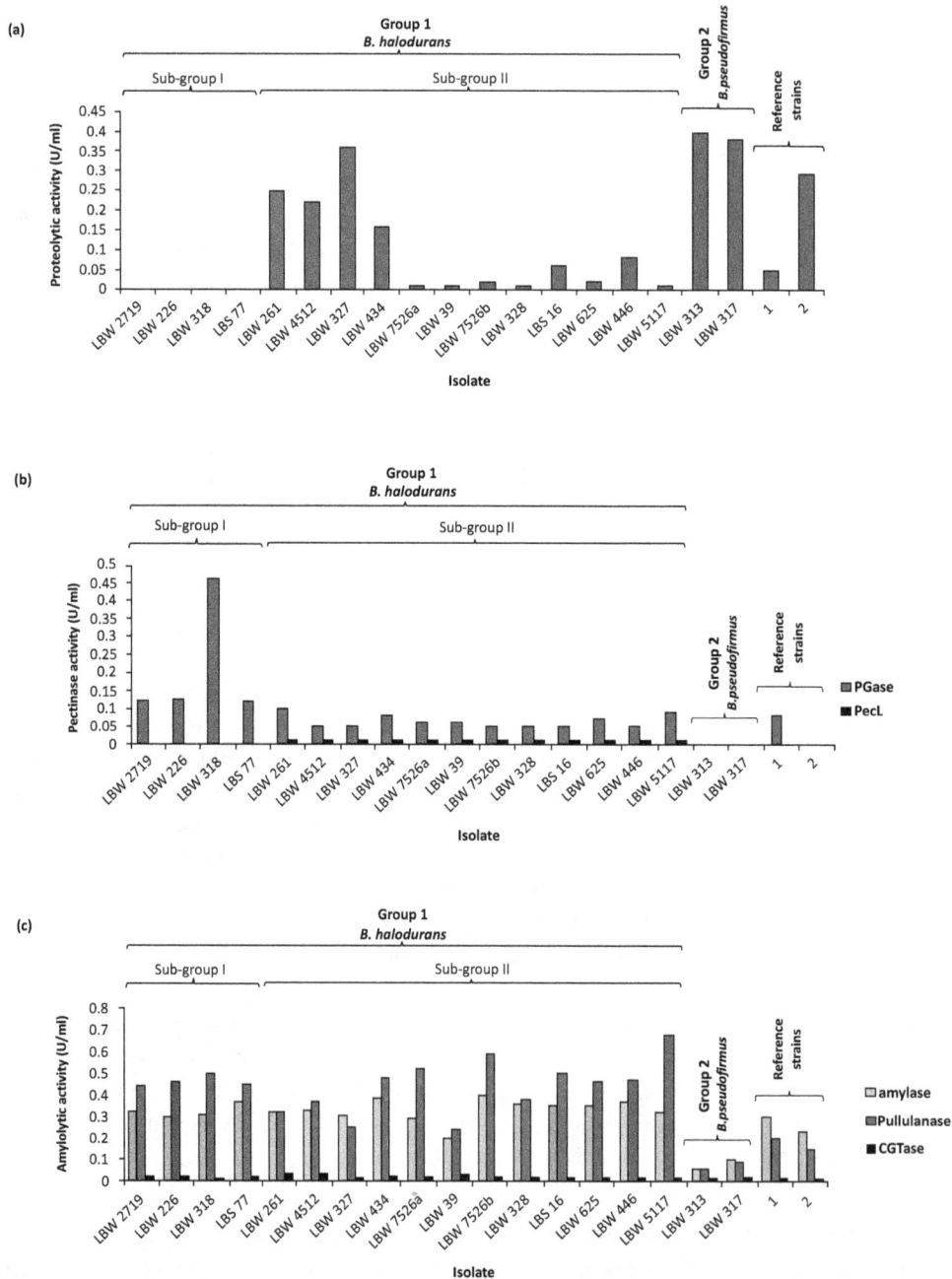

Fig. (4). Proteolytic (a), pectinolytic (b) and amylolytic (c) activities of the alkaliphilic soda lake bacterial isolates and the reference strains (1 = *B. halodurans* DSM 497 and 2 = *B. pseudofirmus* DSM 8715).

CONCLUSION

Twelve facultatively alkaliphilic *B. halodurans*, four phenotypic sub-population variants of this bacterial species and two obligately alkaliphilic *B. pseudofirmus* Lake Bogoria isolates are attractive biotechnological sources of alkaline amylolytic enzymes, with the majority also being sources of alkaline pectinolytic and proteolytic enzymes - all of which can find applications in various industrial technologies for production of high quality products in a safe way. Our current work is targeted at the textile industry where we are evaluating the use of alkaline amylolytic enzymes from *B. halodurans* isolate LBW 5117 in desizing cotton yarn in preparation for its scouring (bioscouring) with PGase from *B. halodurans* isolate LBW 318 in order to improve its performance in further finishing steps. We also intend to evaluate the catalase-positive *B. halodurans* isolate LBK 261 as an immobilized whole cell biocatalyst to degrade hydrogen peroxide in simulated textile bleach wastewaters.

AUTHORS' CONTRIBUTION

Oluoch KR, Okanya PW, Hatti-Kaul R, Mattiasson B and Mulaa FJ collected the samples from Lake Bogoria; Oluoch KR and Okanya PW isolated the bacteria; Oluoch KR and Mulaa FJ designed the study; Oluoch KR performed the experiments, analyzed the data and wrote the manuscript whose final version was approved by Mulaa FJ.

HUMAN AND ANIMAL RIGHTS

No Animals/Humans were used for studies of this research.

ACKNOWLEDGEMENTS

The authors would like to thank the Department of Biochemistry, University of Nairobi, for providing the laboratory space used for this study; the authors would like to express their sincere gratitude to Prof. Wallace D. Bulimo, Dr. Edward K. Muge, Mr. Nelson Khan and Ms Irene K Kateve of the University of Nairobi, for their useful suggestions in the study.

REFERENCES

[1]   Grant WD, Tindall BJ. The alkaline saline environment.Microbes in extreme environments. London: Academic Press 1986; pp. 25-54.

[2]   Grant WD, Horikoshi K. Alkaliphiles: Ecology and biotechnological applications.IEMS Microbiology Letters. 1992; 75: pp. 255-69.
[http://dx.doi.org/10.1007/978-94-011-2274-0_5]

[3]   Horikoshi K. Alkaliphiles: Some applications of their products for biotechnology. Microbiol Mol Biol Rev 1999; 63(4): 735-50.
[PMID: 10585964]

[4]   Grant WD, Jones BE. Alkaline environments.Encyclopedia of microbiology. 2nd ed. New York: Academic Press 2000; Vol. 1: pp. 126-33.

[5]   Fujinami S, Fujisawa M. Industrial applications of alkaliphiles and their enzymes--past, present and future. Environ Technol 2010; 31(8-9): 845-56.
[http://dx.doi.org/10.1080/09593331003762807] [PMID: 20662376]

[6]   Sarethy IP, Saxena Y, Kapoor A, *et al.* Alkaliphilic bacteria: Applications in industrial biotechnology. J Ind Microbiol Biotechnol 2011; 38(7): 769-90.
[http://dx.doi.org/10.1007/s10295-011-0968-x] [PMID: 21479938]

[7]   Sundarram A, Murthy TPK. α-Amylase production and applications: A review. J Appl Environ Microbiol 2014; 2: 166-75.

[8]   Stårnes RL. Industrial potential of cyclodextrin glycosyl transferases. Cereal Foods World 1990; 35: 1094-9.

[9]   Ellaiah P, Srinivasulu B, Adinarayana K. A review on microbial alkaline proteases. J Sci Ind Res (India) 2002; 61: 690-704.

[10]   Kohli P, Gupta R. Alkaline pectinases: A review. Biocatal Agric Biotechnol 2015; 4: 279-85.
[http://dx.doi.org/10.1016/j.bcab.2015.07.001]

[11]   Okanya SWP. Screening for anti-plasmodium activity using extracts from extremophiles found in Lake Bogoria, Kenya. Dissertation, University of Nairobi 2006; pp. 23-25.

[12]   Juwon AD, Emanuel OF. Screening of fungal isolates from Nigerian tar sand and deposit in Ondo state for novel biocatalysts. J Biol Sci 2012; 12: 57-61.
        [http://dx.doi.org/10.3923/jbs.2012.57.61]

[13]   Soares MMCN, Da Silva R, Gomes E. Screening of bacterial strains for pectinolytic activity: Characterization of the polygalacturonase produced by *Bacillus* sp. Rev Microbiol 1999; 30: 299-303.
        [http://dx.doi.org/10.1590/S0001-37141999000400002]

[14]   Kelly CT, Fogarty WM. Production and properties of polygalacturonate lyase by an alkalophilic microorganism *Bacillus* sp. RK9. Can J Microbiol 1978; 24(10): 1164-72.
        [http://dx.doi.org/10.1139/m78-190] [PMID: 31973]

[15]   Horikoshi K. Production of alkaline enzymes by alkalophilic microorganisms Part II. Alkaline amylase produced by *Bacillus* No. A-40-2. Agric Biol Chem 1971; 35: 1783-91.
        [http://dx.doi.org/10.1271/bbb1961.35.1783]

[16]   Dhawale MR, Wilson JJ, Khachatourians GG, Ingledew WM. Improved method for detection of starch hydrolysis. Appl Environ Microbiol 1982; 44(3): 747-50.
        [PMID: 16346102]

[17]   Hitha PK, Girija D. Isolation and screening of native microbial isolates for pectinase activity. Int J Sci Res (Ahmedabad) 2014; 3: 632-4.

[18]   Gerhardt P, Murray RGE, Wood WA, Krieg NR. Methods for general and molecular bacteriology. Washington, DC: American Society for Microbiology 1994.

[19]   Gregersen T. Rapid method for distinction of Gram-negative bacteria from Gram-positive bacteria. Eur J Appl Microbiol Biotechnol 1978; 5: 123-7.
        [http://dx.doi.org/10.1007/BF00498806]

[20]   Martins RF, Davids W, Abu Al-Soud W, Levander F, Rådström P, Hatti-Kaul R. Starch-hydrolyzing bacteria from Ethiopian soda lakes. Extremophiles 2001; 5(2): 135-44.
        [http://dx.doi.org/10.1007/s007920100183] [PMID: 11354457]

[21]   Johnson JL. Similarity analysis of DNAs.Methods for general and molecular bacteriology. Washington, DC: American Society for Microbiology 1994; pp. 655-81.

[22]   Weisburg WG, Barns SM, Pelletier DA, Lane DJ. 16S ribosomal DNA amplification for phylogenetic study. J Bacteriol 1991; 173(2): 697-703.
        [http://dx.doi.org/10.1128/jb.173.2.697-703.1991] [PMID: 1987160]

[23]   Edgar RC. MUSCLE: Multiple sequence alignment with high accuracy and high throughput. Nucleic Acids Res 2004; 32(5): 1792-7.
        [http://dx.doi.org/10.1093/nar/gkh340] [PMID: 15034147]

[24]   Huelsenbeck JP, Larget B, Miller RE, Ronquist F. Potential applications and pitfalls of Bayesian inference of phylogeny. Syst Biol 2002; 51(5): 673-88.
        [http://dx.doi.org/10.1080/10635150290102366] [PMID: 12396583]

[25]   Bakhtiar S, Andersson M, Gessesse A, Mattiasson B, Hatti-Kaul R. Stability characteristics of a calcium- independent alkaline protease from *Nesterenkonia* sp. Enzyme Microb Technol 2002; 32: 525-31.
        [http://dx.doi.org/10.1016/S0141-0229(02)00336-8]

[26]   Wang G, Michailides TJ, Bostock RM. Improved detection of polygalacturonase activity due to *Mucor piriformis* with a modified dinitrosalicylic acid reagent. Phytopathology 1997; 87(2): 161-3.
        [http://dx.doi.org/10.1094/PHYTO.1997.87.2.161] [PMID: 18945136]

[27]   Soriano M, Diaz P, Pastor FIJ. Pectate lyase C from *Bacillus subtilis*: A novel endo-cleaving enzyme with activity on highly methylated pectin. Microbiology 2006; 152(Pt 3): 617-25.
        [http://dx.doi.org/10.1099/mic.0.28562-0] [PMID: 16514142]

[28]   Ash C, Farrow JAE, Wallbanks S, Collins MD. Phylogenetic heterogeneity of the genus *Bacillus* as revealed by comparative analysis of small-subunit-ribosomal RNA sequences. Lett Appl Microbiol 1991; 13: 202-6.
        [http://dx.doi.org/10.1111/j.1472-765X.1991.tb00608.x]

[29]   Takami H, Horikoshi K. Reidentification of facultatively alkaliphilic *Bacillus* sp. C-125 to *Bacillus halodurans*. Biosci Biotechnol Biochem 1999; 63(5): 943-5.
        [http://dx.doi.org/10.1271/bbb.63.943] [PMID: 27385575]

[30]   Takami H, Nogi Y, Horikoshi K. Reidentification of the keratinase-producing facultatively alkaliphilic *Bacillus* sp. AH-101 as *Bacillus halodurans*. Extremophiles 1999; 3(4): 293-6.
        [http://dx.doi.org/10.1007/s007920050130] [PMID: 10591021]

[31]   Mei Y, Chen Y, Zhai R, Liu Y. Cloning, purification and biochemical properties of a thermostable pectinase from *Bacillus halodurans* M29. J Mol Catal, B Enzym 2013; 94: 77-83.
        [http://dx.doi.org/10.1016/j.molcatb.2013.05.004]

[32]   Nielsen P, Fritz D, Priest FG. Phenetic diversity of alkaliphilic *Bacillus strains*: Proposal for nine new species. Microbiol 1995; 141: 1745-61.
        [http://dx.doi.org/10.1099/13500872-141-7-1745]

[33]   Duckworth AW, Grant WD, Jones BE, Steenbergen R. Phylogenetic of soda lake alkaliphiles. FEMS Microbiol Ecol 1996; 19: 181-91.
        [http://dx.doi.org/10.1111/j.1574-6941.1996.tb00211.x]

[34]   Rodriguez H. Colony switching in an alpha-amylase-producing strain of *Bacillus subtilis.* J Ind Microbiol 1995; 15: 112-5.
        [http://dx.doi.org/10.1007/BF01569809]

[35]   Sousa AM, Machado I, Pereira MO. Phenotypic switching: An opportunity to bacteria thrive.Science against microbial pathogens: Communicating current research and technological advances. 2011; pp. 252-62.

[36]   Castro GR, Ferrero MA, Méndez BS, Sineriz F. Screening and selection of bacteria with high amylolytic activity. Acta Biotechnol 1993; 13: 197-201.
        [http://dx.doi.org/10.1002/abio.370130220]

7

Heterologous Expression of Transcription Factor *AtWRKY57* Alleviates Salt Stress-Induced Oxidative Damage

Wei Tang[*]

College of Horticulture and Gardening, Yangtze University, Jingzhou, Hubei 434025, China

Abstract:

Background:

WRKY transcription factors play important roles in the responses to abiotic stresses, seed dormancy, seed germination, developmental processes, secondary metabolism, and senescence in plants. However, molecular mechanisms of WRKY transcription factors-related abiotic stress tolerance have not been fully understood.

Methods:

In this investigation, transcription factor AtWRKY57 was introduced into cell lines of rice (*Oryza sativa* L.), tobacco (*Nicotiana tabacum*), and white pine (*Pinus strobes* L.) for characterization of its function in salt stress tolerance. The purpose of this investigation is to examine the function of AtWRKY in a broad sample of plant species including monocotyledons, dicotyledons, and gymnosperms.

Results:

The experimental results demonstrated that heterologous expression of transcription factor AtWRKY57 improves salt stress tolerance by decreasing Thiobarbituric Acid Reactive Substance (TBARS), increasing Ascorbate Peroxidase (APOX) and Catalase (CAT) activity under salt stress. In rice, overexpression of transcription factor AtWRKY57 enhances expression of Ca^{2+}-dependent protein kinase genes *OsCPk6* and *OsCPk19* to counteract salt stress.

Conclusion:

These results indicated that transcription factor AtWRKY57 might have practical application in genetic engineering of plant salt tolerance throughout the plant kingdom.

Keywords: Abiotic stress, Dicotyledons, Genetic engineering, Gymnosperms, Monocotyledons, *Pinus*.

1. INTRODUCTION

High salinity is one of the main factors limiting plant growth and crop productivity. The responses of plants to salt stress are an important topic for the biotechnological application plant salt tolerance in the field of abiotic stress. Transcription Factors (TFs) ranging from bZIP, AP2/ERF, NAC, zinc finger proteins, and MYB proteins to WRKY highly influence the efficiency of salt stress-induced gene expression under salt stress. Overexpression of these transcription factors will provide new opportunities for the engineering of plant tolerance to salt stress [1]. In cotton, 42 relevant/related genes were induced by the NaCl treatment and they may be candidate genes as potential markers of tolerance to salt stress [2]. In wheat, transcription factor TaWRKY79 enhanced the level of tolerance to salinity stress via reducing the sensitivity to ABA in the TaWRKY79 transgenic plants, indicating that TaWRKY79 operates in an

* Address correspondence to this author at the College of Horticulture and Gardening, Yangtze University, Jingzhou, Hubei 434025, China, E-mail: wt10yu604@gmail.com

ABA-dependent pathway [3]. In *Triticum aestivum* cv. (Yuregir-89), *Triticum turgidum* cv. (Kiziltan-91), and *Triticum monococcum* (Siyez), expression analysis of transcription factors TaWLIP19, TaMBF1, TaWRKY10, TaMYB33 and TaNAC69 indicated that all five selected genes in Kiziltan-91 were induced, suggesting that transcription factors might be used for determination of salinity-tolerant for molecular breeding studies [4].

WRKY transcription factors are plant-specific, zinc finger-type transcription factors that are involved in abiotic stress tolerance in a large number of plant species. Analysis of DNA orthology and gene motif compositions indicated that WRKY members in many plant species generally shared the similar motifs. In *Gossypium aridum*, 28 salt-responsive GarWRKY genes were identified under stress. Overexpression of GarWRKY17 and GarWRKY104 in Arabidopsis demonstrated that these transcription factors would be potential candidates for the genetic improvement of cotton salt stress tolerance [5]. In peanut, transcription factors (NAC, bHLH, WRKY, AP2/ERF) are differentially expressed under salinity stress [6]. In cotton, silencing of C4 (encodes WRKY DNA-binding protein) can significantly enhance cotton susceptibility to salt stress [7]. In foxtail millet (*Setaria italica*), salt stress-induced methylation in WRKY transcription factor genes and modulates the expression of corresponding genes [8]. In Citrus, a total of 1831 differentially expressed genes were identified and a multitude of transcription factors including WRKY, NAC, MYB, AP2/ERF, bZIP, GATA, bHLH, ZFP, SPL, CBF, and CAMTA was related to cell wall loosening and was also involved in salt stress [9].

Transcription factors could differentially regulate salt stress tolerance. Basic region/leucine zipper (bZIP) transcription factors play key roles in plant growth, development, and stress signaling and perform as crucial regulators in ABA-mediated stress response in plants [10]. In ramie, a bZIP transcription factor BnbZIP2 may act as a positive regulator in response to high-salinity stress [11]. The homeodomain leucine zipper (HD-Zip) transcription factors modulate plant growth and response to environmental stresses. In Arabidopsis (*Arabidopsis thaliana*), rice (*Oryza sativa*), maize (*Zea mays*), poplar (*Populus trichocarpa*), soybean, and cucumber (*Cucmis sativus*), HD-Zip transcription factors have important roles in dehydration and salt stress [12]. Trihelix transcription factors, which are characterized by containing a conserved trihelix (helix-loop-helix-loop-helix) domain that binds to GT elements, play roles in abiotic stress responses by regulating the expression of stress tolerance genes that result in reduced reactive oxygen species and lipid peroxidation [13]. NAC (NAM ATAF CUC) transcription factors are involved in ethylene-modulated salt tolerance. In apple, MdNAC047 transcription factor was significantly induced by salt treatment and its overexpression conferred increased tolerance to salt stress [14]. WRKY transcription factors appear as important regulators of abiotic stresses tolerance. In tomato, SlWRKY3 binds to the consensus CGTTGACC/T W box and regulates expression of stress-related genes, indicating that SlWRKY3 is an important regulator of salinity tolerance [15]. In *Reaumuria trigyna*, WRKY transcription factor RtWRKY1 was induced by salt stress and ABA treatment. Overexpression of RtWRKY1 enhanced root length and fresh weight of the transgenic lines under salt stress [16]. In maize, expression of the ZmWRKY17 was up-regulated by salt and ZmWRKY17 acts as a negative regulator involved in the salt stress responses through ABA signaling [17].

NAC transcription factor proteins play important roles in salt stress responses. In *Cucumis melo*, CmNAC14 regulates expression of salt stress-related genes. Overexpression of CmNAC14 increased the sensitivity of transgenic lines to salt stress [18]. In Chickpea (*Cicer arietinum* L.), the transcript levels of CarNAC4 were enhanced in response to abiotic stresses. Over-expression of CarNAC4 improved tolerance to salt stresses [19]. In wheat, TaNAC29 was involved in response to salt. TaNAC29 confers salt stress tolerance through reducing H2O2 accumulation by enhancing the antioxidant system [20]. In tomato, SlNAC4-SlNAC10 genes are involved in the response to salt stress [21]. In *Arabidopsis thaliana*, AtNAC2 is a transcription factor that regulates salt stress and incorporates the environmental and endogenous stimuli into the process of plant lateral root development [22].

WRKY transcription factors play important regulatory roles in plant development and defense response. In *Salvia miltiorrhiza*, a total of 61 SmWRKYs were cloned [23]. In cucumber, a total of 55 WRKY genes were identified [24]. In Arabidopsis, the expression of AtWRKY25, AtWRKY26, and AtWRKY33 was related to NaCl, and osmotic stress [25]. In canola (*Brassica napus* L.), 46 WRKY genes were identified to be involved in stress [26]. In rice (*Oryza sativa*), OsWRKY genes contributing to salt stress tolerance-realted biological processes or signal transduction pathways [27]. In knapweed (*Centaurea maculosa*), differential gene expression was observed for a putative serpin (CmSER-1) and a calmodulin-like (CmCAL-1) protein. In dandelion (*Taraxacum officinale*), differential gene expression was observed for a putative protein phosphatase 2C (ToPP2C-1) and cytochrome P-450 (ToCYP-1) protein. These genes are involved in plant stress responses [28]. In Arabidopsis, 49 of the 72 AtWRKY genes were differentially regulated in the plants infected by an avirulent strain of the bacterial pathogen *Pseudomonas syringae* [29].

Transcription factor WRKY57 is known to function in adaptation to abiotic stresses. In banana (*Musa acuminate*), WRKY57 is necessary for developing drought-resilient plants [30]. In Arabidopsis, WRKY57 directly binds to the promoters of JASMONATE ZIM-DOMAIN1 (JAZ1) and interacts with nuclear-encoded SIGMA FACTOR BINDING PROTEIN1 (SIB1) and SIB2 to allow fine regulation of defense [31]. Overexpression of the Arabidopsis WRKY57 transcription factor AtWRKY57 in rice improved not only drought tolerance but also salt and PEG tolerance, indicating its potential role in crop improvement [32]. In grape (*Vitis vinifera* L.), the WRKY57-like transcription factor has an important role in some biological processes including cell rescue, protein fate, secondary metabolism, and regulation of transcription [33]. Leaf senescence is regulated by diverse developmental and environmental factors. WRKY57 interacts with the AUX/IAA protein IAA29 to regulate leaf senescence process as a common component of the JA- and auxin-mediated signaling pathways [34]. Constitutive expression of WRKY57 conferred drought tolerance by regulating expression of stress-responsive genes (RD29A, NCED3, and ABA3). ChIP assays demonstrated that WRKY57 could directly bind the W-box of RD29A and NCED3 promoter sequences. Increased WRKY57 expression enhanced drought stress tolerance by increasing ABA levels that will increase drought stress tolerance in plants [35]. In this investigation, we demonstrated that transcription factor AtWRKY57 might have practical application in genetic engineering of plants for salt stress tolerance throughout the plant kingdom.

2. MATERIALS AND METHODS

2.1. Plasmid Constructs

The expression vector pBI121 and the cDNA of *WRKY57* were used to construct the pBI-*WRKY57* expression vector. Restrict enzymes *Kpn* I and *Xba* I (Promega, Madison, WI, USA) were used to digest both the pBI121 vector and the *WRKY57* cDNA at 37°C, DNA bands of the pBI121 vector and the *WRKY57* cDNA were purified using QIAquick Gel Extraction Kit (QIAGEN, Valencia, CA, USA). The 864-bp fragment of the *WRKY57* cDNA was ligated with the DNA of digested pBI121 [36] to produce the expression vector pBI-WRKY57. Vector pBI-WRKY57 was introduced into *Agrobacterium tumefaciens* strain LBA4404 by electroporation.

2.2. *Agrobacterium*-Mediated Transformation

Transgenic cell lines of rice (*Oryza sativa* L.), tobacco (*Nicotiana tabacum*), and white pine (*Pinus strobes* L.) were prepared as previously described [37, 38], using *Agrobacterium tumefaciens* strain LBA4404 carrying pBI-WRKY57 to transform cell cultures. Transgenic cells of rice, tobacco, and white pine were sub-cultured on a liquid proliferation medium for six weeks to obtain large quantities of cell cultures for further analysis. After 6 weeks, 50-70 mg of tissue/l cultures can be produced each week, and they were then used molecular analysis including PCR, Southern blot analysis, and northern blotting analysis.

2.3. Polymerase Chain Reaction and Southern Blot Analyses of Transgenic Cultures

Polymerase Chain Reaction (PCR) and Southern blotting analysis of putative transgenic cell lines of rice, tobacco, and white pine were conducted as previously described [37, 38]. 500 mg fresh cells of rice, tobacco, and white pine control and putative transgenic cell lines were used to isolate genomic DNA, using a Genomic DNA Isolation Kit (Sigma) following the manufacturer's protocol. PCR was done in a PTC-100TM Programmable Thermal Controller (MJ Research, San Francisco, CA, USA). The primers used are the nptII forward primer (nfp) 50-ACAAACAGACAATCGGCTGC-30 and the reverse primer (nrp) 50-AAGAACTCGTCAAGAAGGCG-30, as well as the transcription factor WRKY57 forward primer (zfp) 5'-ATGAACGATCCTGATAATCCCGATC-3' the reverse primer (zrp) 5'-TCAAGGGTTGCGCATAGTTTGAGG-3' were used. Southern blot analysis was conducted as previously described [38]. 25 µg DNA was digested overnight at 37 °C. Probes (864 pb fragment of *WRKY57*) were labeled by Digoxigenin (DIG) (Roche Diagnostics, Indianapolis, IN, USA).

2.4. RNA Isolation and Northern Blot Analysis

Total RNA was isolated from 5 g fresh cell cultures of rice, tobacco, and white pine control and putative transgenic cell lines, respectively, using a RNeasy Mini Plant Kit (Germantown, MD, USA) following the manufacturer's protocol. Then, 6 µg RNA from rice, tobacco, and white pine transgenic cells were separated by agarose-gel electrophoresis. Electrophoresis and northern blotting of RNAs were performed as previously described [36]. Digoxigenin (DIG)-labelling WRKY57 DNA (864 pb) (Roche Diagnostics) was used as a hybridization probe. Equal loading of RNA samples of rice, tobacco, and white pine was verified on the control of tobacco 25SrRNA. After PCR,

Southern blotting and northern blotting analyses, four cell lines each containing one copy of the pBI-WRKY57 T-DNA were selected from rice (Os1, Os2, Os3, and Os4), tobacco (Nt1, Nt2, Nt3, andNt6), and white pine (Ps1, Ps2, Ps3, and Ps4) and used for salt-induced oxidative damage experiments.

2.5. Salt Treatment of Transgenic Cell Lines

Salt treatment was applied by adding different concentrations of NaCl (50, 100, 150, 200, 250, and 300 mM) to the media used for transgenic cell cultures, which consisted of TE medium [38] containing 0.5 μM indole-3-butyric acid, 8.9 μM BA. The effects of NaCl on growth of cell cultures of rice (Os1, Os2, Os3, and Os4), tobacco (Nt1, Nt2, Nt3, andNt6), and white pine (Ps1, Ps2, Ps3, and Ps4) were examined by culture of cell cultures on growth medium containing different concentrations of NaCl, as previously described [36 - 38]. The average growth was expressed as mg/g FW/day.

2.6. Thiobarbituric Acid Reactive Substances (TBARS) Determination

The amount of thiobarbituric acid reactive substances (TBARS) was determined using the method of thiobarbituric acid (TBA) reaction as described previously [38]. Cell cultures (1 g) of rice (Os1, Os2, Os3, and Os4), tobacco (Nt1, Nt2, Nt3, andNt6), and white pine (Ps1, Ps2, Ps3, and Ps4) were homogenized in 3 ml of 20% (w/v) trichloroacetic acid (TCA). The homogenate was centrifuged at 5,000 rpm for 20 min and mixed with 20% TCA containing 0.5% (w/v) TBA and100 ll 4% BHT in ethanol at 1:1. The amount of thiobarbituric acid reactive substances was determined using the method previously described [36].

2.7. Antioxidant Enzymes Ascorbate Peroxidase (APOX) and Catalase (CAT) Activity Determination

Determination of activities of APOX and CAT were determined as described previously [36 - 38]. Two grams of non-transgenic control and transgenic cell cultures of rice (Os1, Os2, Os3, and Os4), tobacco (Nt1, Nt2, Nt3, andNt6), and white pine (Ps1, Ps2, Ps3, and Ps4) were homogenized under ice-cold conditions in 3 ml of extraction buffer, containing 50 mM phosphate buffer (pH 7.4), 1 mM EDTA, 1 g PVP, and 0.5% (v/v) Triton X-100 at 4 °C [36 - 38]. The homogenates were centrifuged at 10,000 × g for 20 min and the supernatant fraction was used for the assays. APOX activity was measured immediately in fresh extracts and was assayed as described [36]. CAT activity was detected in a 3 ml 50 mMpotassium phosphate buffer (pH 7.8) containing 3 mMH$_2$O$_2$, as previously described [36].

2.8. Expression Analysis of OsCPK Genes

Expression of OsCPK6 and OsCPK19 was analyzed by northern blotting by the method described previously [37, 38]. Twenty micrograms of total RNA of rice (Os1, Os2, Os3, and Os4) was used. The PCR-amplified fragments of OsCPK6 (amplified by forward primer 5'- atgggcaactactactcgtg -3' and reverse primer 5'- acgtacaggttgtcctcgt -3') and OsCPK19 (amplified by forward primer 5'- ggagcaaacggttatggcta -3' and reverse primer 5'- gcgcttagagatggatttgc -3') were labeled by Digoxigenin (DIG) (Roche Diagnostics Corporation, Roche Applied Science, Indianapolis, IN, USA) and were used as a hybridization probe [37, 38].

2.9. Statistical Analyses

Data on growth rate, amount of TBARS, APOX and CAT activity, amount of OsCPK6 and OsCPK19 obtained from different experiments were analyzed in Graphpad Prism 6 software. The significant differences between mean values derived from three independent biological experiments were determined using the least significant difference test at 5% level of probability.

3. RESULTS

3.1. Molecular Analyses of Transgenic Cell Lines

Transgenic cell lines of rice (*Oryza sativa* L.), tobacco (*Nicotiana tabacum*), and white pine (*Pinus strobes* L.) were obtained through *Agrobacterium tumefaciens*-mediated transformation and were examined by PCR, Southern blotting, and northern blotting (Fig. **1**). A total of 109 transgenic cell lines including 31 cell line from rice, 36 cell line from tobacco, and 42 cell line from white pine, were obtained after PCR analysis using primers for neomycin phosphotransferase II gene and primer for WRKY57. After confirmation by PCR, four cell lines were randomly selected and used for further analysis. The presence of the 717-bp band amplified by primers nrp and nfp from templates obtained from transgenic cells and absence of the 717-bp band amplified by primers nrp and nfp indicate that the T-DNA (Fig. **1A**) is integrated into the genome. This is further confirmed by PCR using primers for WRKy57 that the 864-bp band amplified by primers zf and zr from templates obtained from transgenic cells and absence of the 864-bp band amplified by primers zf and zr (Fig. **1B**). Integration of T-DNA into the genome of rice, tobacco, and white pine was also confirmed by Southern and northern blotting analisys (Figs. **1C, D**).

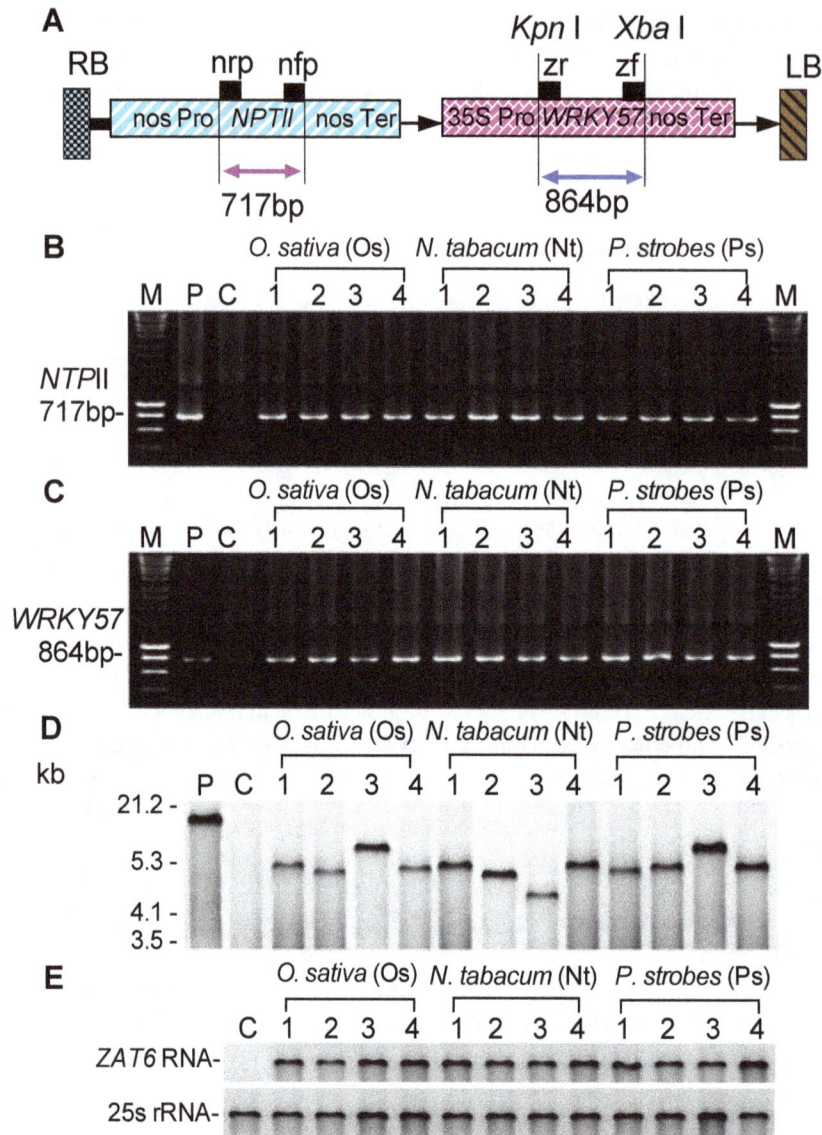

Fig. (1). Plasmid map and molecular analyses of transgenic cell lines. (**A**) A linear plasmid map. (**B**) PCR analysis of WRKY57 transgenic cell lines. (**C**) Southern blot analysis of WRKY57 transgenic cell lines. (**D**) Northern blot analysis of WRKY57 transgenic cell lines.

3.2. Growth of Cell Lines Under Different Concentrations of NaCL

Growth rate (Fig. **2**) of transgenic cell lines of rice (Os1, Os2, Os3, and Os4), tobacco (Nt1, Nt2, Nt3, and Nt4), and white pine (Ps1, Ps2, Ps3, and Ps4) were measured 3 days after transgenic cell cultures were transferred into media containing different concentrations of NaCl (50, 100, 150, 200, 250, and 300 mM). The medium containing no NaCl was used as the control of salt stress. A significant difference in growth rate was obtained when transgenic cell cultures were cultured on media containing 150, 200, 250, or 300 mM NaCl, compared to the control. Similar results were obtained in all 12 transgenic cell lines derived from rice, tobacco, and white pine, indicating that overexpression of WRKY57 enhances salt stress tolerance in plants including monocotyledons (rice, Fig. **2A**), dicotyledons (tobacco,

Fig. **2B**), and gymnosperms (white pine, Fig. **2C**).

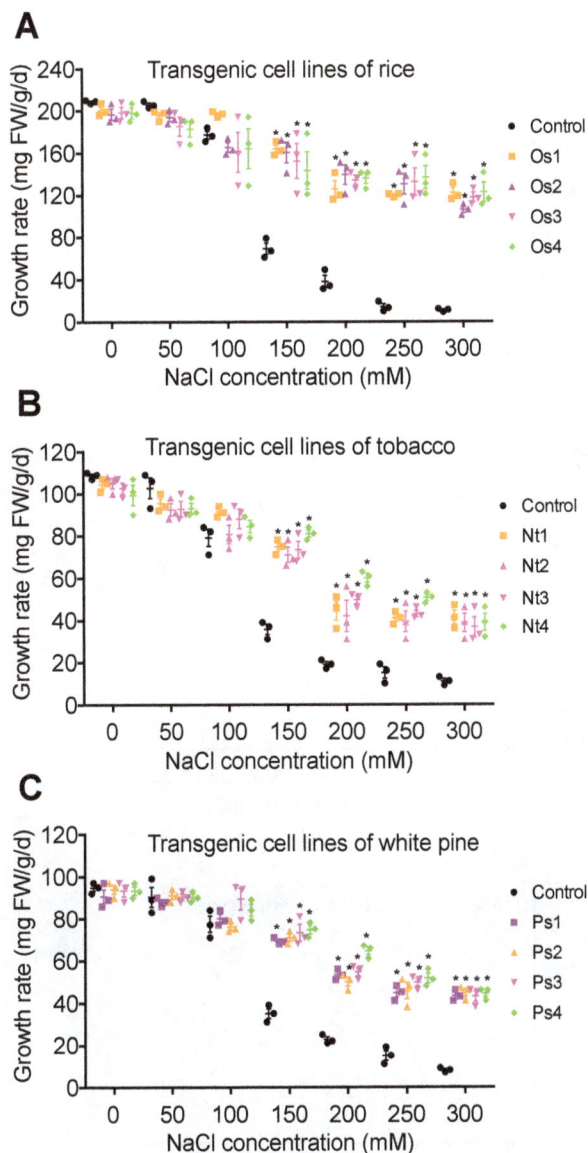

Fig. (2). Growth of cell lines under different concentrations of NaCl. (A-C) Effect of different concentrations of NaCl (50, 100, 150, 200, 250, and 300 mM) on the growth of transgenic cell lines of rice (Os1, Os2, Os3, and Os4), tobacco (Nt1, Nt2, Nt3, and Nt4), and white pine (Ps1, Ps2, Ps3, and Ps4) were measured 3 days after callus transgenic cell cultures were transferred into media containing NaCl. Each experiment was replicated three times, and each replicate consisted of five to ten 250-ml flasks of transgenic cell cultures. Values represent the means ± S.D.

3.3. Thiobarbituric Acid Reactive Substance (TBARS) Changes Under Different Concentrations of NaCl

The salt stress increased the formation of lipid peroxidation via increasing the rate of Reactive Oxygen Species (ROS) formation. Decreasing lipid peroxidation via genetic engineering approaches in cells will increases cells tolerance to salt stress. To determine if salt stress tolerance enhanced by overexpression of WRKY57 is related to the change of Thiobarbituric Acid Reactive Substance (TBARS), TBARS was measured in transgenic cell lines of rice (Os1, Os2, Os3, and Os4), tobacco (Nt1, Nt2, Nt3, and Nt4), and white pine (Ps1, Ps2, Ps3, and Ps4) 3 days after cell cultures were transferred into media containing different concentrations of NaCl (Fig. **3**). A significant decrease in the amount of TBARS was obtained in rice (Fig. **3A**), tobacco (Fig. **3B**), and white pine (Fig. **3C**) when transgenic cell cultures were cultured on media containing 150, 200, 250, or 300 mM NaCl, compared to the control. Similar results were obtained in all 12 transgenic cell lines derived from rice, tobacco, and white pine, indicating that overexpression of

WRKY57 enhances salt stress tolerance by decreasing the amount of TBARS. Compared to the control, no significant decrease in the amount of TBARS was obtained when 50 mM and 100 mM NaCl were applied to the transgenic cells.

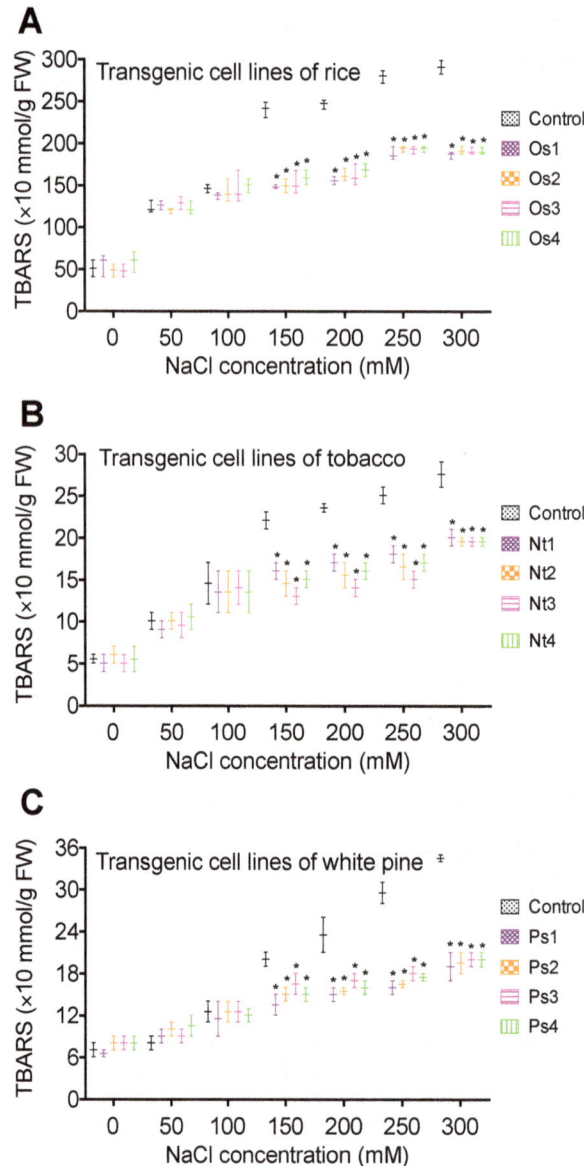

Fig. (3). Thiobarbituric acid reactive substance (TBARS) changes under different concentrations of NaCl. (A-C) TBARS changes in transgenic cell lines of rice (Os1, Os2, Os3, and Os4), tobacco (Nt1, Nt2, Nt3, and Nt4), and white pine (Ps1, Ps2, Ps3, and Ps4). The TBARS were measured 3 days after cell cultures were transferred into media containing different concentrations of NaCl (50, 100, 150, 200, 250, and 300 mM). Each experiment was replicated three times, and each replicate consisted of five to ten 250-ml flasks of transgenic cell cultures. Values represent the means ± S.D.

3.4. Effect of WRKY57 Overexpression on APOX and CAT Activity

To examine if salt stress tolerance enhanced by overexpression of WRKY57 is involved in the change of antioxidant enzymes Ascorbate Peroxidase (APOX) and Catalase (CAT) activity, APOX and CAT activity were measured in transgenic cell lines of rice (Os1, Os2, Os3, and Os4), tobacco (Nt1, Nt2, Nt3, and Nt4), and white pine (Ps1, Ps2, Ps3, and Ps4) 3 days after cell cultures were transferred into media containing different concentrations of NaCl (Fig. **4A**). Increased APOX and CAT activity were observed in transgenic cell lines of rice (Figs. **4A, D**), tobacco (Figs. **4E, 4F**) and white pine (Figs. **4C, F**) indicating that overexpression of WRKY57 enhances salt stress tolerance by increasing the activity of APOX and CAT. Compared to the control, no significant increase in APOX and CAT activity was obtained when 50 mM and 100 mM NaCl were applied to the transgenic cells.

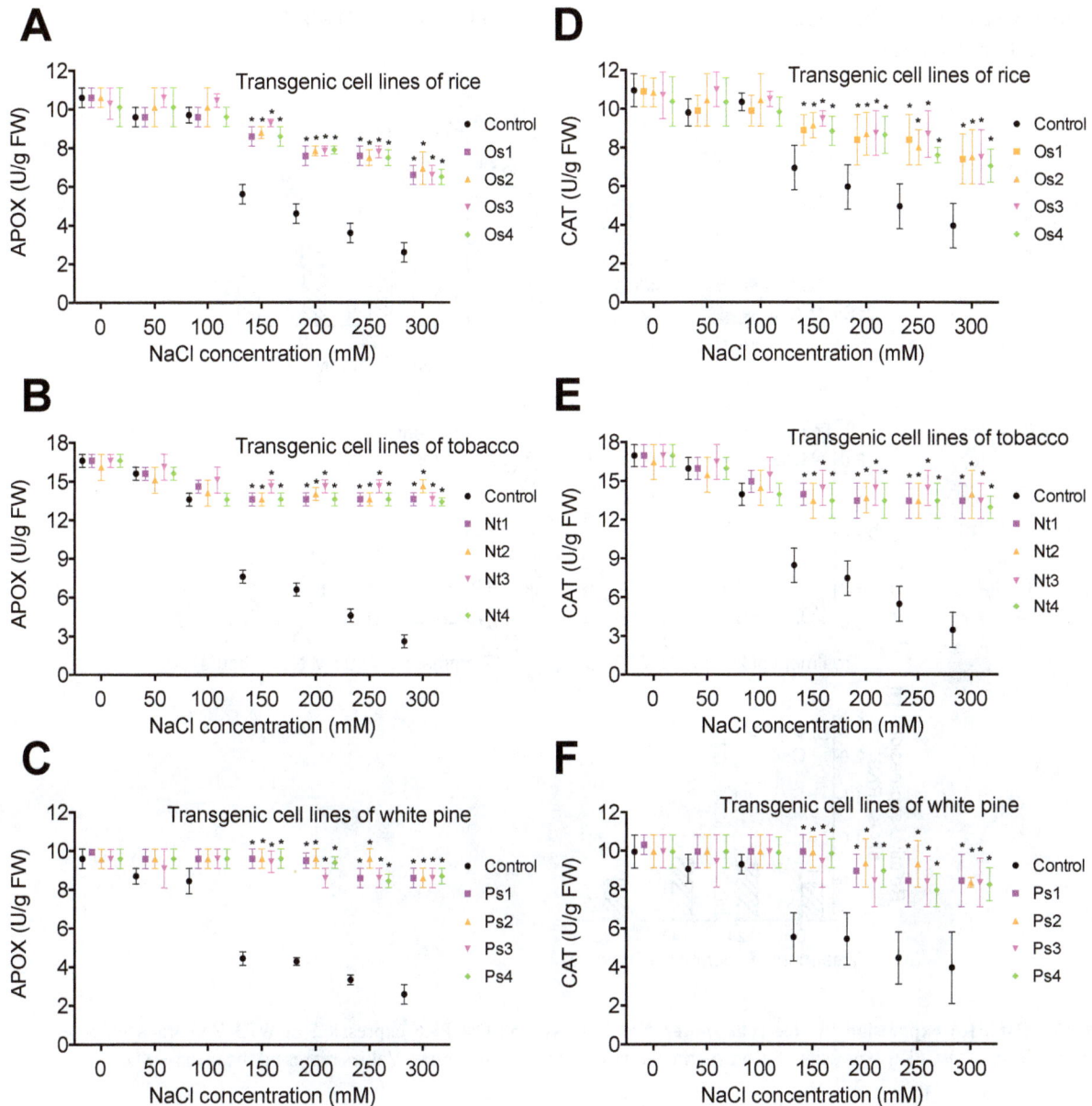

Fig. (4). Effect of WRKY57 overexpression on APOX and CAT activity. APOX and CAT activity changes in transgenic cell lines of rice (Os1, Os2, Os3, and Os4), tobacco (Nt1, Nt2, Nt3, and Nt4), and white pine (Ps1, Ps2, Ps3, and Ps4) were measured 3 days after cell cultures were transferred into media containing different concentrations of NaCl (50, 100, 150, 200, 250, and 300 mM). Each experiment was replicated three times, and each replicate consisted of five to ten 250-ml flasks of transgenic cell cultures. Values represent the means ± S.D.

3.5. Influence of NaCl on Expression of Rice Ca^{2+}-Dependent Protein Kinase Gene Oscpk6 and Oscpk19

To investigate if salt stress tolerance enhanced by overexpression of WRKY57 is related to the expression of Ca^{2+}-dependent protein kinase gene OsCPK6, the amount of OsCPK6 mRNA was measured by northern blotting in transgenic cell lines of rice (Os1, Os2, Os3, and Os4), 24, 48, and 96 hours after cell cultures were transferred into media containing 250 mM NaCl (Fig. **5**). Increase in the amount of OsCPK6 mRNA was obtained in rice 24, 48, and 96 hours after transgenic cell cultures were treated by 250 mM NaCl in all four rice cell lines Quantitative analysis of expression of Ca^{2+}-dependent protein kinase gene OsCPK9 demonstrated that the amount of OsCPK6 mRNA was increased significantly (Figs. **5B**, **E**), compared to the control. Northern blotting analysis of expression of Ca^{2+}-dependent protein kinase gene OsCPK19 (Fig. **6**) in WRKY57 transgenic cell lines Os1, Os2, Os3, and Os4 demonstrated that the amount of OsCPK19 mRNA was increased under 250 mM NaCl treatment at 24, 48, and 96 h. Quantitative analysis of the expression of Ca^{2+}- dependent protein kinase gene OsCPK9 demonstrated that the amount of OsCPK19 mRNA was increased significantly (Figs. **6B**, **E**).

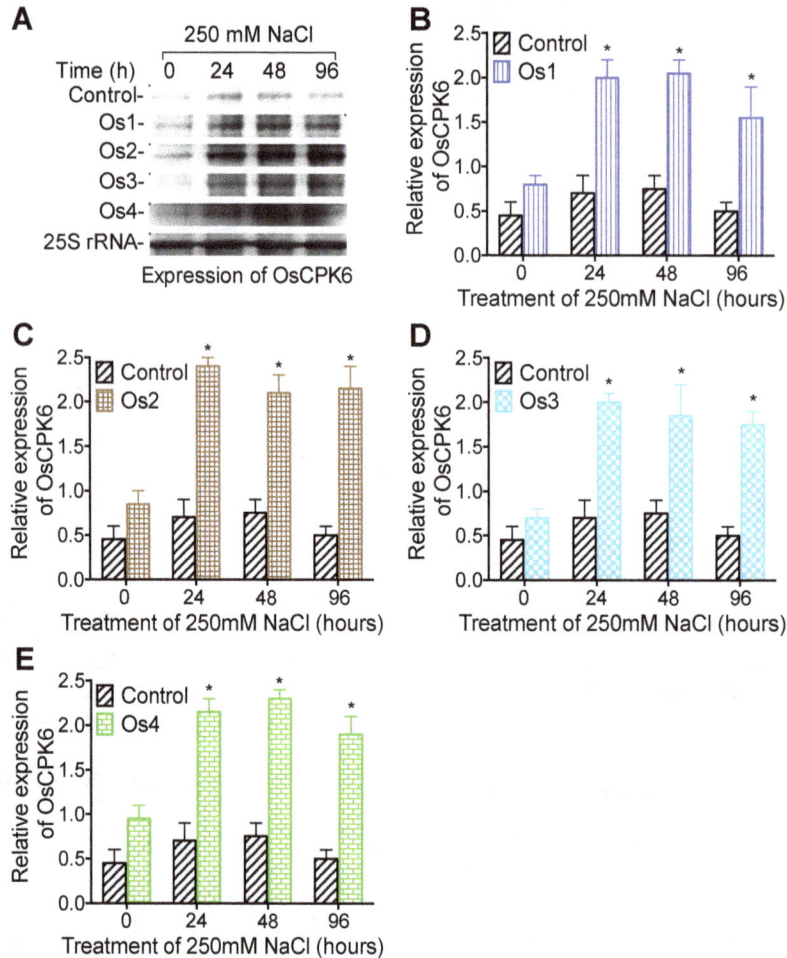

Fig. (5). OsCPK6 expression in rice cells under NaCl stress. (A) OsCPK6 expression in WRKY57 transgenic cells. (B-E) Quantification of OsCPK6 expression. Each experiment was replicated three times. Values represent the mean ± SD.

4. DISCUSSION

The WRKY transcription factors play very important roles in plant growth, some development stages, and plants responses to abiotic stresses. In Sesame (*Sesamum indicum* L.), manipulating WRKYs could improve resistance to waterlogging stress and drought stress [39]. In Arabidopsis, WRKY75 functions as a new component of the GA-mediated signaling pathway to positively regulate flowering in Arabidopsis [40]. WRKY71 activity hastens flowering by providing a means for the plant to complete its life cycle in the presence of salt stress [41]. WRKY36 is a negative regulator of HY5 and UVB represses WRKY36 *via* UVR8 to promote the transcription of HY5 and photomorphogenesis [42]. In oilseed rape (*Brassica napus*), BnWRKY15 overexpression simultaneously increased the susceptibility of *B. napus* to *S. sclerotiorum* and down-regulated BnWRKY33 after different durations of infection [43]. AtWRKY11 and AtWRKY17 are not strictly limited to plant defense responses but are also involved in conferring stress tolerance [44]. Overexpression of WRKY27 displays significantly decreased pollen viability ans is involved in proper plant biomass accumulation and male fertility [45]. In Arabidopsis, WRKY43 acts as a potent modulator of fatty acid desaturation and seed filling, which results in increased tolerance to abiotic stress [46]. WRKY46, WRKY54, and WRKY70, are involved in both BR-regulated plant growth and drought response [47]. Transcription factor WRKY57 enhances salt stress tolerances in plant [32] and is important for developing drought-resilient crops [30]. In grape (*Vitis vinifera* L.), the WRKY57-like protein was strongly upregulated by UV-C irradiation and have great implications for further studies [33 - 35].

Environmental stress can affect the viability of an organism and may affect the genome or the proteome by contributing to protein damage or misfolding [48 - 52]. Plant responses to abiotic stress require two protective measures, one is to reduce stress-inflicted damage to cellular structures, another is the molecular mechanism that efficiently removes damaged and toxic macromolecules [53]. The antioxidant enzyme activities (Superoxide Dismutase (SOD), Ascorbate Peroxidase (APX), and Glutathione Reductase (GR)) and the contents of malondialdehyde (MDA) and H2O2 are important in reactive oxygen scavenging systems that could enhance salinity tolerance and alleviated salinity-induced damage in plants [54]. The Ca^{2+}- dependent protein kinase gene OsCPK4 gene is a member of the complex gene family of calcium-dependent protein kinases in rice (*Oryza sativa*). OsCPK4 is a positive regulator of the salt and drought stress responses in rice that protect cellular membranes and cell signaling transduction from stress-induced oxidative damage [55].

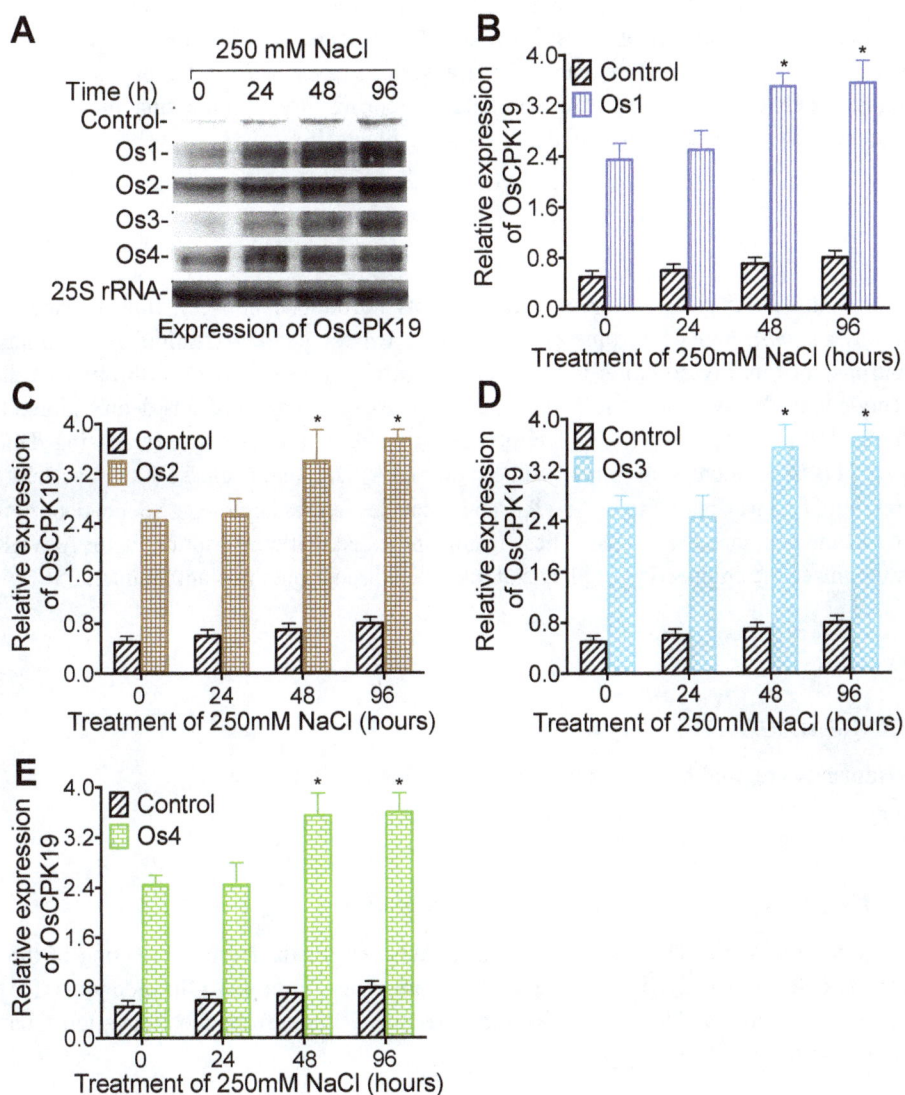

Fig. (6). **OsCPK19 expression in rice cells under NaCl stress.** (**A**) OsCPK19 expression in WRKY57 transgenic cells. (**B-E**) Quantification of OsCPK19 expression. Each experiment was replicated three times. Values represent the mean ± SD.

Inorganic Nanoparticles (NP) have been used in different applications and recent research demonstrated that NP cause adverse effects via induction of an oxidative stress. In vitro methods using oxidative stress biomarkers to measure stress were designed and standardized for conventional NP [56]. It has been reported that a melon Superoxide

Dismutase (SODB) is involved in oxidative stress. SODB could exert its antioxidant properties by inducing the endogenous antioxidant defense [57]. OsCEST and AtCEST were mainly transcribed in photosynthetic tissues. Overexpression of CEST enhanced tolerance not only to salt stress but also to drought stress, and high-temperature stress, which causes photooxidative stress [58]. In potato (*Solanum tuberosum* L. cv. Taedong Valley), a higher level of GSH homeostasis and higher glyoxalase activity inhibit the accumulation of methylglyoxal under salt stress [59]. Investigation of exogenous selenium (Se) in the antioxidant defense demonstrated that exogenous selenium improves stress tolerant to salt stress-induced oxidative damage and enhances their antioxidant defense and MG detoxification systems [60].

Salinity stress-induced production of Reactive Oxygen Species (ROS) and associated oxidative damage is one of the major factors limiting crop production in saline soils [61]. Superoxide Dismutase (SOD), catalase (CAT), ascorbate peroxidase (APX), and glutathione reductase in chickpea plants were up-regulated by salt stress [52]. Transcription factors, such as MYB, WRKY, and zinc finger transcription factors may be regulated by environmental stresses, such as hormones, drought, cold, heat, and pathogens as well as by Pi starvation [62 - 64]. In tomato, high-antioxidant enzyme activities were induced during the response to salt stress [63]. Intricate signaling network responding to the diverse environmental stress could be developed by plants. The transcript expression of SA biosynthetic gene ICS1 and CAT and SOD antioxidative enzymes demonstrated upregulation during stress, indicating biochemical and molecular pathways to maneuver salt stress tolerance of the transgenic plants [64 - 70].

CONCLUSION

In this investigation, transcription factor AtWRKY57 was introduced into cell lines of rice (*Oryza sativa* L.), tobacco (*Nicotiana tabacum*), and white pine (*Pinus strobes* L.) for characterization of its function in salt stress tolerance. The purpose of this investigation is to examine the function of AtWRKY throughout the plant kingdom including monocotyledons, dicotyledons, and gymnosperms. The experimental results demonstrated that heterologous expression of transcription factor AtWRKY57 improves salt stress tolerance by decreasing Thiobarbituric Acid Reactive Substance (TBARS), increasing Ascorbate Peroxidase (APOX) and Catalase (CAT) activity under salt stress. In rice, overexpression of transcription factor AtWRKY57 enhances expression of Ca^{2+}-dependent protein kinase genes *OsCPk6* and *OsCPk19* to counteract salt stress. These results indicated that transcription factor AtWRKY57 might have practical application in genetic engineering of plant salt tolerance throughout the plant kingdom.

HUMAN AND ANIMAL RIGHTS

No Animals/Humans were used for studies that are base of this research.

ACKNOWLEDGEMENTS

The author is grateful to Wells Thompson and Jennifer Whitley for their critical reading of the manuscript. The author appreciates Nicole Franco-Zorrilla, Tinya Backiyarani, and Ambrosia Oluwadahunsi for their support in preparing transgenic cell cultures. This work was supported by a grant from the Education Committee of Hubei Providence of China (Grant No. D20101306).

REFERENCES

[1]   Golldack D, Lüking I, Yang O. Plant tolerance to drought and salinity: Stress regulating transcription factors and their functional significance in the cellular transcriptional network. Plant Cell Rep 2011; 30(8): 1383-91.
        [http://dx.doi.org/10.1007/s00299-011-1068-0] [PMID: 21476089]

[2]   Yao D, Zhang X, Zhao X, *et al.* Transcriptome analysis reveals salt-stress-regulated biological processes and key pathways in roots of cotton

(Gossypium hirsutum L.). Genomics 2011; 98(1): 47-55.
[http://dx.doi.org/10.1016/j.ygeno.2011.04.007] [PMID: 21569837]

[3]   Qin Y, Tian Y, Han L, Yang X. Constitutive expression of a salinity-induced wheat WRKY transcription factor enhances salinity and ionic stress tolerance in transgenic *Arabidopsis thaliana*. Biochem Biophys Res Commun 2013; 441(2): 476-81.
[http://dx.doi.org/10.1016/j.bbrc.2013.10.088] [PMID: 24383079]

[4]   Baloglu MC, Inal B, Kavas M, Unver T. Diverse expression pattern of wheat transcription factors against abiotic stresses in wheat species. Gene 2014; 550(1): 117-22.
[http://dx.doi.org/10.1016/j.gene.2014.08.025] [PMID: 25130909]

[5]   Fan X, Guo Q, Xu P, *et al.* Transcriptome-wide identification of salt-responsive members of the WRKY gene family in *Gossypium aridum*. PLoS One 2015; 10(5): e0126148.
[http://dx.doi.org/10.1371/journal.pone.0126148] [PMID: 25951083]

[6]   Sui J, Jiang D, Zhang D, *et al.* The salinity responsive mechanism of a hydroxyproline-tolerant mutant of peanut based on digital gene expression profiling analysis. PLoS One 2016; 11(9): e0162556.
[http://dx.doi.org/10.1371/journal.pone.0162556] [PMID: 27661086]

[7]   Cai C, Wu S, Niu E, Cheng C, Guo W. Identification of genes related to salt stress tolerance using intron-length polymorphic markers, association mapping and virus-induced gene silencing in cotton. Sci Rep 2017; 7(1): 528.
[http://dx.doi.org/10.1038/s41598-017-00617-7] [PMID: 28373664]

[8]   Pandey G, Yadav CB, Sahu PP, Muthamilarasan M, Prasad M. Salinity induced differential methylation patterns in contrasting cultivars of foxtail millet (*Setaria italica* L.). Plant Cell Rep 2017; 36(5): 759-72.
[http://dx.doi.org/10.1007/s00299-016-2093-9] [PMID: 27999979]

[9]   Xie R, Pan X, Zhang J, *et al.* Effect of salt-stress on gene expression in citrus roots revealed by RNA-seq. Funct Integr Genomics 2018; 18(2): 155-73.
[http://dx.doi.org/10.1007/s10142-017-0582-8] [PMID: 29264749]

[10]   Zhu M, Meng X, Cai J, Li G, Dong T, Li Z. Basic leucine zipper transcription factor SlbZIP1 mediates salt and drought stress tolerance in tomato. BMC Plant Biol 2018; 18(1): 83.
[http://dx.doi.org/10.1186/s12870-018-1299-0] [PMID: 29739325]

[11]   Huang C, Zhou J, Jie Y, *et al.* A ramie bZIP transcription factor bnbzip2 is involved in drought, salt, and heavy metal stress response. DNA Cell Biol 2016; 35(12): 776-86.
[http://dx.doi.org/10.1089/dna.2016.3251] [PMID: 27845851]

[12]   Belamkar V, Weeks NT, Bharti AK, Farmer AD, Graham MA, Cannon SB. Comprehensive characterization and RNA-Seq profiling of the HD-Zip transcription factor family in soybean (glycine max) during dehydration and salt stress. BMC Genomics 2014; 15: 950.
[http://dx.doi.org/10.1186/1471-2164-15-950] [PMID: 25362847]

[13]   Xu H, Shi X, He L, *et al.* *Arabidopsis thaliana* trihelix transcription factor AST1 mediates salt and osmotic stress tolerance by binding to a novel agag-box and some gt motifs. Plant Cell Physiol 2018; 59(5): 946-65.
[http://dx.doi.org/10.1093/pcp/pcy032] [PMID: 29420810]

[14]   An JP, Yao JF, Xu RR, You CX, Wang XF, Hao YJ. An apple NAC transcription factor enhances salt stress tolerance by modulating the ethylene response. Physiol Plant 2018.
[http://dx.doi.org/10.1111/ppl.12724] [PMID: 29527680]

[15]   Hichri I, Muhovski Y, Žižková E, *et al.* The *Solanum lycopersicum* WRKY3 transcription factor slwrky3 is involved in salt stress tolerance in tomato. Front Plant Sci 2017; 8: 1343.
[http://dx.doi.org/10.3389/fpls.2017.01343] [PMID: 28824679]

[16]   Du C, Zhao P, Zhang H, Li N, Zheng L, Wang Y. The reaumuria trigyna transcription factor RtWRKY1 confers tolerance to salt stress in transgenic Arabidopsis. J Plant Physiol 2017; 215: 48-58.
[http://dx.doi.org/10.1016/j.jplph.2017.05.002] [PMID: 28527975]

[17]   Cai R, Dai W, Zhang C, *et al.* The maize WRKY transcription factor ZmWRKY17 negatively regulates salt stress tolerance in transgenic arabidopsis plants. Planta 2017; 246(6): 1215-31.
[http://dx.doi.org/10.1007/s00425-017-2766-9] [PMID: 28861611]

[18]   Wei S, Gao L, Zhang Y, Zhang F, Yang X, Huang D. Genome-wide investigation of the NAC transcription factor family in melon *(Cucumis melo L.)* and their expression analysis under salt stress. Plant Cell Rep 2016; 35(9): 1827-39.
[http://dx.doi.org/10.1007/s00299-016-1997-8] [PMID: 27229006]

[19]   Yu X, Liu Y, Wang S, *et al.* CarNAC4, a NAC-type chickpea transcription factor conferring enhanced drought and salt stress tolerances in Arabidopsis. Plant Cell Rep 2016; 35(3): 613-27.
[http://dx.doi.org/10.1007/s00299-015-1907-5] [PMID: 26650836]

[20]   Xu Z, Gongbuzhaxi C, Wang C, Xue F, Zhang H, Ji W. Wheat NAC transcription factor TaNAC29 is involved in response to salt stress. Plant Physiol Biochem 2015; 96: 356-63.
[http://dx.doi.org/10.1016/j.plaphy.2015.08.013] [PMID: 26352804]

[21]   Zhu M, Chen G, Zhang J, *et al.* The abiotic stress-responsive NAC-type transcription factor SlNAC4 regulates salt and drought tolerance and stress-related genes in tomato (*Solanum lycopersicum*). Plant Cell Rep 2014; 33(11): 1851-63.

[http://dx.doi.org/10.1007/s00299-014-1662-z] [PMID: 25063324]

[22]   He XJ, Mu RL, Cao WH, Zhang ZG, Zhang JS, Chen SY. AtNAC2, a transcription factor downstream of ethylene and auxin signaling
         pathways, is involved in salt stress response and lateral root development. The Plant Journal: For cell and molecular biology 2005; 44(11):
         903-16.

[23]   Li C, Li D, Shao F, Lu S. Molecular cloning and expression analysis of WRKY transcription factor genes in *Salvia miltiorrhiza*. BMC
         Genomics 2015; 16: 200.
         [http://dx.doi.org/10.1186/s12864-015-1411-x] [PMID: 25881056]

[24]   Ling J, Jiang W, Zhang Y, *et al.* Genome-wide analysis of WRKY gene family in *Cucumis sativus*. BMC Genomics 2011; 12: 471.
         [http://dx.doi.org/10.1186/1471-2164-12-471] [PMID: 21955985]

[25]   Fu QT, Yu DQ. Expression profiles of atWRKY25, atWRKY26 and atWRKY33 under abiotic stresses. Yi chuan = Hereditas 2010; 32:
         848-56.

[26]   Yang B, Jiang Y, Rahman MH, Deyholos MK, Kav NN. Identification and expression analysis of WRKY transcription factor genes in canola
         (*Brassica napus* L.) in response to fungal pathogens and hormone treatments. BMC Plant Biol 2009; 9: 68.
         [http://dx.doi.org/10.1186/1471-2229-9-68] [PMID: 19493335]

[27]   Berri S, Abbruscato P, Faivre-Rampant O, *et al.* Characterization of WRKY co-regulatory networks in rice and Arabidopsis. BMC Plant Biol
         2009; 9: 120.
         [http://dx.doi.org/10.1186/1471-2229-9-120] [PMID: 19772648]

[28]   Keates SE, Kostman TA, Anderson JD, Bailey BA. Altered gene expression in three plant species in response to treatment with Nep1, a
         fungal protein that causes necrosis. Plant Physiol 2003; 132(3): 1610-22.
         [http://dx.doi.org/10.1104/pp.102.019836] [PMID: 12857840]

[29]   Dong J, Chen C, Chen Z. Expression profiles of the Arabidopsis WRKY gene superfamily during plant defense response. Plant Mol Biol
         2003; 51(1): 21-37.
         [http://dx.doi.org/10.1023/A:1020780022549] [PMID: 12602888]

[30]   Muthusamy M, Uma S, Backiyarani S, Saraswathi MS, Chandrasekar A. Transcriptomic changes of drought-tolerant and sensitive banana
         cultivars exposed to drought stress. Front Plant Sci 2016; 7: 1609.
         [http://dx.doi.org/10.3389/fpls.2016.01609] [PMID: 27867388]

[31]   Jiang Y, Yu D. The WRKY57 transcription factor affects the expression of jasmonate zim-domain genes transcriptionally to compromise
         botrytis cinerea resistance. Plant Physiol 2016; 171(4): 2771-82.
         [PMID: 27268959]

[32]   Jiang Y, Qiu Y, Hu Y, Yu D. Heterologous expression of AtWRKY57 confers drought tolerance in *Oryza sativa*. Front Plant Sci 2016; 7:
         145.
         [http://dx.doi.org/10.3389/fpls.2016.00145] [PMID: 26904091]

[33]   Xi H, Ma L, Liu G, *et al.* Transcriptomic analysis of grape (*Vitis vinifera L.*) leaves after exposure to ultraviolet C irradiation. PLoS One
         2014; 9(12): e113772.
         [http://dx.doi.org/10.1371/journal.pone.0113772] [PMID: 25464056]

[34]   Jiang Y, Liang G, Yang S, Yu D. Arabidopsis WRKY57 functions as a node of convergence for jasmonic acid- and auxin-mediated signaling
         in jasmonic acid-induced leaf senescence. Plant Cell 2014; 26(1): 230-45.
         [http://dx.doi.org/10.1105/tpc.113.117838] [PMID: 24424094]

[35]   Jiang Y, Liang G, Yu D. Activated expression of WRKY57 confers drought tolerance in arabidopsis. Mol Plant 2012; 5(6): 1375-88.
         [http://dx.doi.org/10.1093/mp/sss080] [PMID: 22930734]

[36]   Tang W, Page M, Fei Y, *et al.* Overexpression of atbZIP60deltaC gene alleviates salt induced oxidative damage in transgenic cell cultures.
         Plant Mol Biol Report 2012; 30: 1183-95.
         [http://dx.doi.org/10.1007/s11105-012-0437-3]

[37]   Tang W, Fei Y, Page M. Elevated tolerance to salt stress in transgenic cells expressing transcription factor AtbZIP60 is associated with the
         increased activities of H+ATPase and acid phosphatase. Plant Biotechnol Rep 2012; 6: 313-25.
         [http://dx.doi.org/10.1007/s11816-012-0226-3]

[38]   Tang W, Page M. Overexpression of the Arabidopsis AtEm6 gene enhances salt tolerance in transgenic rice cell lines. Plant Cell Tissue Organ
         Cult 2013; 114: 339-50. [PCTOC].
         [http://dx.doi.org/10.1007/s11240-013-0329-8]

[39]   Wei W, Hu Y, Han YT, Zhang K, Zhao FL, Feng JY. The WRKY transcription factors in the diploid woodland strawberry *Fragaria vesca*:
         Identification and expression analysis under biotic and abiotic stresses. Plant Physiol Biochem 2016; 105: 129-44.
         [http://dx.doi.org/10.1016/j.plaphy.2016.04.014] [PMID: 27105420]

[40]   Zhang L, Chen L, Yu D. Transcription factor WRKY75 interacts with DELLA proteins to affect flowering. Plant Physiol 2018; 176(1):
         790-803.
         [http://dx.doi.org/10.1104/pp.17.00657] [PMID: 29133369]

[41]   Yu Y, Wang L, Chen J, Liu Z, Park CM, Xiang F. WRKY71 acts antagonistically against salt-delayed flowering in *Arabidopsis thaliana*.
         Plant Cell Physiol 2018; 59(2): 414-22.

[http://dx.doi.org/10.1093/pcp/pcx201] [PMID: 29272465]

[42]   Yang Y, Liang T, Zhang L, et al. UVR8 interacts with WRKY36 to regulate HY5 transcription and hypocotyl elongation in Arabidopsis. Nat Plants 2018; 4(2): 98-107.
         [http://dx.doi.org/10.1038/s41477-017-0099-0] [PMID: 29379156]

[43]   Liu F, Li X, Wang M, et al. Interactions of WRKY15 and WRKY33 transcription factors and their roles in the resistance of oilseed rape to Sclerotinia infection. Plant Biotechnol J 2018; 16(4): 911-25.
         [http://dx.doi.org/10.1111/pbi.12838] [PMID: 28929638]

[44]   Ali MA, Azeem F, Nawaz MA, et al. Transcription factors WRKY11 and WRKY17 are involved in abiotic stress responses in arabidopsis. J Plant Physiol 2018; 226: 12-21.
         [http://dx.doi.org/10.1016/j.jplph.2018.04.007] [PMID: 29689430]

[45]   Mukhtar MS, Liu X, Somssich IE. Elucidating the role of WRKY27 in male sterility in arabidopsis. Plant Signal Behav 2017; 12(9): e1363945.
         [http://dx.doi.org/10.1080/15592324.2017.1363945] [PMID: 28816593]

[46]   Geilen K, Heilmann M, Hillmer S, Böhmer M. WRKY43 regulates polyunsaturated fatty acid content and seed germination under unfavourable growth conditions. Sci Rep 2017; 7(1): 14235.
         [http://dx.doi.org/10.1038/s41598-017-14695-0] [PMID: 29079824]

[47]   Chen J, Nolan TM, Ye H, et al. Arabidopsis WRKY46, WRKY54, and WRKY70 transcription factors are involved in brassinosteroid-regulated plant growth and drought responses. Plant Cell 2017; 29(6): 1425-39.
         [PMID: 28576847]

[48]   Sampuda KM, Riley M, Boyd L. Stress induced nuclear granules form in response to accumulation of misfolded proteins in caenorhabditis elegans. BMC Cell Biol 2017; 18(1): 18.
         [http://dx.doi.org/10.1186/s12860-017-0136-x] [PMID: 28424053]

[49]   Abolaji AO, Olaiya CO, Oluwadahunsi OJ, Farombi EO. Dietary consumption of monosodium L-glutamate induces adaptive response and reduction in the life span of Drosophila melanogaster. Cell Biochem Funct 2017; 35(3): 164-70.
         [http://dx.doi.org/10.1002/cbf.3259] [PMID: 28303588]

[50]   Ranjit SL, Manish P, Penna S. Early osmotic, antioxidant, ionic, and redox responses to salinity in leaves and roots of Indian mustard (Brassica juncea L.). Protoplasma 2016; 253(1): 101-10.
         [http://dx.doi.org/10.1007/s00709-015-0792-7] [PMID: 25786350]

[51]   Penella C, Landi M, Guidi L, et al. Salt-tolerant rootstock increases yield of pepper under salinity through maintenance of photosynthetic performance and sinks strength. J Plant Physiol 2016; 193: 1-11.
         [http://dx.doi.org/10.1016/j.jplph.2016.02.007] [PMID: 26918569]

[52]   Ahmad P, Abdel Latef AA, Hashem A, Abd Allah EF, Gucel S, Tran LS. Nitric oxide mitigates salt stress by regulating levels of osmolytes and antioxidant enzymes in chickpea. Front Plant Sci 2016; 7: 347.
         [http://dx.doi.org/10.3389/fpls.2016.00347] [PMID: 27066020]

[53]   Zhou J, Zhang Y, Qi J, et al. E3 ubiquitin ligase CHIP and NBR1-mediated selective autophagy protect additively against proteotoxicity in plant stress responses. PLoS Genet 2014; 10(1): e1004116.
         [http://dx.doi.org/10.1371/journal.pgen.1004116] [PMID: 24497840]

[54]   Wang R, Liu S, Zhou F, Ding C, Hua C. Exogenous ascorbic acid and glutathione alleviate oxidative stress induced by salt stress in the chloroplasts of oryza sativa L. Z Natforsch C J Biosci 2014; 69(5-6): 226-36.
         [http://dx.doi.org/10.5560/znc.2013-0117] [PMID: 25069161]

[55]   Campo S, Baldrich P, Messeguer J, Lalanne E, Coca M, San Segundo B. Overexpression of a calcium-dependent protein kinase confers salt and drought tolerance in rice by preventing membrane lipid peroxidation. Plant Physiol 2014; 165(2): 688-704.
         [http://dx.doi.org/10.1104/pp.113.230268] [PMID: 24784760]

[56]   Tournebize J, Sapin-Minet A, Bartosz G, Leroy P, Boudier A. Pitfalls of assays devoted to evaluation of oxidative stress induced by inorganic nanoparticles. Talanta 2013; 116: 753-63.
         [http://dx.doi.org/10.1016/j.talanta.2013.07.077] [PMID: 24148470]

[57]   Carillon J, Romain C, Bardy G, et al. Cafeteria diet induces obesity and insulin resistance associated with oxidative stress but not with inflammation: Improvement by dietary supplementation with a melon superoxide dismutase. Free Radic Biol Med 2013; 65: 254-61.
         [http://dx.doi.org/10.1016/j.freeradbiomed.2013.06.022] [PMID: 23792771]

[58]   Yokotani N, Higuchi M, Kondou Y, et al. A novel chloroplast protein, CEST induces tolerance to multiple environmental stresses and reduces photooxidative damage in transgenic Arabidopsis. J Exp Bot 2011; 62(2): 557-69.
         [http://dx.doi.org/10.1093/jxb/erq290] [PMID: 20876334]

[59]   Upadhyaya CP, Venkatesh J, Gururani MA, et al. Transgenic potato overproducing L-ascorbic acid resisted an increase in methylglyoxal under salinity stress via maintaining higher reduced glutathione level and glyoxalase enzyme activity. Biotechnol Lett 2011; 33(11): 2297-307.
         [http://dx.doi.org/10.1007/s10529-011-0684-7] [PMID: 21750996]

[60]   Hasanuzzaman M, Hossain MA, Fujita M. Selenium-induced up-regulation of the antioxidant defense and methylglyoxal detoxification system reduces salinity-induced damage in rapeseed seedlings. Biol Trace Elem Res 2011; 143(3): 1704-21.

[http://dx.doi.org/10.1007/s12011-011-8958-4] [PMID: 21264525]

[61]   Wang H, Shabala L, Zhou M, Shabala S. Hydrogen peroxide-induced root Ca^{2+} and K^+ fluxes correlate with salt tolerance in cereals: Towards the cell-based phenotyping. Int J Mol Sci 2018; 19(3): 19.
        [PMID: 29494514]

[62]   Baek D, Chun HJ, Yun DJ, Kim MC. Cross-talk between phosphate starvation and other environmental stress signaling pathways in plants. Mol Cells 2017; 40(10): 697-705.
        [PMID: 29047263]

[63]   Gharsallah C, Fakhfakh H, Grubb D, Gorsane F. Effect of salt stress on ion concentration, proline content, antioxidant enzyme activities and gene expression in tomato cultivars. AoB Plants 2016; 8: 8.
        [http://dx.doi.org/10.1093/aobpla/plw055] [PMID: 27543452]

[64]   Agarwal P, Dabi M, Sapara KK, Joshi PS, Agarwal PK. Ectopic expression of jcwrky transcription factor confers salinity tolerance *via* salicylic acid signaling. Front Plant Sci 2016; 7: 1541.
        [http://dx.doi.org/10.3389/fpls.2016.01541] [PMID: 27799936]

[65]   Tang W, Page M. Transcription factor AtbZIP60 regulates expression of Ca^{2+} -dependent protein kinase genes in transgenic cells. Mol Biol Rep 2013; 40(3): 2723-32.
        [http://dx.doi.org/10.1007/s11033-012-2362-9] [PMID: 23275191]

[66]   Tang W, Newton RJ, Weidner DA. Genetic transformation and gene silencing mediated by multiple copies of a transgene in eastern white pine. J Exp Bot 2007; 58(3): 545-54.
        [http://dx.doi.org/10.1093/jxb/erl228] [PMID: 17158108]

[67]   Tang W, Weidner DA, Hu BY, Newton RJ, Hu XH. Efficient delivery of small interfering RNA to plant cells by a nanosecond pulsed laser-induced stress wave for posttranscriptional gene silencing. Plant science: An international journal of experimental plant biology 2006; 171: 375-81.

[68]   Tang W, Newton RJ, Lin J, Charles TM. Expression of a transcription factor from Capsicum annuum in pine calli counteracts the inhibitory effects of salt stress on adventitious shoot formation. Mol Genet Genomics 2006; 276(3): 242-53.
        [http://dx.doi.org/10.1007/s00438-006-0137-5] [PMID: 16767459]

[69]   Tang W, Newton RJ, Li C, Charles TM. Enhanced stress tolerance in transgenic pine expressing the pepper CaPF1 gene is associated with the polyamine biosynthesis. Plant Cell Rep 2007; 26(1): 115-24.
        [http://dx.doi.org/10.1007/s00299-006-0228-0] [PMID: 16937149]

[70]   Tang W, Newton RJ. Peroxidase and catalase activities are involved in direct adventitious shoot formation induced by thidiazuron in eastern white pine (*Pinus strobus L.*) zygotic embryos. Plant Physiol Biochem 2005; 43(8): 760-9.
        [http://dx.doi.org/10.1016/j.plaphy.2005.05.008] [PMID: 16129608]

8

Aluminium Adjuvants – A Nanomaterial used as Adjuvants in Human Vaccines for Decades

Ravi Danielsson, Tove Sandberg and Håkan Eriksson[*]

Department of Biomedical Science, Faculty of Health and Society, Malmö University, SE-205 06 Malmö, Sweden

Abstract:

Background:

Aluminium salts have been used for decades in vaccines as adjuvants to facilitate the adaptive immune response against co-administered antigens. Two types of aluminium adjuvant are mostly used, aluminium oxyhydroxide and aluminium hydroxyphosphate. Both types of aluminium adjuvant consist of nanoparticles that form loose, micrometre sized aggregates at circumneutral pH.

Aluminium adjuvants constitute a well-documented example of administration of nanomaterials to humans with infrequent side effects and a safety record generally regarded as excellent. However, despite its prolonged use in human and veterinary medicine, the mechanisms behind the enhanced response and the immune stimulatory effect are still by and large unknown.

Methods:

The present paper reviews existing ideas regarding the immunostimulatory effects of aluminium adjuvants, with a focus on the induction of an inflammatory response by cellular stress. Reviewed information was obtained from peer-reviewed scientific papers published in 1988 to date with one exception, a paper published 1931.

Results:

Cellular stress causes extra cellular signalling of Danger Associated Molecular Patterns (DAMPs) and upon phagocytosis of aluminium adjuvants the cells need to manage the ingested particles.

Conclusion:

A persistent intracellular accumulation of aluminium adjuvants will be a solid depository of sparingly soluble aluminium salts maintaining a constant concentration of Al^{3+} ions in the cytoplasm and this will affect multiple biochemical processes. The cell will be under constant stress and DAMP signalling will occur and we would like to suggest the maintenance of a constant concentration Al^{3+} ions in the cytoplasm as a general underlying feature of the immune stimulation properties of aluminium adjuvants.

Keywords: Aluminium adjuvants, Cellular stress, DAMP, Vaccine, Immunostimulatory effects, Al^{3+} ions.

1. INTRODUCTION

Vaccines, one of the greatest achievements of medicine, have a major impact on public health worldwide, saving millions of people from disease and premature death. In attenuated and inactivated vaccines, such as the tuberculosis and the Salk polio vaccine, the vaccine contains not only the pertinent antigens but also endogenous immune stimulants acting as adjuvants. However, in vaccines containing isolated and purified antigen such as tetanus, diphtheria, and pertussis vaccines, the formulations require addition of exogenous adjuvants and to that end, aluminium salts have been

[*] Address correspondence to this author at the Department of Biomedical Science, Faculty of Health and Society, Malmö University, SE-205 06 Malmö, Sweden, E-mail hakan.eriksson@mah.se

used for decades to facilitate an adaptive immune response against co-administered antigens.

The current and main types of aluminium salts used as adjuvants in vaccine formulations are aluminium oxyhydroxide and aluminium hydroxyphosphate. Both types of aluminium adjuvants consist of nanoparticles that form loose, micrometre sized aggregates at circumneutral pH; the physical and chemical properties are however quite different.

Aluminium oxyhydroxide consists of needle-like crystallites, 4.5 nm x 2.2 nm x 10 nm in size [1], which form micrometre sized aggregates under physiological conditions. The aggregated aluminium oxyhydroxide particles have a large surface area with a positive surface charge at neutral pH.

Aluminium hydroxyphosphate is amorphous, forming plate-like nanoparticles with a diameter of 50 nm [2], which aggregate and form loosely unified micrometre sized particles under physiological conditions. The aggregated aluminium hydroxyphosphate particles have a large surface area at neutral pH, however, the surface charge is negative and thereby opposite in charge compared to aluminium oxyhydroxide particles.

Adsorption of protein antigens onto aluminium adjuvants are mediated through hydrophobic and van der Waals forces, electrostatic interactions, and by ligand exchange [3]. An initial electrostatic interaction will lead to adsorption of the antigen, which upon rearrangements on the adjuvant surface results in an antigen adsorbed onto the adjuvant particles through multi-point attachments, mediated by a mixture of electrostatic as well as hydrophobic and van der Waals interactions. Since, the commonly used aluminium adjuvants have opposite surface charges, aluminium oxyhydroxide is generally used as adjuvant in combination with antigens having an isoelectric point less than 7 whereas aluminium hydroxyphosphate is used with antigens having an isoelectric point greater than 7.

For many years, it was believed that the mechanism of action of aluminium adjuvants was due to adsorption of the antigen to the adjuvant and thereby facilitating antigen presentation and at the same time forming an antigen depot at the injection site, prolonging the exposure of antigen to the immune system. The depot hypothesis was proposed as early as 1931 by Glenny, Buttle, and Stevens [4], however, the depot effect of aluminium adjuvants is today regarded as not being likely since:

- Within hours after administration most of the antigen diffuses away from the injection site [5].
- The half-life of antigen *in situ* is not increased due to administration of antigen adsorbed on aluminium adjuvants [3].
- The magnitude of the immune response is not dependent on the adjuvant remaining at the injection site for more than a few hours [6].
- Fibrin-dependent nodules induced by aluminium adjuvants at the injection site is not a requirement of an immune response [7].

Instead, many reports have been made regarding infiltrating cells at the injection site showing both the accumulation of granulocytes, monocytes, macrophages and dendritic cells [8 - 10] and increased amounts of inflammatory chemokines and cytokines at the injection site [9, 11, 12]. Today the predominant hypothesis is that the aluminium adjuvants induce inflammation at the injection site and several reviews have been published regarding the induction of an inflammatory reaction as the mechanism behind the immune stimulatory properties of aluminium adjuvants [13 - 21]. Induction of an inflammatory response is consistent with Polly Matzinger's proposal in 1994, that immune activation not only needs to be initiated by microbial infections but also by the recognition of molecules associated with danger [22]. Danger molecules such as Pathogen Associated Molecular Patterns (PAMPs), driving inflammation and initiating an adaptive immune response were soon identified [23, 24], and today the danger signals also include alarm signals derived from endogenous molecules expressed at elevated levels upon danger, damage, stress, injury, or necrosis [25, 26].

The involvement of DAMP and alarm molecules in the initiation of inflammation and thereby the induction of an adaptive immune response has to some extent focused the research and lead to the development of adjuvants derived from PAMP structures, inducing an inflammatory response at the injection site [27, 28]. An innate immune response is induced after recognition of PAMP molecules by Toll-Like Receptors (TLR) and agonists against TLRs have been utilized as adjuvants in vaccine formulations. Derivatives of LPS such as Monophosphoryl Lipid A (MPLA) and oligonucleotides as CpG with specificities against TLR4 and TLR9 respectively, have been used in both experimental and approved vaccine formulations.

The initiation of an inflammatory and a subsequent adaptive immune response caused by aluminium adjuvants has been suggested to be mediated by; the activation of caspase -1 and the NLRP3 inflammasome [29]; formation of Reactive Oxygen Species, ROS, either in the phagosomes [30] or by the mitochondrial electron transport chain [31 - 33] and their subsequent release into the cytosol; rupture of phagolysosomes by aluminium adjuvants facilitating release of proteases such as cathepsin B into the cytoplasm [34]; formation of lipid rafts in the cell membranes of antigen presenting cells, APCs [35]. However, none of these suggested mechanisms have been verified as a general mechanism initiating the immune stimulatory features of aluminium adjuvants since no reduction of the immune response has been observed *in vivo* after cellular depletions [36] or in mice lacking NALP3 inflammasomes [11, 37] and cathepsin B [38].

At the injection site, aluminium adjuvants have been shown to induce necrosis and cell injury [39 - 41]. Dead or dying cells at the injection site will release DAMP structures, sometimes referred to as alarmins [42], corroborating the induction of inflammation as a central cause of the adjuvant properties of aluminium adjuvants and indeed, the involvement of DAMP and alarmins such as DNA [39], ATP [43] and HSP70 [44] in the immune response induced by aluminium adjuvants has been reported. Accordingly, cell injury and necrosis caused by aluminium adjuvants will induce a sterile inflammation [45], however, DAMP or alarmin molecules can also be released upon cellular stress [46, 47], in the absence of cell death and necrosis.

Regarding the depot effect of aluminium adjuvants, clearance of the adjuvant is mediated by a complete solubilisation into Al^{3+} ions. Ions are quickly eliminated from the body and using a radioactive isotope of aluminium and a soluble aluminium salt, aluminium citrate, 83% of intravenously injected aluminium was eliminated after 13 days [48]. However, in a vaccine, aluminium adjuvants are injected as aggregates of aluminium salts with a solubility of less than one µg Al^{3+}/ml [49] and in animal experiments injection of aluminium adjuvants have resulted in the clearance of no more than 6% of the aluminium 28 days after administration of aluminium oxyhydroxide, although a somewhat higher elimination was obtained using aluminium hydroxyphosphate showing 28% elimination at day 28 [50]. Information regarding the bio-distribution of aluminium adjuvants after injection and their elimination is limited, however, upon endocytosis and encapsulation in granulomas at the injection site, an elimination rate completely different from the rate obtained using soluble aluminium citrate as tracer can be expected.

Cells of myeloid origin, *i.e.* neutrophils, macrophages, and dendritic cells, are efficient phagocytic cells and a quick uptake of aluminium adjuvants by myeloid cells occurs at the injection site [34, 35, 51, 52]. Several reports have shown long time persistent intracellular aluminium aggregates in macrophages after administration of aluminium adjuvants [53, 54] and in lymphoid organs, cells of myeloid origin have been shown to harbour aluminium aggregates [55, 56]. The cellular machinery will try to handle intracellular aluminium adjuvants and cellular stress can be anticipated in cells holding the challenging aluminium aggregates, and as previously mentioned, cellular stress induces expression of DAMP or alarmin molecules. DAMP molecules are not only released upon stress induced cell death [26, 45, 57 - 59], but also from living, although stressed cells [46, 60 - 62]. In this context, myeloid cells are well known to induce inflammation and to act as Antigen-Presenting Cells (APCs). Release of DAMP and alarmins from these cells fits well with the danger model by Matzinger in which injection of aluminium adjuvants results in a short-term inflammation inducing the adaptive immune system and could concomitantly trigger a long-term inflammation resulting in some of the autoimmune related pathological side effects (ASIA) reported to be caused by aluminium adjuvants [63, 64].

The mechanism behind the immune stimulating properties of aluminium adjuvants have been discussed and debated during decades and this review will focus on possible biochemical mechanisms inducing an initial cellular stress response due to exposure to aluminium adjuvants.

2. CELLULAR IMPACTS DUE TO EXPOSURE AND INTERNALIZATION OF ALUMINIUM ADJUVANTS

Aluminium adjuvants in vaccines consist of aggregated nanomaterial and the impact of nanoparticles on various cellular mechanisms has been reported in several reviews [65 - 68] while Al^{3+} ions have no essential role in any biological mechanism as far as is known. Aluminium salts and ions have only shown toxic effects at very high concentrations and are generally considered as posing low risk of inducing adverse effects. However, studies have suggested that long-term exposure of aluminium has adverse neurological effects [69, 70] and ASIA has been reported in relation to aluminium adjuvants [63, 64].

Inorganic solid state materials have been shown to affect the immune activation due to their physicochemical properties [71], and upon internalization into a cell, aluminium adjuvants may interfere with numerous and diverging biochemical and biological mechanisms. Fig. (1) outlines the various aspects that will be discussed in this review and in which aluminium adjuvants may initiate an inflammatory process and thereby an adaptive immune response.

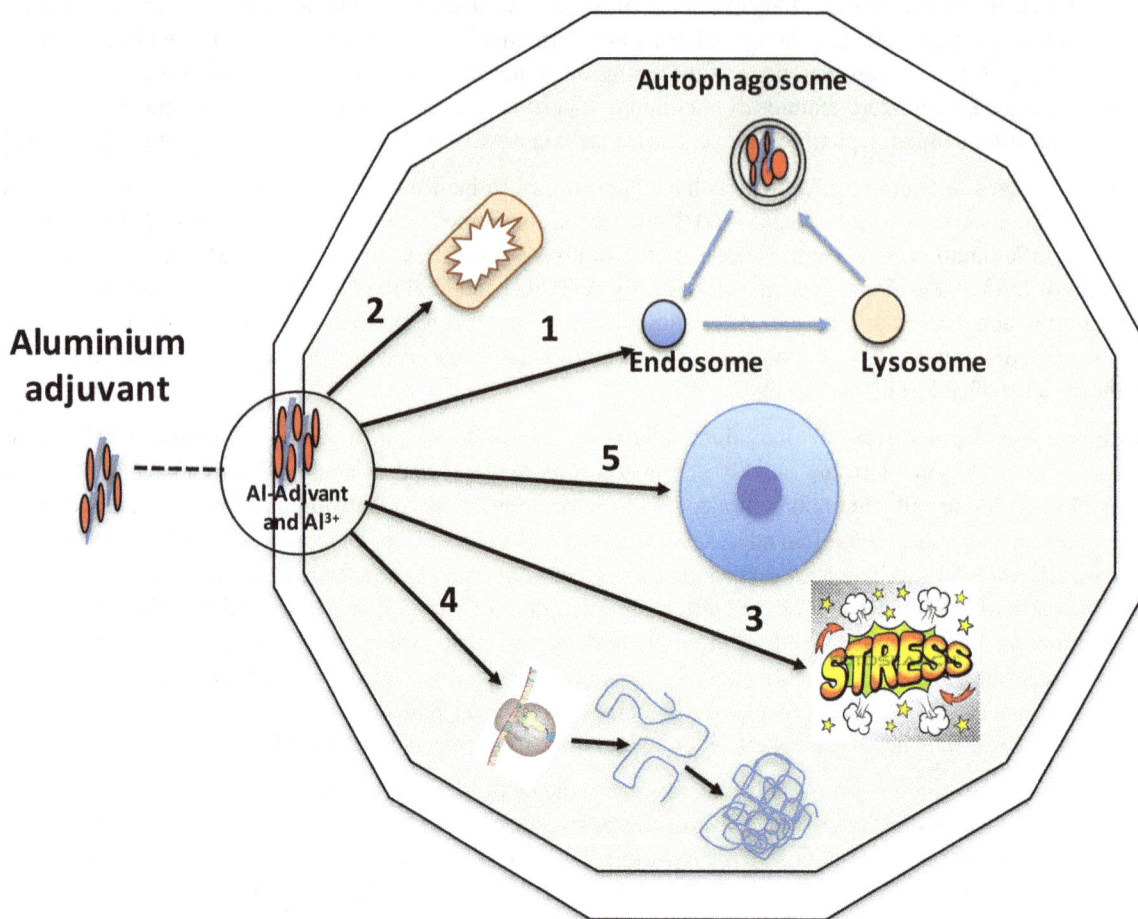

Fig. (1). Schematic picture showing internalization of Al adjuvants into a cell through an early phagosome containing Al-adjuvant aggregate(s) and due to encapsulation of the adjuvant also a saturated solution of A^{13+} ions. Upon entrance into the cell, Al-particles and dissolved Al^{3+} ions can be expected to influence; 1) the endosomal pathway 2) mitochondrial function and activity 3) cellular stress functions 4) protein synthesis and folding 5) nuclear structure and function.

2.1. The Endosomal Pathway, Early Endosome (Phagosome) to Lysosome (Phagolysosome)

Upon injection aluminium adjuvants consist of micrometre sized aggregates, which are quickly recognized by phagocytosing cells and internalized [53, 72, 73]. Phagocytosing cells are mainly cells of the myeloid line which are potential professional APCs and phagocytosis of aluminium adjuvant may promote the survival of the cells [74].

Phagocytosis of aluminium adjuvants will, at the molecular level, give the cell a very high load of both antigen adsorbed on the adjuvant aggregates and of aluminium salts. The aluminium adjuvant consists of aggregated nanoparticles and once inside the phagosome an equilibrium will be established between various levels of aggregated nanoparticles and of soluble Al^{3+} ions. However, upon phagosome maturation into a phagolysosome, the milieu inside the vesicle entrapping the adjuvant aggregate will be altered. Maturation into a phagolysosome will change the protein and peptide content inside the vesicle and due to oxidative burst, change the redox potential in the vesicle. The pH will also drop, reflecting an increase of the H^+ concentration by several orders of magnitude. Taken together this will probably affect the aggregation level of nanoparticles forming the aluminium adjuvant and certainly the Al^{3+} concentration since a decrease in pH increases the solubility of the aluminium salts used as adjuvants. Both aluminium adjuvants and Al^{3+} ions have been reported to interact and destabilize lipid membranes [35, 51, 75, 76], and rupture of phagosomes and release of the content into the cytoplasm has been suggested as possible mechanism of aluminium adjuvants [34]. However, in this context it was only the release of the lysosomal protease cathepsin B that was discussed, not the release of nanoparticles of aluminium adjuvants or Al^{3+} ions. Release of aluminium adjuvants into the cytoplasm constitutes a source of nanoparticles at various degree of aggregation and thus a solid depository of sparingly soluble aluminium salts maintaining a constant concentration of Al^{3+} ions in the cytoplasm. Several reports have verified the presence of small amounts aluminium particles persistent in the cytoplasm [51 - 54, 77] and Mitchell *et al.* [68] have shown a strong upregulation of stress responses as a proactive defence against low loads of intra cellular nanoparticles. Several stress induced proteins have been identified as DAMP molecules and it should be considered that a cell with a low and persistent load of aluminium adjuvants will be a cell with a continuous increased stress response and thereby a potential inducer of a sterile inflammation.

Micrometre sized aggregates of aluminium adjuvants will not be degraded inside a phagolysosome and in a situation with or without disruption of the lysosomal membrane, initiation of an autophagosomal process will be anticipated as an attempt to handle the particles [78 - 80]. The autophagosomal pathway induces a new compartmentalization of the particles, including re-acidification and relocation of the aluminium adjuvant in autophagosomes. Without solubilisation, this process will be repeated and each cycle will drain the cell of energy. Both the phagosomal and the autophagosomal pathways are energy consuming primarily owing to the acidification of the vesicles and increased mitochondrial activity after phagocytosis of aluminium adjuvants has been reported [81].

Solubilisation of aluminium adjuvants occurs at the decreased pH inside the phagosomal/autophagosomal vesicles, however, upon release into the cytoplasm, the solubility will be reduced as the circumneutral pH is re-established. In any event, a constant amount of Al^{3+} ions will be present in the cytoplasm, controlled by the release of ions from solid aluminium salts. As other metal ions, aluminium ions interact with all kinds of biomolecules and can be expected to interfere with numerous biochemical events and mechanisms in the cell.

2.2. Mitochondrial Function and Activity

The energy production of a cell takes place in the mitochondria, and to maintain the function of the mitochondria, the inner mitochondrial membrane potential and the permeability status of the membranes must be sustained. The formation of ATP is driven by a chemiosmotic gradient established by an electron transport chain in the inner mitochondrial membrane and the major generation of ROS occurs at the electron transport chain as a by-product of respiration [82, 83]. At high cellular energy demand, the generation of mitochondrial ROS will increase and several reports have shown that nanoparticles induce and increase mitochondrial ROS generation [84 - 86]. This fits well with the assumption that non- or slowly degradable particles like aluminium adjuvants increase the energy consumption of the cells and reports have shown the accumulation of Al_2O_3 nanoparticles in autophagosomes [87].

Nanoparticles of aluminium adjuvants are more potent adjuvants than their corresponding microparticles and treatment with nanoparticles of aluminium adjuvants establishes a higher intracellular aluminium content [84]. Aluminium adjuvants and Al^{3+} ions impair the stability of lipid membranes [35, 51, 75, 76] and both nano- and microparticles of aluminium adjuvants induce lysosomal rupture and trigger increased levels of intracellular ROS. However, both particle sizes have been shown to be equally efficient in inducing lysosomal rupture and intracellular ROS content [84]. Solubilisation of the adjuvants into aluminium ions is a limiting feature and although treatment with nanoparticles results in a higher level of intra cellular aluminium, the cytoplasmic concentration of aluminium ions can be expected to be the same regardless whether the intracellular delivery is mediated by nano- or microparticles. Instead the time span, during which a saturated level of intracellular Al^{3+} ions remains, will be longer using nanoparticles. The major impact on mitochondria can then be expected to be mediated by Al ions, however, in mammalian systems the effect of intra cellular Al ions is challenging to investigate since they hardly penetrate the cell membrane (unless they are delivered through endocytosis as an aggregated aluminium salt). Instead conclusion can be drawn from experiments using plant materials in which intracellular concentrations of aluminium ions can be maintained. Several biochemical processes can be considered as the same or very similar in plant and mammalian cells and in plants, aluminium ions have been shown to cause ROS bursts and deterioration of the mitochondrial inner membrane potential [88, 89]. Al ions also increased the permeability of the mitochondrial outer membrane, triggered the release of cytochrome c from mitochondria to the cytoplasm, and caused activation of caspase 3-like activity and fragmentation of DNA. Several other reports have shown similar results in plant material with a collapse of the inner mitochondrial membrane potential and the release of cytochrome c from mitochondria to cytoplasm [90].

The function of the mitochondria is affected by Al^{3+} in plants and the same can be expected to be valid in mammalian cells and dysfunctional mitochondria have been detected upon chronic exposure to aluminium ions in mammalians [91]. Damaged and non-functional mitochondria are removed through autophagy, a process called mitophagy [92], and oxidized mitochondrial DNA, released into the cytoplasm due to damaged mitochondria evading mitophagy, activates the inflammasome and the cell becomes a potential inflammatory cell [31, 33]. Mitochondria function and their implication in the induction and continuation of inflammation has received increased attention [93] and the mitochondria are also pronounced sources of potential DAMP molecules and structures such as ATP, N-formyl peptides, and mtDNA [58]. Aluminium adjuvants escaping the endosomal pathway into the cytoplasm, and especially the establishment of a long lasting saturated concentration of aluminium ions in the cytoplasm, will have consequences on the mitochondrial performance and may have a pronounced influence on the manifestation and continuation of an inflammatory response.

2.3. Cellular Stress Functions

Cellular stress responses are physiological responses to an event, involving unfamiliar stimuli that affect the cells in different ways and initiate production and release of various proteins by the cell to contain or overcome the provocation. Polly Matzinger proposed the Danger model 1994 in which she suggests that the immune system is more concerned with damage than with foreignness, and that the immune system is called into action by alarm signals from injured tissues. These alarm signals can be constitutive or inducible, intracellular or secreted, or even a part of the extracellular matrix. Healthy tissues induce tolerance, but when damaged, they stimulate inflammation, and initiate an adaptive immune response [94]. Examples of proteins associated with a stress response in this context are Heat Shock Proteins (HSP) of different molecular size, S100s, calreticulin, High-Mobility Group Box 1 Protein (HMGB1), Glucose-Regulated Proteins (GRP), and mitochondrial components as previously described [60].

Today the use of adjuvants has become vital as additives in vaccine formulations and an awareness of their interactions with PAMP and DAMP biological mechanisms has become apparent [95]. Particles can cause a sterile inflammation, and cells that are not inflammatory become pro-inflammatory when they are dead because of their release of intracellular molecules acting as DAMPs.

Aluminum adjuvants cause cell injury resulting in the release of cellular contents such as uric acid, ATP, and cellular DNA. The thus released cellular contents provide molecular danger signals that cause sterile inflammation and activate dendritic cells by binding to pattern recognition receptors [39, 43, 96]. Stress induced by aluminium adjuvants *in vivo* as a mechanism of adjuvanticity has been proposed. In this respect, subcutaneous administration of aluminium adjuvants induced increased expression of both cell surface and intracellular HSP70 by splenic dendritic cells, and the adjuvant induced immune response was reduced if the expression of HSP70 was inhibited. HSP70 is a hallmark DAMP molecule and these results associate DAMP induction, cellular stress, and aluminium adjuvants [44]. Furthermore, upon oral intake of Al ions, the resulting prolonged exposure to Al ions affects biochemical processes in the cell causing cumulative sub lethal effects that activate different chaperones in the rat kidney and liver. A clear induction of HSP72 was detected in liver and HSP25 in the kidney whereas GRP75 was detected in both liver and kidney [97].

At the administration site of a vaccine, an initial high concentration of various aluminium species can be anticipated, causing necrosis of resident and initially infiltrating phagocytosing cells. Upon diffusion and dilution by interstitial fluid, the concentration of aluminium species will be reduced and the cells will be exposed to sub-toxic concentrations. At this point, phagocytosing cells will accumulate the adjuvant in phagosomes and upon release or escape, micro- and nanometre sized aggregates as well as a constant concentration of Al^{3+} ions will be maintained in the cytoplasm as previously described. Low amounts of intra cellular nanoparticles have been shown to induce a strong upregulation of genes encoding stress response proteins, such as heat shock proteins and other chaperones [68]. However, pro-inflammatory signalling by DAMP molecules can also be mediated by secretion of existing intracellular DAMP molecules as shown by the inflammatory response triggered by HMGB1 [26, 98], and the DAMP signalling is thus not initiated by gene regulation. Consequently, approaches based on proteomics are needed to understand cellular stress responses and DAMP induction underlying the immune stimulating properties of aluminium adjuvants [99].

Cells must survive a variety of stressful conditions, and interference by aluminium ions, and other metal ions, can be expected to involve numerous biochemical events and mechanisms in the cell. Al^{3+} ions have been shown to bind to multifunctional proteins, *e.g.* calmodulin and S100 proteins [100], and it should be noted that upon secretion, proteins belonging to the S100 family show pro-inflammatory features and act as DAMPs [101, 102].

During cellular stress, there is an insistent and peremptory demand on protein synthesis and controlled folding, and several DAMP molecules are involved in these processes. HSPs are divided into seven classes, and exercise chaperone function by stabilizing new proteins to ensure correct folding, and aiding in the refolding of damaged proteins [103].

Calreticulin is another protein located in the endoplasmatic reticulum with chaperon properties and both HSPs and calreticulin are DAMPs with pro-inflammatory abilities [58, 104]. These molecules may very well be targets of aluminium ions resulting in changed conformation, modified function, and cellular re-localization. Hence, as suggested by these results, cytoplasmic nanometre aggregates of aluminium adjuvants in general, and aluminium ions in particularly, impact several biochemical processes, provoking cellular stress and expression of DAMP molecules without causing cell death.

2.4. Protein Synthesis and Folding

The protein content of a cell is regulated by gene expression, synthesis, re-folding, and degradation, and nanoparticles as well as aluminium ions have been shown to affect the conformation of proteins [100, 105, 106]. Conformational changes will increase the demand for cellular protein synthesis and re-folding operations, and consequently, upregulation of the syntheses of proteins and chaperons will result.

Aluminium adjuvants in the cytoplasm will form a solid depository of sparingly soluble aluminium salts, maintaining a constant concentration of Al^{3+} ions over a long period. Protocols designed to study the effects of a constant low concentration of aluminium ions in the cytoplasm of mammalian cells are challenging. The required experimental set up is less challenging when working with plant cells and several reports have shown that aluminium ions affect the cellular content of proteins related to stress/defence, signal transduction, transport, and folding [107, 108]. Chaperones are upregulated and hence, proteins of the HSP family, *e.g.* HSP70 and HSP90, are utilized by plant cells to handle aluminium ion induced stress [109 - 111]. The increased value of chaperone systems in plants to manage intracellular aluminium ions is probably mirrored by mammalian cells and an interesting aspect is that several components of the chaperone systems also function as DAMP molecules in a mammalian context.

2.5. Nuclear Structure and Function

Particulate aluminium adjuvants have not been demonstrated experimentally in the nucleus of adjuvant exposed cells [10, 53 - 56] and nuclear DNA damage has been observed in mammalians only at aluminium ion concentrations several magnitudes higher than those expected from slow dissolution of a solid depository of aluminium adjuvants in the cytoplasm [112, 113]. Increased gene expression of several genes has been monitored upon exposure to aluminium ions. In rat brain cells expression of genes related to mitochondrial respiration, including gene products also encoded by mitochondrial DNA [114] increased, and similar results on mitochondrial genes have been reported in plant cells [88].

Cellular uptake of nanoparticles also results in increased gene expression and the genes expressed are dependent on the surface charge of the particles and the amount nanoparticles accumulated intracellularly [68]. An interesting aspect is that microRNAs (miRNA) have been shown to be involved in the gene regulation of plant cells exposed to Al_2O_3 nanoparticles [115] and to our knowledge, there are no reports regarding miRNA based gene regulation upon cellular exposure to aluminium adjuvants.

CONCLUSION

Aluminium adjuvants have been used in vaccines during decades and yet no consensus has been reached regarding the mechanism of immune stimulation by aluminium salts. Several mechanisms have been proposed but none of these cover all the aspects of the immune response induced by aluminium adjuvants.

This review has focused on the possible biological impacts on a cell of an intra cellular depository of sparingly soluble aluminium salts, maintaining a constant concentration of Al^{3+} ions in the cytoplasm. A general cellular stress response resulting in the release of DAMP molecules can be expected due to the intracellular presence of particles and aluminium ions. The release of DAMP molecules will initially be mediated by necrotic cells that have succumbed to too high amounts intra cellular aluminium adjuvants, and later by livings cells that are able to handle the intra cellular amounts of adjuvants. Upon administration of vaccine myeloid cells, *e.g.* monocytes, macrophages, and dendritic cells accumulate aluminium adjuvants and since those cells are all potential APCs, they will become efficient inducers of an adjuvant based immune response. Aluminium adjuvants will upon phagocytosis deliver an antigen and at the same time convert the cells into DAMP expressing cells. On this basis, we would like to suggest the maintenance of a constant concentration of Al^{3+} ions in the cytoplasm as a general underlying feature of the immune stimulation properties of aluminium adjuvants. The presence of a constant intracellular concentration of aluminium ions will affect multiple biochemical processes in the cell, efficiently inducing DAMP molecules in viable cells that otherwise manage to handle the presence of aluminium ions (Fig. 1). Other metal ions may also have the same intracellular cellular effect as Al^{3+}

ions and thus mediate the induction of an immune response, and correlations between the immune stimulating properties and the physicochemical properties of inorganic crystalline materials containing metal ions have been shown [71]. However, in this report by Williams *et.al*, no information was given regarding the elimination, or lack of elimination, of the crystallites once phagocytosed, and no information was given regarding the expected metal ion concentration in the cytoplasm of a cell.

According to regulations, the maximal amount of aluminium adjuvant in the administered dose of a vaccine is 1.25 mg Al per dose [116]. In general, a vaccine contains 0.5 – 1 mg aluminium adjuvant per dose and the amount of aluminium adjuvant is designed to provide an excessive load of antigen. An inflammation is the inducer of an immune response and upon an adaptive immune activation, the inflammation is supposed to subside. However, if the inflammation persists, it may turn into a chronic inflammation with all the negative implications [117]. Side effects such as ASIA [63, 64] caused by aluminium adjuvants have raised concerns on the use of aluminium salts as adjuvants in vaccine formulations and a chronic inflammation induced by a constant intracellular concentration of Al^{3+} ions could be the core of the autoimmune related side effect observed. By lowering the amounts of aluminium adjuvants in the vaccine formulation, an efficient immune response may still be achieved, however, the amounts of aluminium adjuvants persisting intracellularly will be reduced and thus mitigate the induction of a chronic inflammation and the occurrence of adverse reactions.

ACKNOWLEDGEMENTS

We thank Prof. Zoltan Blum for helpful comments and discussions.

REFERENCES

[1]   Johnston CT, Wang SL, Hem SL. Measuring the surface area of aluminum hydroxide adjuvant. J Pharm Sci 2002; 91(7): 1702-6.
        [http://dx.doi.org/10.1002/jps.10166] [PMID: 12115832]

[2]   Burrell LS, Johnston CT, Schulze D, Klein J, White JL, Hem SL. Aluminium phosphate adjuvants prepared by precipitation at constant pH. Part I: Composition and structure. Vaccine 2000; 19(2-3): 275-81.
        [http://dx.doi.org/10.1016/S0264-410X(00)00160-2] [PMID: 10930682]

[3]   Hem SL, Hogenesch H. Relationship between physical and chemical properties of aluminum-containing adjuvants and immunopotentiation. Expert Rev Vaccines 2007; 6(5): 685-98.
        [http://dx.doi.org/10.1586/14760584.6.5.685] [PMID: 17931150]

[4]   Glenny AT, Buttle GAH, Stevens MF. Rate of disappearance of diphtheria toxoid injected into rabbits and guinea - pigs: Toxoid precipitated with alum. J Pathol Bacteriol 1931; 34(2): 267-75.
        [http://dx.doi.org/10.1002/path.1700340214]

[5]   Gupta RK, Chang A-C, Griffin P, Rivera R, Siber GR. *In vivo* distribution of radioactivity in mice after injection of biodegradable polymer microspheres containing 14C-labeled tetanus toxoid. Vaccine 1996; 14(15): 1412-6.
        [http://dx.doi.org/10.1016/S0264-410X(96)00073-4] [PMID: 8994315]

[6]   Hutchison S, Benson RA, Gibson VB, Pollock AH, Garside P, Brewer JM. Antigen depot is not required for alum adjuvanticity. FASEB J 2012; 26(3): 1272-9.
        [http://dx.doi.org/10.1096/fj.11-184556] [PMID: 22106367]

[7]   Munks MW, McKee AS, Macleod MK, *et al.* Aluminum adjuvants elicit fibrin-dependent extracellular traps *in vivo*. Blood 2010; 116(24): 5191-9.
        [http://dx.doi.org/10.1182/blood-2010-03-275529] [PMID: 20876456]

[8]   Mosca F, Tritto E, Muzzi A, *et al.* Molecular and cellular signatures of human vaccine adjuvants. Proc Natl Acad Sci USA 2008; 105(30): 10501-6.
        [http://dx.doi.org/10.1073/pnas.0804699105] [PMID: 18650390]

[9]   Kool M, Pétrilli V, De Smedt T, *et al.* Cutting edge: Alum adjuvant stimulates inflammatory dendritic cells through activation of the NALP3 inflammasome. J Immunol 2008; 181(6): 3755-9.
        [http://dx.doi.org/10.4049/jimmunol.181.6.3755] [PMID: 18768827]

[10]   Lu F, Hogenesch H. Kinetics of the inflammatory response following intramuscular injection of aluminum adjuvant. Vaccine 2013; 31(37): 3979-86.
[http://dx.doi.org/10.1016/j.vaccine.2013.05.107] [PMID: 23770306]

[11]   McKee AS, Munks MW, MacLeod MK, et al. Alum induces innate immune responses through macrophage and mast cell sensors, but these sensors are not required for alum to act as an adjuvant for specific immunity. J Immunol 2009; 183(7): 4403-14.
[http://dx.doi.org/10.4049/jimmunol.0900164] [PMID: 19734227]

[12]   Korsholm KS, Petersen RV, Agger EM, Andersen P. T-helper 1 and T-helper 2 adjuvants induce distinct differences in the magnitude, quality and kinetics of the early inflammatory response at the site of injection. Immunology 2010; 129(1): 75-86.
[http://dx.doi.org/10.1111/j.1365-2567.2009.03164.x] [PMID: 19824919]

[13]   Brewer JM. (How) do aluminium adjuvants work? Immunol Lett 2006; 102(1): 10-5.
[http://dx.doi.org/10.1016/j.imlet.2005.08.002] [PMID: 16188325]

[14]   Marrack P, McKee AS, Munks MW. Towards an understanding of the adjuvant action of aluminium. Nat Rev Immunol 2009; 9(4): 287-93.
[http://dx.doi.org/10.1038/nri2510] [PMID: 19247370]

[15]   Reed SG, Bertholet S, Coler RN, Friede M. New horizons in adjuvants for vaccine development. Trends Immunol 2009; 30(1): 23-32.
[http://dx.doi.org/10.1016/j.it.2008.09.006] [PMID: 19059004]

[16]   Mbow ML, De Gregorio E, Valiante NM, Rappuoli R. New adjuvants for human vaccines. Curr Opin Immunol 2010; 22(3): 411-6.
[http://dx.doi.org/10.1016/j.coi.2010.04.004] [PMID: 20466528]

[17]   Hogenesch H. Mechanism of immunopotentiation and safety of aluminum adjuvants. Front Immunol 2013; 3: 406.
[http://dx.doi.org/10.3389/fimmu.2012.00406] [PMID: 23335921]

[18]   Kool M, Fierens K, Lambrecht BN. Alum adjuvant: Some of the tricks of the oldest adjuvant. J Med Microbiol 2012; 61(Pt 7): 927-34.
[http://dx.doi.org/10.1099/jmm.0.038943-0] [PMID: 22174375]

[19]   Awate S, Babiuk LA, Mutwiri G. Mechanisms of action of adjuvants. Front Immunol 2013; 4: 114.
[http://dx.doi.org/10.3389/fimmu.2013.00114] [PMID: 23720661]

[20]   Reed SG, Orr MT, Fox CB. Key roles of adjuvants in modern vaccines. Nat Med 2013; 19(12): 1597-608.
[http://dx.doi.org/10.1038/nm.3409] [PMID: 24309663]

[21]   Ghimire TR. The mechanisms of action of vaccines containing aluminum adjuvants: An in vitro vs in vivo paradigm. Springerplus 2015; 4: 181.
[http://dx.doi.org/10.1186/s40064-015-0972-0] [PMID: 25932368]

[22]   Matzinger P. Tolerance, danger, and the extended family. Annu Rev Immunol 1994; 12: 991-1045.
[http://dx.doi.org/10.1146/annurev.iy.12.040194.005015] [PMID: 8011301]

[23]   Medzhitov R, Preston-Hurlburt P, Janeway CA Jr. A human homologue of the drosophila toll protein signals activation of adaptive immunity. Nature 1997; 388(6640): 394-7.
[http://dx.doi.org/10.1038/41131] [PMID: 9237759]

[24]   Medzhitov R, Janeway CA Jr. An ancient system of host defense. Curr Opin Immunol 1998; 10(1): 12-5.
[http://dx.doi.org/10.1016/S0952-7915(98)80024-1] [PMID: 9523104]

[25]   Pradeu T, Cooper EL. The danger theory: 20 years later. Front Immunol 2012; 3: 287.
[http://dx.doi.org/10.3389/fimmu.2012.00287] [PMID: 23060876]

[26]   Sangiuliano B, Pérez NM, Moreira DF, Belizário JE. Cell death-associated molecular-pattern molecules: Inflammatory signaling and control. Mediators Inflamm 2014; 2014: 821043.
[http://dx.doi.org/10.1155/2014/821043] [PMID: 25140116]

[27]   Di Pasquale A, Preiss S, Tavares Da Silva F, Garçon N. Vaccine adjuvants: From 1920 to 2015 and beyond. Vaccines (Basel) 2015; 3(2): 320-43.
[http://dx.doi.org/10.3390/vaccines3020320] [PMID: 26343190]

[28]   De Gregorio E, Caproni E, Ulmer JB. Vaccine adjuvants: Mode of action. Front Immunol 2013; 4: 214.
[http://dx.doi.org/10.3389/fimmu.2013.00214] [PMID: 23914187]

[29]   Li H, Nookala S, Re F. Aluminum hydroxide adjuvants activate caspase-1 and induce IL-1beta and IL-18 release. J Immunol 2007; 178(8): 5271-6.
[http://dx.doi.org/10.4049/jimmunol.178.8.5271] [PMID: 17404311]

[30]   Pollock JD, Williams DA, Gifford MA, et al. Mouse model of X-linked chronic granulomatous disease, an inherited defect in phagocyte superoxide production. Nat Genet 1995; 9(2): 202-9.
[http://dx.doi.org/10.1038/ng0295-202] [PMID: 7719350]

[31]   Nakahira K, Haspel JA, Rathinam VA, et al. Autophagy proteins regulate innate immune responses by inhibiting the release of mitochondrial DNA mediated by the NALP3 inflammasome. Nat Immunol 2011; 12(3): 222-30.
[http://dx.doi.org/10.1038/ni.1980] [PMID: 21151103]

[32]   Zhou R, Yazdi AS, Menu P, Tschopp J. A role for mitochondria in NLRP3 inflammasome activation. Nature 2011; 469(7329): 221-5.
[http://dx.doi.org/10.1038/nature09663] [PMID: 21124315]

[33]   Shimada K, Crother TR, Karlin J, *et al.* Oxidized mitochondrial DNA activates the NLRP3 inflammasome during apoptosis. Immunity 2012; 36(3): 401-14.
        [http://dx.doi.org/10.1016/j.immuni.2012.01.009] [PMID: 22342844]

[34]   Hornung V, Bauernfeind F, Halle A, *et al.* Silica crystals and aluminum salts activate the NALP3 inflammasome through phagosomal destabilization. Nat Immunol 2008; 9(8): 847-56.
        [http://dx.doi.org/10.1038/ni.1631] [PMID: 18604214]

[35]   Flach TL, Ng G, Hari A, *et al.* Alum interaction with dendritic cell membrane lipids is essential for its adjuvanticity. Nat Med 2011; 17(4): 479-87.
        [http://dx.doi.org/10.1038/nm.2306] [PMID: 21399646]

[36]   Mitchell LA, Henderson AJ, Dow SW. Suppression of vaccine immunity by inflammatory monocytes. J Immunol 2012; 189(12): 5612-21.
        [http://dx.doi.org/10.4049/jimmunol.1202151] [PMID: 23136203]

[37]   Franchi L, Núñez G. The Nlrp3 inflammasome is critical for aluminium hydroxide-mediated IL-1beta secretion but dispensable for adjuvant activity. Eur J Immunol 2008; 38(8): 2085-9.
        [http://dx.doi.org/10.1002/eji.200838549] [PMID: 18624356]

[38]   Dostert C, Pétrilli V, Van Bruggen R, Steele C, Mossman BT, Tschopp J. Innate immune activation through Nalp3 inflammasome sensing of asbestos and silica. Science 2008; 320(5876): 674-7.
        [http://dx.doi.org/10.1126/science.1156995] [PMID: 18403674]

[39]   Marichal T, Ohata K, Bedoret D, *et al.* DNA released from dying host cells mediates aluminum adjuvant activity. Nat Med 2011; 17(8): 996-1002.
        [http://dx.doi.org/10.1038/nm.2403] [PMID: 21765404]

[40]   McKee AS, Burchill MA, Munks MW, *et al.* Host DNA released in response to aluminum adjuvant enhances MHC class II-mediated antigen presentation and prolongs CD4 T-cell interactions with dendritic cells. Proc Natl Acad Sci USA 2013; 110(12): E1122-31.
        [http://dx.doi.org/10.1073/pnas.1300392110] [PMID: 23447566]

[41]   Cain DW, Sanders SE, Cunningham MM, Kelsoe G. Disparate adjuvant properties among three formulations of "alum". Vaccine 2013; 31(4): 653-60.
        [http://dx.doi.org/10.1016/j.vaccine.2012.11.044] [PMID: 23200935]

[42]   Yang D, de la Rosa G, Tewary P, Oppenheim JJ. Alarmins link neutrophils and dendritic cells. Trends Immunol 2009; 30(11): 531-7.
        [http://dx.doi.org/10.1016/j.it.2009.07.004] [PMID: 19699678]

[43]   Riteau N, Baron L, Villeret B, *et al.* ATP release and purinergic signaling: A common pathway for particle-mediated inflammasome activation. Cell Death Dis 2012; 3: e403.
        [http://dx.doi.org/10.1038/cddis.2012.144] [PMID: 23059822]

[44]   Wang Y, Rahman D, Lehner T. A comparative study of stress-mediated immunological functions with the adjuvanticity of alum. J Biol Chem 2012; 287(21): 17152-60.
        [http://dx.doi.org/10.1074/jbc.M112.347179] [PMID: 22474329]

[45]   Rock KL, Latz E, Ontiveros F, Kono H. The sterile inflammatory response. Annu Rev Immunol 2010; 28: 321-42.
        [http://dx.doi.org/10.1146/annurev-immunol-030409-101311] [PMID: 20307211]

[46]   Muralidharan S, Mandrekar P. Cellular stress response and innate immune signaling: Integrating pathways in host defense and inflammation. J Leukoc Biol 2013; 94(6): 1167-84.
        [http://dx.doi.org/10.1189/jlb.0313153] [PMID: 23990626]

[47]   Rider P, Voronov E, Dinarello CA, Apte RN, Cohen I. Alarmins: feel the stress. J Immunol 2017; 198(4): 1395-402.
        [http://dx.doi.org/10.4049/jimmunol.1601342] [PMID: 28167650]

[48]   Priest ND, Talbot RJ, Austin JG, *et al.* The bioavailability of 26Al-labelled aluminium citrate and aluminium hydroxide in volunteers. Biometals 1996; 9(3): 221-8.
        [http://dx.doi.org/10.1007/BF00817919] [PMID: 8696074]

[49]   Rinella JV, White JL, Hem SL. Effect of pH on the elution of model antigens from aluminum-containing adjuvants. J Colloid Interface Sci 1998; 205(1): 161-5.
        [http://dx.doi.org/10.1006/jcis.1998.5648] [PMID: 9710509]

[50]   Flarend RE, Hem SL, White JL, *et al. In vivo* absorption of aluminium-containing vaccine adjuvants using 26Al. Vaccine 1997; 15(12-13): 1314-8.
        [http://dx.doi.org/10.1016/S0264-410X(97)00041-8] [PMID: 9302736]

[51]   Rimaniol AC, Gras G, Verdier F, *et al.* Aluminum hydroxide adjuvant induces macrophage differentiation towards a specialized antigen-presenting cell type. Vaccine 2004; 22(23-24): 3127-35.
        [http://dx.doi.org/10.1016/j.vaccine.2004.01.061] [PMID: 15297065]

[52]   Morefield GL, Tammariello RF, Purcell BK, *et al.* An alternative approach to combination vaccines: Intradermal administration of isolated components for control of anthrax, botulism, plague and staphylococcal toxic shock. J Immune Based Ther Vaccines 2008; 6: 5.
        [http://dx.doi.org/10.1186/1476-8518-6-5] [PMID: 18768085]

[53]   Gherardi RK, Coquet M, Cherin P, *et al.* Macrophagic myofasciitis lesions assess long-term persistence of vaccine-derived aluminium hydroxide in muscle. Brain 2001; 124(Pt 9): 1821-31.
[http://dx.doi.org/10.1093/brain/124.9.1821] [PMID: 11522584]

[54]   Verdier F, Burnett R, Michelet-Habchi C, Moretto P, Fievet-Groyne F, Sauzeat E. Aluminium assay and evaluation of the local reaction at several time points after intramuscular administration of aluminium containing vaccines in the Cynomolgus monkey. Vaccine 2005; 23(11): 1359-67.
[http://dx.doi.org/10.1016/j.vaccine.2004.09.012] [PMID: 15661384]

[55]   Crépeaux G, Eidi H, David MO, *et al.* Highly delayed systemic translocation of aluminum-based adjuvant in CD1 mice following intramuscular injections. J Inorg Biochem 2015; 152: 199-205.
[http://dx.doi.org/10.1016/j.jinorgbio.2015.07.004] [PMID: 26384437]

[56]   Eidi H, David MO, Crépeaux G, *et al.* Fluorescent nanodiamonds as a relevant tag for the assessment of alum adjuvant particle biodisposition. BMC Med 2015; 13: 144.
[http://dx.doi.org/10.1186/s12916-015-0388-2] [PMID: 26082187]

[57]   Chen GY, Nuñez G. Sterile inflammation: Sensing and reacting to damage. Nat Rev Immunol 2010; 10(12): 826-37.
[http://dx.doi.org/10.1038/nri2873] [PMID: 21088683]

[58]   Krysko DV, Agostinis P, Krysko O, *et al.* Emerging role of damage-associated molecular patterns derived from mitochondria in inflammation. Trends Immunol 2011; 32(4): 157-64.
[http://dx.doi.org/10.1016/j.it.2011.01.005] [PMID: 21334975]

[59]   Zelenay S, Reis e Sousa C. Adaptive immunity after cell death. Trends Immunol 2013; 34(7): 329-35.
[http://dx.doi.org/10.1016/j.it.2013.03.005] [PMID: 23608152]

[60]   Vénéreau E, Ceriotti C, Bianchi ME. DAMPs from Cell Death to New Life. Front Immunol 2015; 6: 422.
[http://dx.doi.org/10.3389/fimmu.2015.00422] [PMID: 26347745]

[61]   Srikrishna G, Freeze HH. Endogenous damage-associated molecular pattern molecules at the crossroads of inflammation and cancer. Neoplasia 2009; 11(7): 615-28.
[http://dx.doi.org/10.1593/neo.09284] [PMID: 19568407]

[62]   Garg AD, Martin S, Golab J, Agostinis P. Danger signalling during cancer cell death: Origins, plasticity and regulation. Cell Death Differ 2014; 21(1): 26-38.
[http://dx.doi.org/10.1038/cdd.2013.48] [PMID: 23686135]

[63]   Agmon-Levin N, Paz Z, Israeli E, Shoenfeld Y. Vaccines and autoimmunity. Nat Rev Rheumatol 2009; 5(11): 648-52.
[http://dx.doi.org/10.1038/nrrheum.2009.196] [PMID: 19865091]

[64]   Shoenfeld Y, Agmon-Levin N. 'ASIA' - autoimmune/inflammatory syndrome induced by adjuvants. J Autoimmun 2011; 36(1): 4-8.
[http://dx.doi.org/10.1016/j.jaut.2010.07.003] [PMID: 20708902]

[65]   Gregory AE, Titball R, Williamson D. Vaccine delivery using nanoparticles. Front Cell Infect Microbiol 2013; 3: 13.
[http://dx.doi.org/10.3389/fcimb.2013.00013] [PMID: 23532930]

[66]   Weissleder R, Nahrendorf M, Pittet MJ. Imaging macrophages with nanoparticles. Nat Mater 2014; 13(2): 125-38.
[http://dx.doi.org/10.1038/nmat3780] [PMID: 24452356]

[67]   Zhao L, Seth A, Wibowo N, *et al.* Nanoparticle vaccines. Vaccine 2014; 32(3): 327-37.
[http://dx.doi.org/10.1016/j.vaccine.2013.11.069] [PMID: 24295808]

[68]   Mitchell HD, Markillie LM, Chrisler WB, *et al.* Cells respond to distinct nanoparticle properties with multiple strategies as revealed by single-cell RNA-seq. ACS Nano 2016; 10(11): 10173-85.
[http://dx.doi.org/10.1021/acsnano.6b05452] [PMID: 27788331]

[69]   Willhite CC, Karyakina NA, Yokel RA, *et al.* Systematic review of potential health risks posed by pharmaceutical, occupational and consumer exposures to metallic and nanoscale aluminum, aluminum oxides, aluminum hydroxide and its soluble salts. Crit Rev Toxicol 2014; 44(Suppl. 4): 1-80.
[http://dx.doi.org/10.3109/10408444.2014.934439] [PMID: 25233067]

[70]   Krewski D, Yokel RA, Nieboer E, *et al.* Human health risk assessment for aluminium, aluminium oxide, and aluminium hydroxide. J Toxicol Environ Health B Crit Rev 2007; 10(Suppl. 1): 1-269.
[http://dx.doi.org/10.1080/10937400701597766] [PMID: 18085482]

[71]   Williams GR, Fierens K, Preston SG, *et al.* Immunity induced by a broad class of inorganic crystalline materials is directly controlled by their chemistry. J Exp Med 2014; 211(6): 1019-25.
[http://dx.doi.org/10.1084/jem.20131768] [PMID: 24799501]

[72]   Morefield GL, Sokolovska A, Jiang D, HogenEsch H, Robinson JP, Hem SL. Role of aluminum-containing adjuvants in antigen internalization by dendritic cells *in vitro*. Vaccine 2005; 23(13): 1588-95.
[http://dx.doi.org/10.1016/j.vaccine.2004.07.050] [PMID: 15694511]

[73]   Mold M, Eriksson H, Siesjö P, Darabi A, Shardlow E, Exley C. Unequivocal identification of intracellular aluminium adjuvant in a monocytic THP-1 cell line. Sci Rep 2014; 4: 6287.
[http://dx.doi.org/10.1038/srep06287] [PMID: 25190321]

[74]   Hamilton JA, Byrne R, Whitty G. Particulate adjuvants can induce macrophage survival, DNA synthesis, and a synergistic proliferative
        response to GM-CSF and CSF-1. J Leukoc Biol 2000; 67(2): 226-32.
        [http://dx.doi.org/10.1002/jlb.67.2.226] [PMID: 10670584]

[75]   Goto N, Kato H, Maeyama J, Eto K, Yoshihara S. Studies on the toxicities of aluminium hydroxide and calcium phosphate as immunological
        adjuvants for vaccines. Vaccine 1993; 11(9): 914-8.
        [http://dx.doi.org/10.1016/0264-410X(93)90377-A] [PMID: 8212836]

[76]   Bazzoni GB, Bollini AN, Hernández GN, Contini MdelC, Chiarotto MM, Rasia ML. *In vivo* effect of aluminium upon the physical properties
        of the erythrocyte membrane. J Inorg Biochem 2005; 99(3): 822-7.
        [http://dx.doi.org/10.1016/j.jinorgbio.2004.12.012] [PMID: 15708804]

[77]   Mile I, Svensson A, Darabi A, Mold M, Siesjö P, Eriksson H. Al adjuvants can be tracked in viable cells by lumogallion staining. J Immunol
        Methods 2015; 422: 87-94.
        [http://dx.doi.org/10.1016/j.jim.2015.04.008] [PMID: 25896212]

[78]   Stern ST, Adiseshaiah PP, Crist RM. Autophagy and lysosomal dysfunction as emerging mechanisms of nanomaterial toxicity. Part Fibre
        Toxicol 2012; 9: 20.
        [http://dx.doi.org/10.1186/1743-8977-9-20] [PMID: 22697169]

[79]   Wirawan E, Vanden Berghe T, Lippens S, Agostinis P, Vandenabeele P. Autophagy: For better or for worse. Cell Res 2012; 22(1): 43-61.
        [http://dx.doi.org/10.1038/cr.2011.152] [PMID: 21912435]

[80]   Deretic V. Autophagy as an innate immunity paradigm: expanding the scope and repertoire of pattern recognition receptors. Curr Opin
        Immunol 2012; 24(1): 21-31.
        [http://dx.doi.org/10.1016/j.coi.2011.10.006] [PMID: 22118953]

[81]   Ohlsson L, Exley C, Darabi A, Sandén E, Siesjö P, Eriksson H. Aluminium based adjuvants and their effects on mitochondria and lysosomes
        of phagocytosing cells. J Inorg Biochem 2013; 128: 229-36.
        [http://dx.doi.org/10.1016/j.jinorgbio.2013.08.003] [PMID: 23992993]

[82]   Kowaltowski AJ, de Souza-Pinto NC, Castilho RF, Vercesi AE. Mitochondria and reactive oxygen species. Free Radic Biol Med 2009; 47(4):
        333-43.
        [http://dx.doi.org/10.1016/j.freeradbiomed.2009.05.004] [PMID: 19427899]

[83]   Murphy MP. How mitochondria produce reactive oxygen species. Biochem J 2009; 417(1): 1-13.
        [http://dx.doi.org/10.1042/BJ20081386] [PMID: 19061483]

[84]   Ruwona TB, Xu H, Li X, Taylor AN, Shi YC, Cui Z. Toward understanding the mechanism underlying the strong adjuvant activity of
        aluminum salt nanoparticles. Vaccine 2016; 34(27): 3059-67.
        [http://dx.doi.org/10.1016/j.vaccine.2016.04.081] [PMID: 27155490]

[85]   Poborilova Z, Opatrilova R, Babula P. Toxicity of aluminium oxide nanoparticles demonstrated using a BY-2 plant cell suspension culture
        model. Environ Exp Bot 2013; 91: 1-11.
        [http://dx.doi.org/10.1016/j.envexpbot.2013.03.002]

[86]   Yamamoto Y, Kobayashi Y, Devi SR, Rikiishi S, Matsumoto H. Aluminum toxicity is associated with mitochondrial dysfunction and the
        production of reactive oxygen species in plant cells. Plant Physiol 2002; 128(1): 63-72.
        [http://dx.doi.org/10.1104/pp.010417] [PMID: 11788753]

[87]   Li H, Li Y, Jiao J, Hu HM. Alpha-alumina nanoparticles induce efficient autophagy-dependent cross-presentation and potent antitumour
        response. Nat Nanotechnol 2011; 6(10): 645-50.
        [http://dx.doi.org/10.1038/nnano.2011.153] [PMID: 21926980]

[88]   Huang W, Yang X, Yao S, *et al.* Reactive oxygen species burst induced by aluminum stress triggers mitochondria-dependent programmed
        cell death in peanut root tip cells. Plant Physiol Biochem 2014; 82: 76-84.
        [http://dx.doi.org/10.1016/j.plaphy.2014.03.037] [PMID: 24907527]

[89]   Matsumoto H, Motoda H. Aluminum toxicity recovery processes in root apices. Possible association with oxidative stress. Plant Sci 2012;
        185-186: 1-8.
        [http://dx.doi.org/10.1016/j.plantsci.2011.07.019] [PMID: 22325861]

[90]   Zhan J, Li W, He HY, Li CZ, He LF. Mitochondrial alterations during Al-induced PCD in peanut root tips. Plant Physiol Biochem 2014; 75:
        105-13.
        [http://dx.doi.org/10.1016/j.plaphy.2013.12.010] [PMID: 24398246]

[91]   Kumar V, Bal A, Gill KD. Impairment of mitochondrial energy metabolism in different regions of rat brain following chronic exposure to
        aluminium. Brain Res 2008; 1232: 94-103.
        [http://dx.doi.org/10.1016/j.brainres.2008.07.028] [PMID: 18691561]

[92]   Ding WX, Yin XM. Mitophagy: Mechanisms, pathophysiological roles, and analysis. Biol Chem 2012; 393(7): 547-64.
        [http://dx.doi.org/10.1515/hsz-2012-0119] [PMID: 22944659]

[93]   Tschopp J. Mitochondria: Sovereign of inflammation? Eur J Immunol 2011; 41(5): 1196-202.
        [http://dx.doi.org/10.1002/eji.201141436] [PMID: 21469137]

[94]   Matzinger P. The danger model: A renewed sense of self. Science 2002; 296(5566): 301-5.
        [http://dx.doi.org/10.1126/science.1071059] [PMID: 11951032]

[95]   Powell BS, Andrianov AK, Fusco PC. Polyionic vaccine adjuvants: Another look at aluminum salts and polyelectrolytes. Clin Exp Vaccine
        Res 2015; 4(1): 23-45.
        [http://dx.doi.org/10.7774/cevr.2015.4.1.23] [PMID: 25648619]

[96]   Kool M, Soullié T, van Nimwegen M, et al. Alum adjuvant boosts adaptive immunity by inducing uric acid and activating inflammatory
        dendritic cells. J Exp Med 2008; 205(4): 869-82.
        [http://dx.doi.org/10.1084/jem.20071087] [PMID: 18362170]

[97]   Stacchiotti A, Rodella LF, Ricci F, Rezzani R, Lavazza A, Bianchi R. Stress proteins expression in rat kidney and liver chronically exposed to
        aluminium sulphate. Histol Histopathol 2006; 21(2): 131-40.
        [PMID: 16329037]

[98]   Gardella S, Andrei C, Ferrera D, et al. The nuclear protein HMGB1 is secreted by monocytes via a non-classical, vesicle-mediated secretory
        pathway. EMBO Rep 2002; 3(10): 995-1001.
        [http://dx.doi.org/10.1093/embo-reports/kvf198] [PMID: 12231511]

[99]   Le Naour F, Hohenkirk L, Grolleau A, et al. Profiling changes in gene expression during differentiation and maturation of monocyte-derived
        dendritic cells using both oligonucleotide microarrays and proteomics. J Biol Chem 2001; 276(21): 17920-31.
        [http://dx.doi.org/10.1074/jbc.M100156200] [PMID: 11279020]

[100]   Levy R, Shohat L, Solomon B. Specificity of an anti-aluminium monoclonal antibody toward free and protein-bound aluminium. J Inorg
        Biochem 1998; 69(3): 159-63.
        [http://dx.doi.org/10.1016/S0162-0134(97)10013-7] [PMID: 9629674]

[101]   Roth J, Vogl T, Sorg C, Sunderkötter C. Phagocyte-specific S100 proteins: A novel group of proinflammatory molecules. Trends Immunol
        2003; 24(4): 155-8.
        [http://dx.doi.org/10.1016/S1471-4906(03)00062-0] [PMID: 12697438]

[102]   Sorci G, Bianchi R, Riuzzi F, et al. S100B Protein, A Damage-Associated Molecular Pattern Protein in the Brain and Heart, and Beyond.
        Cardiovasc Psychiatry Neurol 2010.

[103]   Lindquist S, Craig EA. The heat-shock proteins. Annu Rev Genet 1988; 22: 631-77.
        [http://dx.doi.org/10.1146/annurev.ge.22.120188.003215] [PMID: 2853609]

[104]   Jiang Y, Dey S, Matsunami H. Calreticulin: Roles in cell-surface protein expression. Membranes (Basel) 2014; 4(3): 630-41.
        [http://dx.doi.org/10.3390/membranes4030630] [PMID: 25230046]

[105]   Saptarshi SR, Duschl A, Lopata AL. Interaction of nanoparticles with proteins: Relation to bio-reactivity of the nanoparticle. J
        Nanobiotechnology 2013; 11: 26.
        [http://dx.doi.org/10.1186/1477-3155-11-26] [PMID: 23870291]

[106]   Levi R, Wolf T, Fleminger G, Solomon B. Immuno-detection of aluminium and aluminium induced conformational changes in calmodulin--
        implications in Alzheimer's disease. Mol Cell Biochem 1998; 189(1-2): 41-6.
        [http://dx.doi.org/10.1023/A:1006887809463] [PMID: 9879652]

[107]   Zhen Y, Qi JL, Wang SS, et al. Comparative proteome analysis of differentially expressed proteins induced by Al toxicity in soybean. Physiol
        Plant 2007; 131(4): 542-54.
        [http://dx.doi.org/10.1111/j.1399-3054.2007.00979.x] [PMID: 18251846]

[108]   Oh MW, Roy SK, Kamal AH, et al. Proteome analysis of roots of wheat seedlings under aluminum stress. Mol Biol Rep 2014; 41(2): 671-81.
        [http://dx.doi.org/10.1007/s11033-013-2905-8] [PMID: 24357239]

[109]   Zheng L, Lan P, Shen RF, Li WF. Proteomics of aluminum tolerance in plants. Proteomics 2014; 14(4-5): 566-78.
        [http://dx.doi.org/10.1002/pmic.201300252] [PMID: 24339160]

[110]   Wang W, Vinocur B, Shoseyov O, Altman A. Role of plant heat-shock proteins and molecular chaperones in the abiotic stress response.
        Trends Plant Sci 2004; 9(5): 244-52.
        [http://dx.doi.org/10.1016/j.tplants.2004.03.006] [PMID: 15130550]

[111]   Wang CY, Shen RF, Wang C, Wang W. Root protein profile changes induced by Al exposure in two rice cultivars differing in Al tolerance. J
        Proteomics 2013; 78: 281-93.
        [http://dx.doi.org/10.1016/j.jprot.2012.09.035] [PMID: 23059537]

[112]   Lima PD, Leite DS, Vasconcellos MC, et al. Genotoxic effects of aluminum chloride in cultured human lymphocytes treated in different
        phases of cell cycle. Food Chem Toxicol 2007; 45(7): 1154-9.
        [http://dx.doi.org/10.1016/j.fct.2006.12.022] [PMID: 17321660]

[113]   Mihaljević Z, Ternjej I, Stanković I, Kerovec M, Kopjar N. Application of the comet assay and detection of DNA damage in haemocytes of
        medicinal leech affected by aluminium pollution: A case study. Environ Pollut 2009; 157(5): 1565-72.
        [http://dx.doi.org/10.1016/j.envpol.2009.01.002] [PMID: 19200629]

[114]   Sharma DR, Sunkaria A, Wani WY, et al. Aluminium induced oxidative stress results in decreased mitochondrial biogenesis via modulation
        of PGC-1α expression. Toxicol Appl Pharmacol 2013; 273(2): 365-80.
        [http://dx.doi.org/10.1016/j.taap.2013.09.012] [PMID: 24084166]

[115]   Burklew CE, Ashlock J, Winfrey WB, Zhang B. Effects of aluminum oxide nanoparticles on the growth, development, and microRNA expression of tobacco (*Nicotiana tabacum*). PLoS One 2012; 7(5): e34783.
[http://dx.doi.org/10.1371/journal.pone.0034783] [PMID: 22606225]

[116]   Vecchi S, Bufali S, Skibinski DA, O'Hagan DT, Singh M. Aluminum adjuvant dose guidelines in vaccine formulation for preclinical evaluations. J Pharm Sci 2012; 101(1): 17-20.
[http://dx.doi.org/10.1002/jps.22759] [PMID: 21918987]

[117]   Murakami M, Hirano T. The molecular mechanisms of chronic inflammation development. Front Immunol 2012; 3: 323.
[http://dx.doi.org/10.3389/fimmu.2012.00323] [PMID: 23162547]

9

Polyphenols as Suitable Control for Obesity and Diabetes

Dasha Mihaylova[1], Aneta Popova[2], Iordanka Alexieva[2], Albert Krastanov[1] and Anna Lante[3,*]

[1]*Department of Biotechnology, University of Food Technologies, 26 Maritza Blvd., Plovdiv 4002, Bulgaria*
[2]*Department of Catering and Tourism, University of Food Technologies, 26 Maritza Blvd., Plovdiv 4002, Bulgaria*
[3]*Department of Agronomy, Food, Natural Resources, Animals and Environment – DAFNAE, University of Padova, Viale Università 16, Agripolis 35020, Italy*

Abstract:

Modern life is characterized by physical inactivity and poor food choices, which is often a prerequisite for unhealthy weight gain and overweight/obesity. These factors unlock the emergence of a number of diseases including diabetes, cardiovascular problems, different types of cancer, *etc.*

The pursuit of scientists to seek strategies to prevent, relieve and cure the patient leads to the usage of natural compounds of potential beneficial effect. Polyphenols are a large group of naturally occurring secondary metabolites mainly found in plants and beverages. The presence of these secondary metabolites seems to decrease the manifestation of miscellaneous disease-causing symptoms.

The purpose of this review is to synthesize information about polyphenols and their potential in controlling obesity and diabetes. Polyphenols are considered as health-beneficial sources and thus could be involved in novel strategies for preventing diabetes and obesity complications.

Keywords: Polyphenols, Diabetes, Obesity, Nutrition, Functional foods, Foods choices.

1. INTRODUCTION

Nutrition is an important health factor, as the main recommendations for achieving it include the consumption of a variety of foods. Nutritional knowledge is gained in the early years of development, thus promoting healthier choices at an early stage of habitual development may lead to minimizing the display of 21st century diseases *i.e.* diabetes type 2, obesity, heart complications, teeth decay, *etc.* with the help of adequate dietary assessment and food choices relevant to nutritional requirements.

A well-balanced diet usually means the provision of the required nutrients for the different physiological groups. People need a wide variety of nutrients in order to attain the anthropometric reference standards and lead an active life. Nutrient balance most often means an equilibrium of the input and output calories. In order to make recommendations, nutrient data is needed to propose a customary intake.

International experience has shown that food safety issues, as well as providing full and healthy food available to the general population, reducing nutrient deficiencies, morbidity and mortality from a number of Noncommunicable Diseases (NCDs) - cardiovascular disease, obesity, diabetes, hyperlipoproteinemias, certain cancers, *etc.*, can be best addressed by implementing a unified national food and nutrition policy with coordinated cross-sectoral activities [1].

According to the WHO, noncommunicable diseases kill 40 million people each year, equivalent to 70% of all deaths globally. People of all age groups, regions and countries are affected by NCDs but raised blood pressure; overweight/obesity; hyperglycemia (high blood glucose levels); and hyperlipidemia (high levels of fat in the blood) are

* Address correspondence to this author at the Department of Agronomy, Food, Natural Resources, Animals and Environment – DAFNAE, University of Padova, Viale Università 16, 35020, Agripolis, Italy; anna.lante@unipd.it

the four key metabolic changes that increase the risk of NCDs [2]. Therefore the European Commission established a coherent and comprehensive Community Strategy to address the issues of overweight and obesity and related health issues [3] as well as the EU Action Plan on Childhood Obesity 2014-2020 [4].

2. DIABETES AND OBESITY

Diabetes (Diabetes *mellitus*) is a serious, lifelong, progressive endocrine disorder, which can influence the entire body. It is characterized by hyperglycemia (high blood sugar (glucose) levels in the body) due to absolute or relative insulin production dysfunction. Diabetes leads to serious consequences *i.e.* limb amputation, kidney failure, heart attack, stroke and blindness. In 2016, there are 3.7 million registered deaths due to diabetes. The three main types of diabetes are type 1 (10% of all diabetes), type 2 (85% of all diabetes) and gestational diabetes. 6th April 2016 was announced World Health Day: Beat diabetes, and according to the WHO global diabetes info gram, 422 million adults have diabetes, which means that every 11th person has diabetes [5].

Type 1 diabetes is an auto-immune condition in which the cells that produce insulin are destroyed so lifelong treatment with insulin is required to prevent death. Type 2 diabetes, which may remain undetected for many years, occurs when the body either stops producing enough insulin for its needs or becomes resistant to the effect of insulin produced. The condition is progressive requiring lifestyle management (diet and exercise) at all stages. In this respect, functional foods may be a promising strategy to address this issue as reported by Tessari and Lante [6]. These authors studied the metabolic effects of a specifically designed functional bread, low in starch and rich in fibers in people with type 2 diabetes mellitus. Over time, most people with type 2 diabetes will require oral drugs and/or insulin. Obesity is only associated with type 2 diabetes and according to the WHO report every 3rd person is overweight and every 10th is obese. The physical inactivity and unhealthy diet are the main contributors to being overweight and obese.

Women with gestational diabetes are at an increased risk of complications during pregnancy and at delivery. It can occur at any stage of pregnancy, and does not usually cause any symptoms, but is more likely at 20 weeks or later. Screening for gestational diabetes is between weeks 8 to 12 when a doctor evaluates the risk of gestational diabetes. Women and their children are also at increased risk of type 2 diabetes in the future.

The likelihood and severity of type 2 diabetes are closely linked to Body Mass Index (BMI). A BMI above 30 is an alarming symptom. The epidemic of diabetes has major health and socioeconomic impacts, especially in developing countries. Obese people are seven times more likely to have diabetes, while overweight people are at only three times greater risk compared to those of healthy weight [7].

There are several theories of why obesity is linked to diabetes type 2. One of them states that abdominal obesity may cause fat cells to release pro-inflammatory chemicals, which lower the body sensitivity to insulin and disrupt its ability to properly respond to insulin [8]. Scientists claim that obesity is most likely to trigger changes to the body's metabolism that cause fat tissue to release increased amounts of fatty acids, glycerol, hormones, pro-inflammatory cytokines and other factors that are involved in the development of insulin resistance [9]. Studies with the skeletal muscle of type 2 diabetic humans demonstrate impaired insulin activation of the IRS-1/PI3K/Akt signaling pathway, which is a critical step in the regulation of glucose transport in response to insulin [10].

There is a clear association between increasing age and greater diabetes prevalence. In the UK, less than 2% of people aged 16-34 years are estimated to have diabetes compared to 5.1% of people aged 35-54 years, 14.3% of people aged 55-74 years and 16.5% of those aged over 75 years [11].

Both obesity and type 2 diabetes are strongly associated with an unhealthy diet and physical inactivity. Physical and social environments are important influences on diet and physical activity behavior along with interrelated economic, psychological and cultural factors [12]. Obesity and diabetes are characterized by both insulin resistance and endothelial dysfunction as increased artery intima-media thickness, and increased vascular stiffness leading to substantial increases in cardiovascular morbidity and mortality [13, 14].

Obesity is due primarily to an imbalance between caloric intake and activity. Obesity has rapidly become a serious public health concern. Obesity and overweight are different stages of unhealthy weight. Overweight people are those with BMI between 25 and 30, while obese ones are those with BMI over 30. This means that obese people have accumulated excessive fat tissues in their body. Social inequalities in overweight and obesity are strong, especially among women [15]. Absolute social inequalities were largest in Hungary (11.6% obese rates in men and 18.3% in women) and Spain (10% obese rates in men and 18.9% in women) across the education spectrum. Relative inequalities with poor education (largest in France and Sweden) showed that poorly educated men are 3.2 and 2.8 times more likely to be obese than men with the higher education. According to Devaux and Sassi [15], USA and England had the highest rates of obesity and overweight. Obesity levels are expected to be particularly high in the United States (47%), Mexico (39%) and England (35%) by the year 2030.

On the contrary, the WHO foresees a weaker increase in obesity and overweight in Italy and Korea, with obesity rates projected to be 13% and 9% in 2030, respectively. The level of obesity in France is projected to nearly match that of Spain, at 21% in 2030. Obesity rates are projected to increase at a faster pace in Korea and Switzerland where rates have been historically low [16].

3. POLYPHENOLS AND HUMAN HEALTH

Polyphenols are a large group of naturally occurring secondary metabolites mainly found in plants (fruits, vegetables, cereals *etc.*) and beverages. They have a wide variety of diverse structures, which belong to two main classes: non-flavonoids (particularly phenolic acids, stilbenes and lignans) and flavonoids, which are characterized by the basic C_6-C_3-C_6 skeleton (Fig. 1). The two aromatic rings within the flavonoid structure are linked by a heterocyclic ring, which differs in the degree of oxidation and leads to the following sub-classifications: flavones, flavonols, isoflavones, flavanones, anthocyanins and flavanols, usually called catechins. Some of the widespread representatives of natural phenolic compounds are kaempherol (flavonol), quercetin (flavonol), luteolin (flavone) and resveratrol (stilbenoid) (Fig 1). Both luteolin and quercetin are associated with their protective effects on diabetic nephropathy and retinopathy [17, 18]. Treatment with quercetin has the ability to abrogate hypertension progression induced by diabetes together with amelioration of the exaggerated contractile responses of aorta. Studies have confirmed the efficacy of resveratrol in type 2 diabetes. Furthermore, *in vitro* and *in vivo* studies have described resveratrol as a potent activator of histone deacetylase Sirtuin1 (Sirt1) [19].

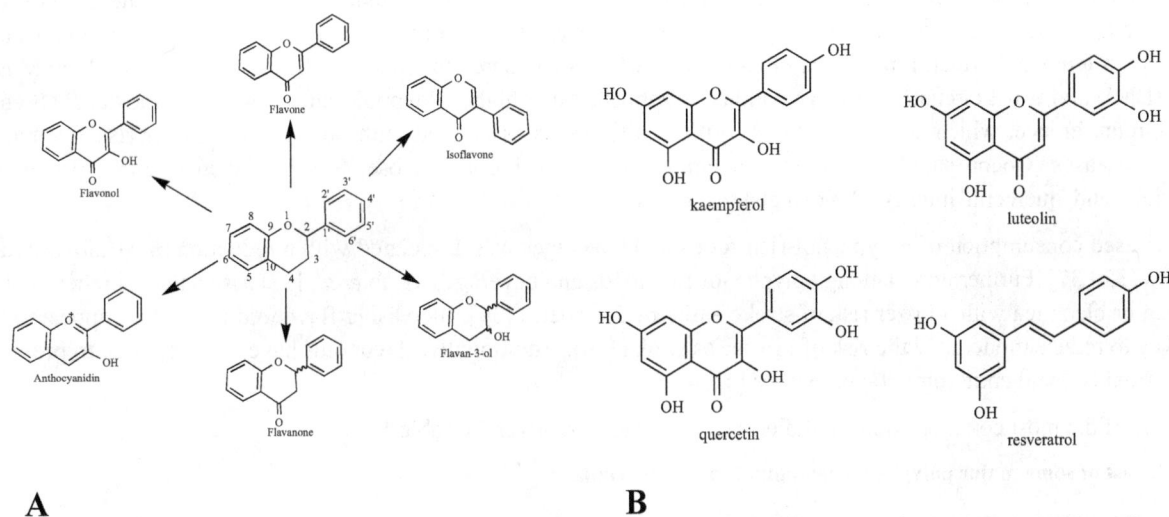

Fig. (1). Structure of flavonoid: **A** – structure of flavonoid skeleton; **B** – chemical structures of some dietary flavonoids.

Polyphenols are generally involved in defense against ultraviolet radiation or aggression by pathogens. In food, they may contribute to the bitterness, astringency, color, flavor, odor and oxidative stability. Because of their possible beneficial effects on human health, polyphenols and other food phenolics are still increasing their scientific interest. Studying the dietary polyphenol intake in Europe, Zamora *et al.* [20] reported a large number of dietary individual polyphenols consumed and high variability of their intakes between European populations, in particular between Mediterranean (MED) and non-MED countries. The main polyphenol contributors were established to be phenolic acids (52.5-56.9%), except in men from MED countries and in the UK health-conscious group where they were flavonoids (49.1-61.7%). Coffee, tea, and fruit were the most important food sources of total polyphenols. Zamora *et al.* [20] reported that the consumption of 437 different individual polyphenols (including 94 consumed at 1 mg/day) had health-promoting properties. The most abundant ones according to the above-mentioned authors were caffeoylquinic acids and the proanthocyanidin olygomers and polymers [20].

The polyphenol content of foods is strongly influenced by the methods of culinary preparation. The process of peeling of fruits and vegetables, for example, can significantly reduce polyphenol content, not only because these substances are often present in high concentrations in the outer parts, but also due to enzymatic browning which occurs after the breakdown of plant cell structure and the subsequent interaction between enzyme Polyphenol Oxidase (PPO) and substrate during post-harvest stages leading to color alteration and antioxidant degradation as a consequence of the phenolic oxidation. For this reason, the research of new eco-friendly systems for controlling PPO activity is focused on innovative non-thermal technologies and bioactive compounds to replace the conventional thermal treatments and traditional additives which could impair the sensory, nutritional and health properties of food products [21 - 26].

The process of heating, also, has an outstanding effect. About 75% of the initial quercetin content in tomatoes and onions is lost after boiling for 15 min, 65% after cooking in a microwave oven, and 30% after frying. Potatoes contain up to 190 mg chlorogenic acid/kg mainly in the skin [27]; so, an extensive loss occurs during culinary processes, and as a result, no remaining phenolic acids are found in French fries [28]. Also, extraction by heating or boiling of green tea leaves seems to increase the epimerization of tea catechins and therefore a lower recovery of bioactive compounds [29].

Polyphenols, of natural origin, are potential sources of various beneficial effects - anti-obesity, anti-diabetic, antihypertensive, anti-hyperlipidemic and anti-inflammatory effects [30]. Commonly consumed polyphenols such as green tea catechins, especially epigallocatechin gallates, as well as resveratrol and curcumin are considered to impact obesity and obesity-related inflammation. Dietary polyphenols are demonstrated to reduce viability of adipocytes and proliferation of preadipocytes, suppress adipocyte differentiation and triglyceride accumulation, stimulate lipolysis and fatty acid β-oxidation, and reduce inflammation. Furthermore, polyphenols can modulate signaling pathways including the adenosine monophosphate-activated protein kinase, peroxisome proliferator-activated receptor γ, CCAAT/enhancer binding protein α, peroxisome proliferator activator receptor gamma activator 1-alpha, sirtuin 1, sterol regulatory element binding protein-1c, uncoupling proteins 1 and 2, and nuclear factor-κB that regulate adipogenesis, antioxidant and anti-inflammatory responses [31]. Black tea polyphenols inhibit the emulsion of droplets and the activity of pancreatic lipase, α-amylase and glucosidases [32]. Green tea was reported to reduce significantly body mass index and waist circumference. Furthermore, Vernarelli and Lambert [33] compared the intake of flavonoids with the Body Mass Index (BMI) and waist circumference. The authors established that higher flavonoid intake influenced lower BMI and waist circumference, which could contribute to the health issues associated with obesity as a higher risk for many chronic diseases. Cocoa supplementation and cinnamon reduce blood glucose. Soy isoflavones, citrus products, hesperidin and quercetin improve lipid metabolism [34].

Increased consumption of polyphenol-rich foods and beverages was associated with a reduction of cardiovascular diseases [35 - 37]. Furthermore, among polyphenol-rich foods and beverages, Arab et al. [38] associated a higher intake of green or black tea with a lower risk of stroke and type 2 diabetes [39]. Increasing flavonoid intake also appeared to be a way to reduce moderately the risk of disease as stroke [40]. Additionally, flavonoids have been reported to provide both antioxidant and antithrombotic properties [41, 42].

Some of the most common sources of dietary polyphenols are given in Table 1.

Table 1. List of some major polyphenol compounds of plant origin.

Food Group	Some Major Sources of Polyphenols
Fruit	oranges, apples, grapes, peaches, grapefruit juice, cherries, blueberries, pomegranate juice, raspberries, cranberries, black elderberries, blackcurrants, plums, blackberries, strawberries, apricots
Vegetables	spinach, onions, shallots, potatoes, black and green olives, globe artichoke heads, broccoli, asparagus, carrots
Whole grains	whole grain wheat, rye, and oat flours
Nuts, seeds, and legumes	roasted soybeans, black beans, white beans, chestnuts, hazelnuts, pecans, almonds, walnuts, flaxseed
Beverages	coffee, tea, red wine
Fats	dark chocolate, virgin olive oil, sesame seed oil
Spices and seasonings	cocoa powder, capers, saffron, dried oregano, dried rosemary, soy sauce, cloves, dried peppermint, star anise, celery seed, dried sage, dried spearmint, dried thyme, dried basil, curry powder, dried ginger, cumin, cinnamon

Numerous scientific reports give evidence regarding the intake of polyphenols (and their food sources) and the influence on related-diabetes risk factors. Several studies highlight the anti-obesity effects of polyphenol-rich diets affecting the ability of polyphenols to interact, directly or indirectly, with adipose tissues (preadipocytes, adipose stem cells, and immune cells). However, obesity and diabetes nowadays are among the major diseases of health concern, which provoke the scientific interest. Obese individuals are at a greater risk for the development of several pathologies including diabetes *etc.* [43], which show the relationship between both.

Polyphenols (p-coumaric acid, m-coumaric acid, ferulic acid and hydroxyhippuric acid) could boost insulin sensitivity, slow down the rate, the digestion, and absorption of sugar [44]. Recent studies associate the decrease of the insulin resistance to type of flavonoid (flavan-3-ols) and revealed the flavonoids as the type of polyphenols most often associated with a lower risk for type 2 diabetes [45, 46]. Unfortunately, in order to state concrete flavonoids conducting large-scale, well-designed, and population-based studies is required in the future [47].

Lambert *et al.* [48] evaluated a synergistic effect of grape polyphenols supplementation combined with the exercise by increasing muscle lipid oxidation and sparing glycogen utilization, which resulted in enhancement of endurance capacity based on the intensification of their individual metabolic effects. The reported, underlined the importance of both dietary and physical training recommendations in insulin resistance condition. Furthermore, the use of a combination of polyphenols is proposed to treat diabesity complications in view of reported synergisms of resveratrol when combined with quercetin or resveratrol plus quercetin and genistein in *in vitro* studies [49, 50].

However, it is difficult such beneficial effects to be proven in *in vivo* experiments. Bruckbauer *et al.* [51] stated synergisms during *in vitro* and *in vivo* approaches with other compounds, such as a leucine metabolite, methylxanthines and metformin as potential therapeutic agents in obesity and diabetes prevention and management.

Resveratrol, commonly found in grapes, berries, and some nuts, is one of the most commonly used nutritional supplement polyphenolic compound with potent antioxidant activity. Resveratrol is particularly interesting to the scientific field because of its health benefits associated with diseases such as cancer, type 2 diabetes, cardiovascular disease, and neurological conditions [52]. This polyphenol gains worldwide interest when reports of its cardio protective effects originating from red wine become available [53]. Resveratrol was reported as a potential obesity treatment as it can lead to lowering body weight [54]; reducing the fat mass by inhibiting adipogenesis. However, the data on long-term resveratrol toxicity is contradictious [55, 56] even though clinical trials proved a well-tolerated and pharmacologically safe dosage of up to 5 g/day [57]. Yet, the research of its bioavailability and effectiveness *in vivo* is essential thus, the current data are still insufficient.

Retinopathy, nephropathy and neuropathy are only some of the serious health concern problems diabetes can cause in the long term. Various anthocyanins, flavonoids and other polyphenolic compounds have been found with potential preventive effects for improving the quality of life in the patients. Anthocyanins and anthocyanins-rich extracts have been reported as potential to alleviate some pathologic conditions because of diabetes [58]. Flavanols, on the other hand, were assumed as potential to improve cognitive disorders and cholinergic dysfunction related to diabetes and other secondary consequence of changes in the nervous system induced by hyperglycemia and diabetes oxidative stress and, as for example, quercetin was reported to improve mental function and memory in rats with diabetes [59].

Numerous literature data support the natural phenolic compounds as health-beneficial sources, furthermore, functional foods and dietary supplements containing phenolic compounds gain extensive publicity and interest as a contemporary inexpensive therapeutic approach [60]. Based on the increased scientific interest, various research studies are involved in the evaluation of the polyphenol dietary sources potential of fruit, vegetables, herbs and spices [61 - 64]. All this is a prerequisite polyphenols to be involved in novel strategies for preventing diabetes and obesity complications [65].

Certain dietary components and over 800 plants were claimed to help prevent or moderate metabolic syndrome by assisting the body homeostasis mechanism [66]. Phenolic beneficial effects are exacted from plants belonging to taxonomic families such as Moraceae, Fabaceae, Asteraceae, Pyrus, Lauraceae, Lythraceae, and Malvaceae [67].

Possible pathways of polyphenol metabolism by intestinal bacteria [68] have been investigated in several research projects, highlighting the significant evidence of the diet-derived bioactive metabolites produced by gut microbiota, with a particular emphasis on polyphenols and their potential impact on human health. In this regard, the characterization of microbial strains, suitable for the fermentation process of plants historically recognized by folk medicine, could be used to design new functional food with added beneficial attributes [69, 70]. Furthermore, other health beneficial functions associated with obesity symptoms improvement have been associated with maintaining normal gut microbiota [71 - 73].

Microbial engineering is one of the alternative approaches, which can meet the global demand for natural products in an eco-friendly manner [74]. Metabolic engineering of microorganisms could provoke microbial synthesis of many plant polyphenols. However, recent technological innovations such as the development of biosensor-driven directed evolution approaches could enable the rapid engineering of microbial host strains for providing more precursor molecules to increase polyphenol synthesis [75, 76]. New molecular techniques, as well as approaches and concepts from synthetic biology, could be also employed to harmonize endogenous and heterologous pathways for maximizing product titers and host could improve rerouting of carbon fluxes to and through polyphenol-providing pathways of interest [77, 78].

In particular, nanotechnological approaches could help to overcome the pharmacokinetic issues of bioactive compounds by providing improved bioavailability, overcoming the first-pass metabolism and trounce enterohepatic recirculation, by protection against degradation, enhancement in intracellular penetration and control delivery, and by reducing potential toxicity [79]. Stilbenes can be therefore used in a natural form for prevention or in their pure form of therapy, for which large doses or nanoformulations are recommended [80].

4. FUTURE INSIGHT AND PERSPECTIVE

An intake of above 600 mg/d of polyphenols within a healthy dietary pattern, rich in fruits and vegetables has been recommended. Future clinical and epidemiological studies are warranted to replicate these associations, especially using biomarkers, in other populations. The advances in polyphenols research will be important to make dietary recommendations for developing effective public health policies and for improving the autonomy and quality of life of older people [81].

In addition, epidemiological studies and associated meta-analyses recommended the long-term consumption of rich in plant polyphenols diets as significant protection against the development of cancers, cardiovascular diseases, diabetes, osteoporosis and neurodegenerative diseases [34, 82].

The measurement of phenolic metabolome in humans would point out which phenolic molecules are present in the plasma/urine of humans followed the intake of polyphenol-rich foods (green tea, cocoa, citrus, apple) and should facilitate the provision of clearer evidence on the relations between food composition and risk of major chronic diseases such as cancer, cardiovascular diseases or diabetes and shed new light on the causes of such diseases [83]. The future research in this field should focus on the actual contribution of dietary polyphenols, new cellular and molecular targets and on clarification of *in vivo* biotransformation processes, including analyses of the biological effects of each single metabolite.

Applying good practices for health promotion and disease prevention through conscious nutritious choices and fostering a positive attitude towards nutrition. Sustainable diets are a key factor to long-term wellbeing and healthy lifestyle. Deepening consumer's knowledge will create healthy eating patterns and therefore promote personalized nutrition.

CONCLUSION

Promoting healthier choices at an early stage of human development may lead to minimizing the display of 21st-century diseases (*i.e.* diabetes type 2, obesity, heart complications, teeth decay, *etc.*) based on adequate dietary assessment and food choices relevant to nutritional requirements.

Diabetes and obesity are issues among serious health concern. Both conditions are strongly associated with an unhealthy diet and physical inactivity. Several recent studies on plant deriving polyphenols revealed their possible beneficial effects on human health. Developing efficient public health policies and improving people's quality of life by personalized nutrition, based on the recommendation of polyphenol intake, would enhance consumer's trust in sustainable dietary choices lying on a sound scientific basis.

ACKNOWLEDGEMENTS

Declared none.

REFERENCES

[1]   Weichselbaum E, Hooper B, Buttriss J, et al. Behaviour change initiatives to promote a healthy diet and physical activity in European countries. Brit Nutr Found Nutr Bull 2013; 38: 85-99. [https://doi.org/10.1111/nbu.12011].
        [http://dx.doi.org/10.1111/nbu.12011]

[2]   World Health Organization. Noncommunicable Diseases Progress Monitor, 2017. Geneva 2017. Licence: CC BY-NC-SA 3.0 IGO.

[3]   Commission of the European communities. White Paper on a Strategy for Europe on Nutrition, Overweight and Obesity related health issues. 2007.

[4]   European communities. EU Action Plan on Childhood Obesity 2014-2020. 2014.

[5]   World Health Organization. Global report on diabetes 2016. ISBN 978 92 4 156525 7. http:// apps.who.int/ iris/ bitstream/ handle/ 10665/204871/9789241565257_eng.pdf;jsessionid=201B2F321B5D3ECAA4B4886C6E915E70?sequence=1

[6]   Tessari P, Lante A. Multifunctional bread rich in beta glucans and low in starch improves metabolic control in type 2 diabetes: A controlled trial. Nutrients 2017; 9(3): E297.
        [http://dx.doi.org/10.3390/nu9030297] [PMID: 28304350]

[7]   Abdullah A, Peeters A, de Courten M, Stoelwinder J. The magnitude of association between overweight and obesity and the risk of diabetes: A meta-analysis of prospective cohort studies. Diabetes Res Clin Pract 2010; 89(3): 309-19.
        [http://dx.doi.org/10.1016/j.diabres.2010.04.012] [PMID: 20493574]

[8]   Després JP. Body fat distribution and risk of cardiovascular disease: An update. Circulation 2012; 126(10): 1301-13.
        [http://dx.doi.org/10.1161/CIRCULATIONAHA.111.067264] [PMID: 22949540]

[9]   Kahn SE, Hull RL, Utzschneider KM. Mechanisms linking obesity to insulin resistance and type 2 diabetes. Nature 2006; 444(7121): 840-6.
        [http://dx.doi.org/10.1038/nature05482] [PMID: 17167471]

[10]   Choi K, Kim Y-B. Molecular mechanism of insulin resistance in obesity and type 2 diabetes. Korean J Intern Med (Korean Assoc Intern Med) 2010; 25(2): 119-29.
        [http://dx.doi.org/10.3904/kjim.2010.25.2.119] [PMID: 20526383]

[11]   National Cardiovascular Intelligence Intelligence network. Diabetes Prevalence Model for Local Authorities and CCGs. Secondary National cardiovascular intelligence intelligence network. Diabetes Prevalence Model for Local Authorities and CCGs. 2015.

[12]   Roberts K, Cavill N, Hancock C, et al. Social and economic inequalities in diet and physical activity Oxford: Public Health England Obesity Knowledge and Intelligence, Crown copyright, 2013.

[13]   Beckman JA, Creager MA, Libby P. Diabetes and atherosclerosis: Epidemiology, pathophysiology, and management. JAMA 2002; 287(19): 2570-81.
        [http://dx.doi.org/10.1001/jama.287.19.2570] [PMID: 12020339]

[14]   Herouvi D, Karanasios E, Karayianni C, Karavanaki K. Cardiovascular disease in childhood: The role of obesity. Eur J Pediatr 2013; 172(6): 721-32.
        [http://dx.doi.org/10.1007/s00431-013-1932-8] [PMID: 23340698]

[15]   Devaux M, Sassi F. Social inequalities in obesity and overweight in 11 OECD countries. Eur J Public Health 2013; 23(3): 464-9.
        [http://dx.doi.org/10.1093/eurpub/ckr058] [PMID: 21646363]

[16]   FAO, IFAD, UNICEF, WFP and WHO. 2017. The State of Food Security and Nutrition in the World 2017. Building resilience for peace and food security. Rome, FAO.

[17]   Wang GG, Lu XH, Li W, Zhao X, Zhang C. Protective effects of luteolin on diabetic nephropathy in STZ-induced diabetic rats. Evid Based Complement Alternat Med 2011; 2011: 323171.
        [http://dx.doi.org/10.1155/2011/323171] [PMID: 21584231]

[18]   Ola MS, Ahmed MM, Shams S, Al-Rejaie SS. Neuroprotective effects of quercetin in diabetic rat retina. Saudi J Biol Sci 2017; 24(6): 1186-94.
        [http://dx.doi.org/10.1016/j.sjbs.2016.11.017] [PMID: 28855811]

[19]   Yar AS, Menevse S, Alp E. The effects of resveratrol on cyclooxygenase-1 and -2, nuclear factor kappa beta, matrix metalloproteinase-9, and sirtuin 1 mRNA expression in hearts of streptozotocin-induced diabetic rats. Genet Mol Res 2011; 10(4): 2962-75.
        [http://dx.doi.org/10.4238/2011.November.29.7] [PMID: 22179968]

[20]   Zamora-Ros R, Knaze V, Rothwell JA, et al. Dietary polyphenol intake in Europe: The European Prospective Investigation into Cancer and nutrition (EPIC) study. Eur J Nutr 2016; 55(4): 1359-75.
        [http://dx.doi.org/10.1007/s00394-015-0950-x] [PMID: 26081647]

[21]   Lante A, Tinello F. Citrus hydrosols as useful by-products for tyrosinase inhibition. Innov Food Sci Emerg 2015; 27: 154-9.
        [http://dx.doi.org/10.1016/j.ifset.2014.11.001]

[22]   Lante A, Tinello F, Lomolino G. The use of polyphenol oxidase activity to identify a potential raisin variety. Food Biotechnol 2016; 30: 98-109.
        [http://dx.doi.org/10.1080/08905436.2016.1166125]

[23]   Lante A, Tinello F, Nicoletto M. UV-A light treatment for controlling enzymatic browning of fresh-cut fruits. Innov Food Sci Emerg 2016;

34: 141-7.
[http://dx.doi.org/10.1016/j.ifset.2015.12.029]

[24]   Tinello F, Lante A. Evaluation of antibrowning and antioxidant activities in unripe grapes recovered during bunch thinning. Aust J Grape
        Wine Res 2017; 23: 33-41.
        [http://dx.doi.org/10.1111/ajgw.12256]

[25]   Zocca F, Lomolino G, Lante A. Antibrowning potential of *Brassicacaea* processing water. Bioresour Technol 2010; 101(10): 3791-5.
        [http://dx.doi.org/10.1016/j.biortech.2009.12.126] [PMID: 20116236]

[26]   Zocca F, Lomolino G, Lante A. Dog rose and pomegranate extracts as agents to control enzymatic browning. Food Res Int 2011; 44: 957-63.
        [http://dx.doi.org/10.1016/j.foodres.2011.02.010]

[27]   Friedman M. Chemistry, biochemistry, and dietary role of potato polyphenols. A review. J Agric Food Chem 1997; 45: 1523-40.
        [http://dx.doi.org/10.1021/jf960900s]

[28]   Clifford MN. Chlorogenic acids and other cinnamates. Nature, occurrence, dietary burden, absorption and metabolism. J Sci Food Agric 2000;
        80: 1033-43.
        [http://dx.doi.org/10.1002/(SICI)1097-0010(20000515)80:7<1033::AID-JSFA595>3.0.CO;2-T]

[29]   Lante A, Friso D. Oxidative stability and rheological properties of nanoemulsions with ultrasonic extracted green tea infusion. Food Res Int
        2013; 54: 269-76.
        [http://dx.doi.org/10.1016/j.foodres.2013.07.009]

[30]   Cherniack EP. Polyphenols: Planting the seeds of treatment for the metabolic syndrome. Nutrition 2011; 27(6): 617-23.
        [http://dx.doi.org/10.1016/j.nut.2010.10.013] [PMID: 21367579]

[31]   Wang S, Moustaid-Moussa N, Chen L, *et al.* Novel insights of dietary polyphenols and obesity. J Nutr Biochem 2014; 25(1): 1-18.
        [http://dx.doi.org/10.1016/j.jnutbio.2013.09.001] [PMID: 24314860]

[32]   Pan H, Gao Y, Tu Y. Mechanisms of body weight reduction by black tea polyphenols. Molecules 2016; 21(12): 1659.
        [http://dx.doi.org/10.3390/molecules21121659] [PMID: 27941615]

[33]   Vernarelli JA, Lambert JD. Flavonoid intake is inversely associated with obesity and C-reactive protein, a marker for inflammation, in US
        adults. Nutr Diabetes 2017; 7(5): e276.
        [http://dx.doi.org/10.1038/nutd.2017.22] [PMID: 28504712]

[34]   Amiot MJ, Riva C, Vinet A. Effects of dietary polyphenols on metabolic syndrome features in humans: A systematic review. Obes Rev 2016;
        17(573): 586.

[35]   Arts ICW, Hollman PCH. Polyphenols and disease risk in epidemiologic studies. Am J Clin Nutr 2005; 81(1): 317S-25S.
        [http://dx.doi.org/10.1093/ajcn/81.1.317S] [PMID: 15640497]

[36]   Hooper L, Kroon PA, Rimm EB, *et al.* Flavonoids, flavonoid-rich foods, and cardiovascular risk: A meta-analysis of randomized controlled
        trials. Am J Clin Nutr 2008; 88(1): 38-50.
        [http://dx.doi.org/10.1093/ajcn/88.1.38] [PMID: 18614722]

[37]   Hollman PCH, Geelen A, Kromhout D. Dietary flavonol intake may lower stroke risk in men and women. J Nutr 2010; 140(3): 600-4.
        [http://dx.doi.org/10.3945/jn.109.116632] [PMID: 20089788]

[38]   Arab L, Liu W, Elashoff D. Green and black tea consumption and risk of stroke: A meta-analysis. Stroke 2009; 40(5): 1786-92.
        [http://dx.doi.org/10.1161/STROKEAHA.108.538470] [PMID: 19228856]

[39]   Jing Y, Han G, Hu Y, Bi Y, Li L, Zhu D. Tea consumption and risk of type 2 diabetes: A meta-analysis of cohort studies. J Gen Intern Med
        2009; 24(5): 557-62.
        [http://dx.doi.org/10.1007/s11606-009-0929-5] [PMID: 19308337]

[40]   Tang Z, Li M, Zhang X, Hou W. Dietary flavonoid intake and the risk of stroke: A dose-response meta-analysis of prospective cohort studies.
        BMJ Open 2016; 6(6): e008680.
        [http://dx.doi.org/10.1136/bmjopen-2015-008680] [PMID: 27279473]

[41]   Frei B, Higdon JV. Antioxidant activity of tea polyphenols *in vivo*: Evidence from animal studies. J Nutr 2003; 133(10): 3275S-84S.
        [http://dx.doi.org/10.1093/jn/133.10.3275S] [PMID: 14519826]

[42]   Sheng R, Gu ZL, Xie ML, Zhou WX, Guo CY. EGCG inhibits proliferation of cardiac fibroblasts in rats with cardiac hypertrophy. Planta
        Med 2009; 75(2): 113-20.
        [http://dx.doi.org/10.1055/s-0028-1088387] [PMID: 19096994]

[43]   Moon H-S, Lee H-G, Choi Y-J, Kim T-G, Cho C-S. Proposed mechanisms of (-)-epigallocatechin-3-gallate for anti-obesity. Chem Biol
        Interact 2007; 167(2): 85-98.
        [http://dx.doi.org/10.1016/j.cbi.2007.02.008] [PMID: 17368440]

[44]   Paquette M, Medina Larqué AS, Weisnagel SJ, *et al.* Strawberry and cranberry polyphenols improve insulin sensitivity in insulin-resistant,
        non-diabetic adults: A parallel, double-blind, controlled and randomised clinical trial. Br J Nutr 2017; 117(4): 519-31.
        [http://dx.doi.org/10.1017/S0007114517000393] [PMID: 28290272]

[45]   Liu Y-J, Zhan J, Liu XL, Wang Y, Ji J, He QQ. Dietary flavonoids intake and risk of type 2 diabetes: A meta-analysis of prospective cohort
        studies. Clin Nutr 2014; 33(1): 59-63.

[http://dx.doi.org/10.1016/j.clnu.2013.03.011] [PMID: 23591151]

[46]   Guasch-Ferré M, Merino J, Sun Q, Fitó M, Salas-Salvadó J. Dietary polyphenols, mediterranean diet, prediabetes, and type 2 diabetes: A
        narrative review of the evidence. Oxid Med Cell Longev 2017; 2017: 6723931.
        [http://dx.doi.org/10.1155/2017/6723931] [PMID: 28883903]

[47]   Bird SR, Hawley JA. Update on the effects of physical activity on insulin sensitivity in humans. BMJ Open Sport Exerc Med 2017; 2(1):
        e000143.
        [http://dx.doi.org/10.1136/bmjsem-2016-000143] [PMID: 28879026]

[48]   Lambert K, Hokayem M, Thomas C, et al. Combination of nutritional polyphenols supplementation with exercise training counteracts insulin
        resistance and improves endurance in high-fat diet-induced obese rats. Sci Rep 2018; 8(1): 2885.
        [http://dx.doi.org/10.1038/s41598-018-21287-z] [PMID: 29440695]

[49]   Yang JY, Della-Fera MA, Rayalam S, et al. Enhanced inhibition of adipogenesis and induction of apoptosis in 3T3-L1 adipocytes with
        combinations of resveratrol and quercetin. Life Sci 2008; 82(19-20): 1032-9.
        [http://dx.doi.org/10.1016/j.lfs.2008.03.003] [PMID: 18433793]

[50]   Park HJ, Yang JY, Ambati S, et al. Combined effects of genistein, quercetin, and resveratrol in human and 3T3-L1 adipocytes. J Med Food
        2008; 11(4): 773-83.
        [http://dx.doi.org/10.1089/jmf.2008.0077] [PMID: 19053873]

[51]   Bruckbauer A, Zemel MB. Synergistic effects of polyphenols and methylxanthines with Leucine on AMPK/Sirtuin-mediated metabolism in
        muscle cells and adipocytes. PLoS One 2014; 9(2): e89166.
        [http://dx.doi.org/10.1371/journal.pone.0089166] [PMID: 24551237]

[52]   Marques FZ, Markus MA, Morris BJ. Resveratrol: Cellular actions of a potent natural chemical that confers a diversity of health benefits. Int J
        Biochem Cell Biol 2009; 41(11): 2125-8.
        [http://dx.doi.org/10.1016/j.biocel.2009.06.003] [PMID: 19527796]

[53]   Wu JM, Wang ZR, Hsieh TC, Bruder JL, Zou JG, Huang YZ. Mechanism of cardioprotection by resveratrol, a phenolic antioxidant present in
        red wine (Review). Int J Mol Med 2001; 8(1): 3-17.
        [PMID: 11408943]

[54]   Rayalam S, Yang J-Y, Ambati S, Della-Fera MA, Baile CA. Resveratrol induces apoptosis and inhibits adipogenesis in 3T3-L1 adipocytes.
        Phytother Res 2008; 22(10): 1367-71.
        [http://dx.doi.org/10.1002/ptr.2503] [PMID: 18688788]

[55]   Zamora-Ros R, Urpí-Sardà M, Lamuela-Raventós RM, et al. Diagnostic performance of urinary resveratrol metabolites as a biomarker of
        moderate wine consumption. Clin Chem 2006; 52(7): 1373-80.
        [http://dx.doi.org/10.1373/clinchem.2005.065870] [PMID: 16675507]

[56]   Chow HH, Garland LL, Hsu CH, et al. Resveratrol modulates drug- and carcinogen-metabolizing enzymes in a healthy volunteer study.
        Cancer Prev Res (Phila) 2010; 3(9): 1168-75.
        [http://dx.doi.org/10.1158/1940-6207.CAPR-09-0155] [PMID: 20716633]

[57]   Patel KR, Scott E, Brown VA, Gescher AJ, Steward WP, Brown K. Clinical trials of resveratrol. Ann N Y Acad Sci 2011; 1215: 161-9.
        [http://dx.doi.org/10.1111/j.1749-6632.2010.05853.x] [PMID: 21261655]

[58]   Ghosh D, Konishi T. Anthocyanins and anthocyanin-rich extracts: Role in diabetes and eye function. Asia Pac J Clin Nutr 2007; 16(2): 200-8.
        [PMID: 17468073]

[59]   Bhutada P, Mundhada Y, Bansod K, et al. Ameliorative effect of quercetin on memory dysfunction in streptozotocin-induced diabetic rats.
        Neurobiol Learn Mem 2010; 94(3): 293-302.
        [http://dx.doi.org/10.1016/j.nlm.2010.06.008] [PMID: 20620214]

[60]   Redan BW, Buhman KK, Novotny JA, Ferruzzi MG. Altered transport and metabolism of phenolic compounds in obesity and diabetes:
        Implications for functional food development and assessment. Adv Nutr 2016; 7(6): 1090-104.
        [http://dx.doi.org/10.3945/an.116.013029] [PMID: 28140326]

[61]   Denev P, Lojek A, Ciz M, Kratchanova M. Antioxidant activity and polyphenol content of Bulgarian fruits. Bulg J Agric Sci 2013; 19: 22-7.

[62]   Mihaylova DS, Lante A, Tinello F, Krastanov AI. Study on the antioxidant and antimicrobial activities of Allium ursinum L. pressurised-
        liquid extract. Nat Prod Res 2014; 28(22): 2000-5.
        [http://dx.doi.org/10.1080/14786419.2014.923422] [PMID: 24895887]

[63]   Georgieva L, Mihaylova D. Screening of total phenolic content and radical scavenging capacity of Bulgarian plant species. Int Food Res J
        2015; 22(1): 240-5.

[64]   Shahidi F, Ambigaipalan P. Phenolics and polyphenolics in foods, beverages and spices: Antioxidant activity and health effects - A review. J
        Funct Foods 2015; 18: 820-97.
        [http://dx.doi.org/10.1016/j.jff.2015.06.018]

[65]   Carpéné C, Gomez-Zorita S, Deleruyelle S, Carpéné MA. Novel strategies for preventing diabetes and obesity complications with natural
        polyphenols. Curr Med Chem 2015; 22(1): 150-64.
        [http://dx.doi.org/10.2174/0929867321666140815124052] [PMID: 25139462]

[66]   Suhaila M. Functional foods against metabolic syndrome (obesity, diabetes, hypertension and dyslipidemia) and cardiovasular disease. Trends Food Sci Technol 2014; 35(2): 114-28.
        [http://dx.doi.org/10.1016/j.tifs.2013.11.001]

[67]   Lin D, Xiao M, Zhao J, et al. An overview of plant phenolic compounds and their importance in human nutrition and management of type 2 diabetes. Molecules 2016; 21(10): 1374.
        [http://dx.doi.org/10.3390/molecules21101374] [PMID: 27754463]

[68]   Duda-Chodak A, Tarko T, Satora P, Sroka P. Interaction of dietary compounds, especially polyphenols, with the intestinal microbiota: A review. Eur J Nutr 2015; 54(3): 325-41.
        [http://dx.doi.org/10.1007/s00394-015-0852-y] [PMID: 25672526]

[69]   Kagkli DM, Corich V, Bovo B, Lante A, Giacomini A. Antiradical and antimicrobial properties of fermented red chicory (Cichorium intybus L.) by-products. Ann Microbiol 2016; 66: 1377-86.
        [http://dx.doi.org/10.1007/s13213-016-1225-3]

[70]   Tinello F, Vendramin V, Divino VB, et al. Co-fermentation of onion and whey: A promising synbiotic combination. J Funct Foods 2017; 39: 233-7.
        [http://dx.doi.org/10.1016/j.jff.2017.10.018]

[71]   Hervert-Hernández D, Goñi I. Dietary polyphenols and human gut microbiota: A review. Food Rev Int 2011; 27(2): 154-69.
        [http://dx.doi.org/10.1080/87559129.2010.535233]

[72]   Yuan JP, Wang JH, Liu X. Metabolism of dietary soy isoflavones to equol by human intestinal microflora-implications for health. Mol Nutr Food Res 2007; 51(7): 765-81.
        [http://dx.doi.org/10.1002/mnfr.200600262] [PMID: 17579894]

[73]   Ley RE, Turnbaugh PJ, Klein S, Gordon JI. Microbial ecology: Human gut microbes associated with obesity. Nature 2006; 444(7122): 1022-3.
        [http://dx.doi.org/10.1038/4441022a] [PMID: 17183309]

[74]   Chouhan S, Sharma K, Zha J, Guleria S, Koffas MAG. Recent advances in the recombinant biosynthesis of polyphenols. Front Microbiol 2017; 8: 2259.
        [http://dx.doi.org/10.3389/fmicb.2017.02259] [PMID: 29201020]

[75]   Schallmey M, Frunzke J, Eggeling L, Marienhagen J. Looking for the pick of the bunch: High-throughput screening of producing microorganisms with biosensors. Curr Opin Biotechnol 2014; 26: 148-54.
        [http://dx.doi.org/10.1016/j.copbio.2014.01.005] [PMID: 24480185]

[76]   Siedler S, Khatri NK, Zsohár A, et al. Development of a bacterial biosensor for rapid screening of yeast p-coumaric acid production. ACS Synth Biol 2017; 6(10): 1860-9.
        [http://dx.doi.org/10.1021/acssynbio.7b00009] [PMID: 28532147]

[77]   van Summeren-Wesenhagen PV, Marienhagen J. Metabolic engineering of Escherichia coli for the synthesis of the plant polyphenol pinosylvin. Appl Environ Microbiol 2015; 81(3): 840-9.
        [http://dx.doi.org/10.1128/AEM.02966-14] [PMID: 25398870]

[78]   Xu P, Li L, Zhang F, Stephanopoulos G, Koffas M. Improving fatty acids production by engineering dynamic pathway regulation and metabolic control. Proc Natl Acad Sci USA 2014; 111(31): 11299-304.
        [http://dx.doi.org/10.1073/pnas.1406401111] [PMID: 25049420]

[79]   Siddiqui IA, Sanna V, Ahmad N, Sechi M, Mukhtar H. Resveratrol nanoformulation for cancer prevention and therapy. Ann N Y Acad Sci 2015; 1348(1): 20-31.
        [http://dx.doi.org/10.1111/nyas.12811] [PMID: 26109073]

[80]   Seyed MA, Jantan I, Bukhari SN, Vijayaraghavan K. Comprehensive review on the chemotherapeutic potential of piceatannol for cancer treatment, with mechanistic insights. J Agric Food Chem 2016; 64(4): 725-37.
        [http://dx.doi.org/10.1021/acs.jafc.5b05993] [PMID: 26758628]

[81]   Rabassa M, Cherubini A, Zamora-Ros R, et al. Low levels of a urinary biomarker of dietary polyphenol are associated with substantial cognitive decline over a 3-year period in older adults: The invecchiare in CHIANTI study. J Am Geriatr Soc 2015; 63(5): 938-46.
        [http://dx.doi.org/10.1111/jgs.13379] [PMID: 25919574]

[82]   Graf BA, Milbury PE, Blumberg JB. Flavonols, flavones, flavanones, and human health: Epidemiological evidence. J Med Food 2005; 8(3): 281-90.
        [http://dx.doi.org/10.1089/jmf.2005.8.281] [PMID: 16176136]

[83]   Mursu J, Virtanen JK, Tuomainen T-P, Nurmi T, Voutilainen S. Intake of fruit, berries, and vegetables and risk of type 2 diabetes in Finnish men: The Kuopio Ischaemic Heart Disease Risk Factor Study. Am J Clin Nutr 2014; 99(2): 328-33.
        [http://dx.doi.org/10.3945/ajcn.113.069641] [PMID: 24257723]

10

Study of Optimum Condition for Rapid Preparation of Thrombin using Russell's Viper Venom Factor X Activator

Narin Kijkriengkraikul[1,*] and Issarang Nuchprayoon[2]

[1]*Technopreneurship and Innovation Management Program, Graduate School, Chulalongkorn University, Bangkok 10330, Thailand*
[2]*Department of Pediatrics, Faculty of Medicine, Chulalongkorn University, Bangkok 10330, Thailand*

Abstract:

Background:

The purpose of this study is to investigate a simple method with the optimum condition for rapid thrombin preparation from Cryoprecipitate-depleted Plasma (CDP) using RVV-X in the process.

Methods:

Thrombin preparation from human CDP was studied with the presence of different factors in batch condition including: 1) RVV-X; 2) volume of calcium chloride solution; 3) volume of sodium chloride solution for final extraction; and 4) incubation time. The properties of the prepared sample were analyzed for fibrin clot formation, total protein by Kjeldahl method, thrombin time, molecular weight and protein patterns by SDS-PAGE, and thrombin concentration by coagulation analyzer. The method and process of preparing thrombin and the study of optimum condition for rapidly preparing the highest yield of thrombin from starting CDP 100 ml were introduced.

Results:

The best four conditions were concluded: 1) RVV-X 50 mcg should be present in the process; 2) volume of 0.25 M calcium chloride should be 3 ml; 3) volume of 0.85% sodium chloride for the final protein precipitate extraction should be 10 ml and; 4) no incubation time needed for prothrombin activation process. A solution prepared from the optimum condition showed an obvious band on SDS-PAGE at a molecular weight about 36,000 Da which is our target protein thrombin. The prepared solution had a total protein content of 0.065 g/dl and gave satisfactory results of thrombin time (9 seconds) and fibrin clot formation. The test results of thrombin concentration between the method with and without incubation time were 269.4 and 295.2 IU/ml, respectively.

Conclusion:

This result showed that the method with RVV-X but without incubation time for prothrombin activation (optimum condition) gave the highest yield of thrombin.

Keywords: Preparation of Thrombin, Russell's Viper Venom Factor X Activator, Fibrinogen, Blood coagulation, Fibrin clot, Plasma.

1. INTRODUCTION

Thrombin, a serine protease that converts fibrinogen into fibrin in blood coagulation, has played a crucial role in hemostasis and thrombosis. Thrombin is activated from its zymogen, prothrombin, at the site of tissue injury by Factor Xa and its cofactor Factor Va in the presence of phospholipid membrane and calcium [1, 2]. Thrombin is a common

---

* Address correspondence to this author at the Technopreneurship and Innovation Management Program, Graduate School, Chulalongkorn University, 1405-1409 14th Floor, Chamchuri Square Building, Phayathai Road, Bangkok 10330, Thailand; E-mail: narin4k@yahoo.com

hemostatic drug used in surgical practice for over 100 years because of its simplicity and efficacy [3]. Thrombin is present in fibrin glue (also known as fibrin sealant or fibrin tissue adhesive), a plasma-derived hemostatic agent that is composed of thrombin, fibrinogen, and sometimes factor XIII and antifibrinolytic agents. This agent mimics the final steps of the physiological coagulation cascade to form a fibrin clot. Fibrin glue is an effective tissue adhesive used for reducing blood loss in a variety of surgical specialties, and it is biocompatible and biodegradable [4 - 9].

In addition, thrombin in the form of test reagents has been used for two types of blood coagulation tests: 1) fibrinogen level test; and 2) thrombin time test. Fibrinogen is measured in plasma most commonly using the Clauss method, based on the comparison of thrombin clotting times of dilutions of plasma against a reference plasma with a known level of fibrinogen. Thrombin time (or thrombin clotting time) is a widely performed coagulation assay, which evaluates the ability of fibrinogen to be converted into fibrin after the addition of bovine or human thrombin reagent to the citrated plasma [10 - 12].

Thrombin is isolated from plasma obtained from bovine or human sources [13]. The plasma is processed through a series of separation and filtration steps followed by incubation of the solution with calcium chloride to isolate and activate prothrombin to thrombin [13], it is also well known that prothrombin can be activated by some components of snake venom to yield thrombin [8]. The solution subsequently undergoes ultrafiltration, vapor heat treatment, solvent-detergent treatment, sterile filtration and freeze-drying [14]. The large-scale production of thrombin from blood plasma (batch size 1200 liters) with a high degree of virus safety can be carried out by these following steps: 1) Isolating prothrombin by the following separation techniques: cryoprecipitation, ion-exchange chromatography (diethyl amino ethyl, DEAE-IEX), heparin affinity chromatography, a second DEAE-IEX step, and Immobilized Metal-Affinity Chromatography (IMAC); 2) Activating prothrombin to thrombin, purifying by Hydrophobic Interaction Chromatography (HIC) and concentrated by ultrafiltration; 3) Reducing viral activity by using substantially different techniques, namely: solvent/detergent (S/D) treatment, pasteurization, and virus filtration (nanofiltration) [15].

Russell's viper venom factor X activator (RVV-X) is a major procoagulant in Russell's viper venom. It is a glycoprotein containing 13% carbohydrate with a molecular mass of approximately 93,000 Da, and is composed of a heavy chain (RVV-XH) of molecular mass 58,000 Da and two light chains (RVV-XL) of heterogenous molecular mass 19,000 and 16,000 Da. It directly activates factor X in the final common coagulation pathway, which leads to rapid formation of blood clots [16].

Since purified RVV-X from Russell viper (*Daboia russelli siamensis)* venom can be prepared at the Snake Bite and Venom Research Unit, Faculty of Medicine, Chulalongkorn University, Thailand [17] and it can be used for prothrombin activation. This study aims to investigate a simple method with the optimum condition for rapid thrombin preparation from human CDP using RVV-X to activate prothrombin in the process.

2. MATERIALS AND METHODS

2.1. Cryoprecipitate-depleted Plasma (CDP)

CDP, the plasma from which cryoprecipitate has been removed, was obtained from Blood Components Production Section, National Blood Centre, Thai Red Cross Society (NBC-TRCS). For the preparation of CDP, a unit of citrated whole blood was spun using a heavy spin in the centrifuge, and the plasma was removed within 6 hours after collection. Fresh Frozen Plasma (FFP) was then prepared by snap freezing the plasma unit at -70 °C and stored at below -30 °C. FFP was slowly thawed overnight at 4°C and centrifuged to separate the cryo-supernatant from the insoluble cryoprecipitate. The insoluble cryoprecipitate was removed and the remaining plasma (CDP) was refrozen.

2.2. Russell's Viper Venom Factor X Activator (RVV-X)

RVV-X, obtained from Snake Bite and Venom Research Unit, Faculty of Medicine, Chulalongkorn University, Thailand, was purified from crude *Daboia russelli siamensis* venom by a modification of the procedure of Kisiel *et al.* (1976) [18] using sequential column chromatography. The specific Factor Xa activity was 1.240 nkat/ng.

2.3. Thrombin Preparation

Thrombin was prepared by a modification procedure of Biggs R. (1972) [19]. The starting volume 100 ml of CDP was diluted to 1000 ml with distilled water, the pH was adjusted to 5.3 with 2% acetic acid and followed by centrifuge. The precipitate was dissolved in 25 ml of 0.85% sodium chloride and the pH was adjusted to 7.0 with 2% sodium carbonate. This was followed by the addition of 0.25 M calcium chloride with or without RVV-X, and incubated for full thrombin formation or without incubation. The coagulated fibrin was removed and acetone was added to the thrombin crude solution (volume 1:1) at room temperature to eliminate electrolytes and to denature some of the protein impurities [20]. The solution was centrifuged to separate the precipitate. The precipitate was finally extracted with 0.85% sodium chloride and followed by centrifuge. The supernatant was collected as thrombin solution and stored in a freezer at a temperature below -20 °C for its stability before the properties were tested.

Thrombin was prepared using the procedure described above in different conditions varying 4 related factors including: 1) RVV-X; 2) 0.25 M calcium chloride volume; 3) incubation time; and 4) 0.85% sodium chloride final extraction volume. The study compared different conditions step by step to determine the optimum preparing condition as the following methods: M1.0 vs. M2.0; M2.0 vs. M2.1; M2.0 vs. M2.2; and M2.0 vs. M2.3 (Table 1). The prepared thrombin was then further analyzed by different methods for the following properties: 1) fibrin clot formation; 2) total protein; and 3) Thrombin Clotting Time (TT). The prepared thrombin from two optimum conditions was analyzed and compared for: 1) Molecular Weight (MW) and protein patterns; and 2) thrombin concentration.

Table 1. Conditions used for thrombin preparation in the study.

Test Factors	Method	Conditions			
-	-	0.25 M CaCl₂ Volume (ml)	RVV-X (mcg)	0.85% NaCl Final Extraction Volume (ml)	Incubation Time (hours)
1) RVV-X	M1.0	3	-	25	2
	M2.0	3	50	25	2
2) 0.25 M CaCl₂	M2.0	3	50	25	2
Volume	M2.1	1.5	50	25	2
3) 0.85% NaCl Final	M2.0	3	50	25	2
Extraction Volume	M2.2	3	50	10	2
4) Incubation Time	M2.2	3	50	10	2
-	M2.3	3	50	10	-

2.4. Fibrin Clot Formation

The prepared thrombin was preliminarily tested for its activity by observation of fibrin clot formation after mixing with cryoprecipitate (volume 1:1) obtained from Blood Components Production Section, NBC-TRCS.

2.5. Total Protein Assay

Kjeldahl method (determination of nitrogen by sulfuric acid digestion) was used for total protein assay of prepared thrombin according to the method described by European Pharmacopoeia 8.0 (2014) [21].

2.6. Thrombin Clotting Time

The prepared thrombin was diluted to 1:50 with distilled water. One hundred microliters of diluted thrombin was added to 100 microliters of fresh plasma in the test tube at 37 °C. The tube was gently shaken and tilted back and forth. The time taken for the first appearance of a fibrin clot was recorded.

2.7. Molecular Weight and Protein Patterns

The Molecular Weight (MW) of the prepared thrombin was determined by sodium dodecyl sulfate-polyacrylamide gel electrophoresis (SDS-PAGE) using standard protein molecular weight markers.

2.8. Thrombin Concentration

The prepared thrombin was determined by the automated coagulation analyzer (Sysmex® CA-560) compared to Japanese Pharmacopoeia Reference Standard Thrombin (690 units/ampoule, Control: THR0104) from the Society of Japanese Pharmacopoeia, for its activity in the International Units (IU/ml).

3. RESULTS AND DISCUSSION

Thrombin solution was prepared in different conditions and the properties of produced thrombin were compared step by step, in order to study the effects of four related factors in the thrombin preparation process.

3.1. RVV-X

Prothrombin is the coagulation proenzyme present in the highest concentration blood (0.07-0.1 mg/ml) and was recognized very early as a prime contributor to the blood coagulation process [22]. By the common pathway, activated factor X (factor Xa) along with its cofactor (factor V), tissue phospholipids, platelet phospholipids and calcium forms the prothrombinase complex which converts prothrombin into thrombin [23]. In our study, factor X has been enzymatically activated by a coagulant protein (heterotrimeric metalloproteinase) [24] in Russell's viper venom (RVV-X) to help make the conversion of prothrombin more complete than adding calcium to the plasma alone.

The first study was conducted on the preparation of thrombin using Method M 1.0 and M 2.0 to compare the presence and absence of RVV-X in the preparation process. Thrombin was prepared from CDP (or cryo-supernatant), which is rich in prothrombin complex and has a low level of fibrinogen [25, 26], by the procedure described in 2.3. From starting volume 100 ml of CDP, we obtained about 25 ml of thrombin solution in the preparation batch. The solution was slightly turbid with some small particles due to no filtration step in the preparation process.

The preliminary test confirmed that there was thrombin in the prepared solutions, which can be observed from the reactions between thrombin and fibrinogen in cryoprecipitate to make the fibrin clot formation (Fig. 1).

Fig. (1). The appearance of fibrin clot after mixed the prepared thrombin with cryoprecipitate.

Total protein content analyzed by Kjeldahl method showed that the thrombin solution of Method M2.0 had higher protein content than of Method M1.0, as well as the shorter thrombin time of Method M2.0 than of Method M1.0. At this step, we concluded that the use of RVV-X in Method M2.0 resulted in a greater amount of thrombin than that of Method M1.0 without RVV-X (Table 2).

Table 2. Properties of thrombin prepared by condition M1.0 and M2.0.

Condition	Fibrin Clot Formation	Total Protein (g/dl)	Thrombin Time (s)
M1.0 (no RVV-X)	Fibrin clot formed	0.036	18.0
M2.0 (used RVV-X)	Fibrin clot formed	0.052	14.0

3.2. Calcium Chloride Volume

The reaction of coagulant protein on factor X is proteolytic, has an absolute requirement for Ca (II), and proceeds at physiological pH [27]. In this study, 0.25 M calcium chloride was reduced by half to 1.5 ml to see how it affects the preparation of thrombin.

At the half-dose of 0.25 M calcium chloride (Method M2.1), the total protein content of thrombin solution was lower than Method M2.0, and the thrombin time of Method M2.1 was twice longer than Method M2.0, indicating that the amount of thrombin prepared by Method M2.1 was much lower than of Method M2.0. In this step, it was suggested that the thrombin preparation process should have a sufficient amount of calcium in order to maximize the conversion of prothrombin (Table 3). This result corresponded to the previous study which indicated that prothrombin concentrates incubated with near-physiological levels of calcium appeared to correct the abnormal APTT to an increasing degree as the calcium concentration was increased [28].

Table 3. Properties of thrombin prepared by condition M2.0 and M2.1.

Condition	Fibrin Clot Formation	Total Protein (g/dl)	Thrombin Time (s)
M2.0 (CaCl$_2$ = 3 ml)	Fibrin clot formed	0.052	14.0
M2.1 (CaCl$_2$ = 1.5 ml)	Fibrin clot formed	0.022	29.0

3.3. Sodium Chloride Final Extraction Volume

For the next study, the different volume of 0.85% sodium chloride solution used to dissolve the final protein precipitate was compared between 25 ml (100%) in Method M2.0 and 10 ml (40%) in Method M2.2, to determine whether the sodium chloride extraction volume affected the properties of prepared thrombin or not.

The total protein content of Method M2.2 thrombin solution was higher than that of Method M2.0, and the thrombin time of Method M2.2 thrombin solution was nearly twice shorter than that of Method M2.0. These results indicated that thrombin concentration of Method M2.2 was highly concentrated, therefore resulted in higher activity than Method M2.0 (Table 4).

Table 4. Properties of thrombin prepared by condition M2.0 and M2.2.

Condition	Fibrin Clot Formation	Total Protein (g/dl)	Thrombin Time (s)
M2.0 (NaCl = 25 ml)	Fibrin clot formed	0.052	14.0
M2.2 (NaCl = 10 ml)	Fibrin clot formed	0.063	8.0

3.4. Incubation Time

After the first three studies on RVV-X, calcium and sodium chloride final extraction volume, the most appropriate method for producing thrombin was Method M2.2 (using RVV-X 50 mcg, 0.25 M calcium chloride 3 ml, 0.85% sodium chloride volume 10 ml for final extraction, and incubation time 2 hours) as it gave the highest total protein and lowest thrombin time. The next step was to compare the incubation time factor between Method M2.2 (incubation time 2 hours) and M2.3 (no incubation time) to determine whether thrombin can be generated without incubation time. Theoretically, the RVV-X, a factor X activator, is capable of promptly converting prothrombin into thrombin [29]. Therefore, it was expected that Method M2.3 would result in thrombin amount not different from that of M2.2.

The results showed that the total protein content of thrombin solutions including the thrombin time of Method M2.2 and M2.3 was very similar. However, the preparation process of Method M2.3 was more convenient as it did not require incubation time up to 2 hours (Table 5).

Table 5. Properties of thrombin prepared by condition M2.2 and M2.3.

Condition	Fibrin Clot Formation	Total Protein (g/dl)	Thrombin Time (s)
M2.2 (incubation time 2 hours)	Fibrin clot formed	0.063	8.0
M2.3 (no incubation time)	Fibrin clot formed	0.065	9.0

The MW of the prepared thrombin from two optimum conditions (Method M2.2 and M2.3) and diluted CDP were determined by SDS-PAGE using standard protein molecular weight markers (Fig. 2). From SDS-PAGE results, CDP contained several proteins including 72,000 Da prothrombin (lane C). Upon activation with calcium and RVV-X, 36,000 Da protein band which is the size of thrombin protein, appeared (lane A and B) [30, 31]. Thrombin bands on the digest of Method M2.2 (lane A) and Method M2.3 (lane B) had a similar intensity even without incubation process in Method M2.3 due to the ability of RVV-X in activated prothrombin into thrombin promptly.

At this stage, thrombin concentration has been tested using a coagulation analyzer by comparison with the standard thrombin, and reported as a thrombin unit. It was found that the thrombin concentration of Method M2.2 and M2.3 was 269.4 and 295.2 IU/ml, respectively, which indicated that Method M2.3 was the most rapid preparation method to produce thrombin and gave the highest yield of thrombin as well (Table 6).

Table 6. Properties of thrombin prepared by condition M2.3 and M2.4.

Condition	SDS-PAGE	Thrombin Concentration (IU/ml)
M2.2 (incubation time 2 hours)	Thrombin band appeared at about 36,000 Da	269.4
M2.3 (no incubation time)	Thrombin band appeared at about 36,000 Da	295.2

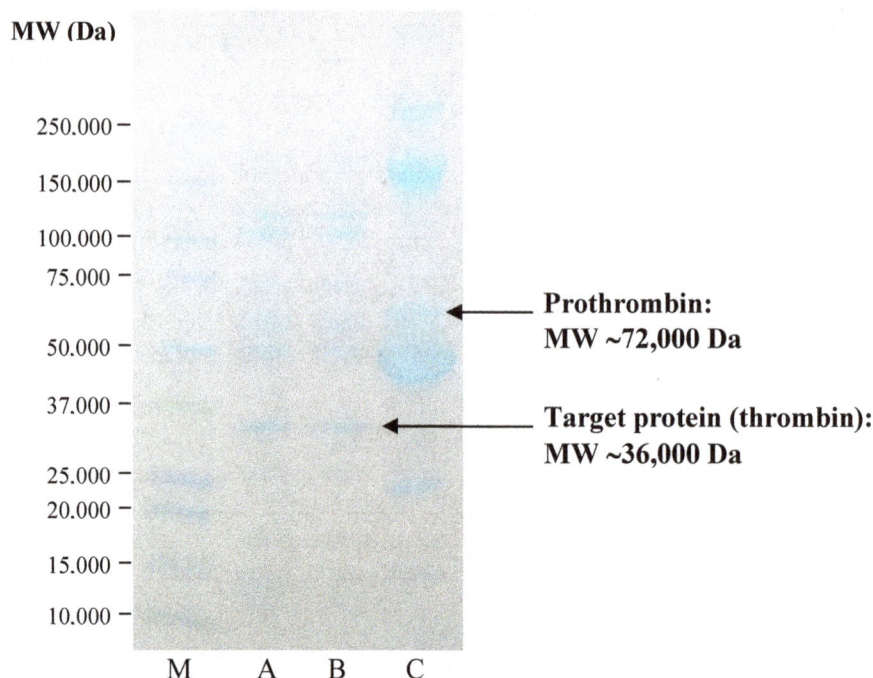

Fig. (2). SDS-PAGE of produced thrombin prepared from two optimum conditions. Lane M: molecular weight markers; Lane A: condition A (method M2.2); Lane B: condition B (method M2.3); Lane D: diluted CDP (10x).

CONCLUSION AND RECOMMENDATIONS

The method and process of preparing thrombin and the study of optimum condition for rapid preparing the highest yield of thrombin from starting CDP 100 ml were introduced. The best four conditions were concluded: 1) RVV-X 50 mcg should be present in the process; 2) volume of 0.25 M calcium chloride should be 3 ml; 3) volume of 0.85% sodium chloride for the final protein precipitate extraction should be 10 ml and 4) no incubation time needed for prothrombin activation process. The solution prepared from the optimum condition showed an obvious band on SDS-PAGE at a molecular weight about 36,000 Da which is our target protein thrombin. The prepared solution had a total protein content of 0.065 g/dl and gave satisfactory results of thrombin time (9 seconds) and fibrin clot formation. Testing comparison of thrombin concentration between the method with and without incubation time was carried out on a coagulation analyzer, and the results were 269.4 and 295.2 IU/ml, respectively. This result showed that the method with RVV-X but without incubation time for prothrombin activation (optimum condition) gave the highest yield of thrombin.

The prepared thrombin should be further purified with suitable affinity chromatography for higher purity and potency in the next study.

LIST OF ABBREVIATIONS

APTT	=	Activated Partial Thromboplastin Time
Ca	=	Calcium
CaCl$_2$	=	Calcium Chloride
CDP	=	Cryoprecipitate-depleted Plasma

Da	=	Dalton (s)
DEAE-IEX	=	Diethyl Amino Ethyl Ion-Exchange Chromatography
dl	=	Deciliter (s)
FFP	=	Fresh Frozen Plasma
g	=	Gram (s)
HIC	=	Hydrophobic Interaction Chromatography
IMAC	=	Immobilized Metal-Affinity Chromatography
IU	=	International Unit (s)
M	=	Molar
ml	=	Milliliter (s)
MW	=	Molecular Weight
NaCl	=	Sodium Chloride
NBC-TRCS	=	National Blood Centre, Thai Red Cross Society
RVV-X	=	Russell's Viper Venom Factor X Activator
s	=	Second (s)
S/D	=	Solvent/Detergent
SDS-PAGE	=	Sodium Dodecyl Sulfate Polyacrylamide Gel Electrophoresis
vs.	=	Versus

HUMAN AND ANIMAL RIGHTS

No animals/humans were used for studies that are base of this research.

ACKNOWLEDGEMENTS

The authors would like to acknowledge the Antiserum and Standard Cells Production Section of NBC-TRCS for providing equipment and instruments for our research with excellent facilities.

We also would like to extend our sincere thanks to the Quality Control Section of NBC-TRCS for testing results in this study.

REFERENCES

[1]   Fenton JW II. Thrombin specificity. Ann N Y Acad Sci 1981; 370: 468-95.
       [http://dx.doi.org/10.1111/j.1749-6632.1981.tb29757.x] [PMID: 7023326]

[2]   Suttie JW, Jackson CM. Prothrombin structure, activation, and biosynthesis. Physiol Rev 1977; 57(1): 1-70.
       [http://dx.doi.org/10.1152/physrev.1977.57.1.1] [PMID: 319462]

[3]   Diesen DL, Lawson JH. Bovine thrombin: History, use, and risk in the surgical patient. Vascular 2008; 16(Suppl 1): S29-36.
       [PMID: 18544303]

[4]   Radosevich M, Goubran HI, Burnouf T. Fibrin sealant: Scientific rationale, production methods, properties, and current clinical use. Vox Sang 1997; 72(3): 133-43.
       [http://dx.doi.org/10.1159/000461980] [PMID: 9145483]

[5]   Jackson MR, MacPhee MJ, Drohan WN, Alving BM. Fibrin sealant: Current and potential clinical applications. Blood Coagul Fibrinolysis 1996; 7(8): 737-46.
       [http://dx.doi.org/10.1097/00001721-199611000-00001] [PMID: 9034553]

[6]   Martinowitz U, Saltz R. Fibrin sealant. Curr Opin Hematol 1996; 3(5): 395-402. [PMID: 9372108]
       [http://dx.doi.org/10.1097/00062752-199603050-00011]

[7]   Tock B, Drohan W, Hess J, Pusateri A, Holcomb J, MacPhee M. Haemophilia and advanced fibrin sealant technologies. Haemophilia 1998; 4(4): 449-55.
       [http://dx.doi.org/10.1046/j.1365-2516.1998.440449.x] [PMID: 9873774]

[8]   Ngai PK, Chang JY. A novel one-step purification of human alpha-thrombin after direct activation of crude prothrombin enriched from plasma. Biochem J 1991; 280(Pt 3): 805-8.
        [http://dx.doi.org/10.1042/bj2800805] [PMID: 1764042]

[9]   Jackson MR. Fibrin sealants in surgical practice: An overview. Am J Surg 2001; 182(2)(Suppl.): 1S-7S.
        [http://dx.doi.org/10.1016/S0002-9610(01)00770-X] [PMID: 11566470]

[10]   Undas A. Determination of fibrinogen and thrombin time (TT). Methods Mol Biol 2017; 1646: 105-10.
        [http://dx.doi.org/10.1007/978-1-4939-7196-1_8] [PMID: 28804822]

[11]   Flanders MM, Crist R, Rodgers GM. Comparison of five thrombin time reagents. Clin Chem 2003; 49(1): 169-72.
        [http://dx.doi.org/10.1373/49.1.169] [PMID: 12507975]

[12]   Ignjatovic V. In: Monagle P, Ed. Thrombin Clotting Time. Haemostasis. Methods in Molecular Biology (Methods and Protocols). Humana Press: Totowa, NJ 2013; pp. 992.
        [http://dx.doi.org/10.1007/978-1-62703-339-8_10]

[13]   Norton JA, Barie PS, Bolinger RR, Chang AE, Eds. Surgery: Basic Science and Clinical Evidence, 2nd ed. Springer, New York 2008; pp. 2334.
        [http://dx.doi.org/10.1007/978-0-387-68113-9]

[14]   Ham SW, Lew WK, Weaver FA. Thrombin use in surgery: An evidence-based review of its clinical use. J Blood Med 2010; 1: 135-42.
        [https://doi.org/10.2147/JBM.S6622].
        [PMID: 22282693]

[15]   Aizawa P, Winge S, Karlsson G. Large-scale preparation of thrombin from human plasma. Thromb Res 2008; 122(4): 560-7.
        [http://dx.doi.org/10.1016/j.thromres.2007.12.027] [PMID: 18329699]

[16]   Suntravat M, Nuchprayoon I. Recombinant Russell's viper venom-factor X activator (RVV-X)-specific antibody: Neutralization and crossreactivity with *Cryptelytrops albolabris* and *Calloselasma rhodostoma* venoms. Asian Biomed 2011; 5(3): 371-9.
        [https://doi.org/10.5372/1905-7415.0503.048].

[17]   Suntravat M, Nuchprayoon I, Pérez JC. Comparative study of anticoagulant and procoagulant properties of 28 snake venoms from families Elapidae, Viperidae, and purified Russell's viper venom-factor X activator (RVV-X). Toxicon 2010; 56(4): 544-53.
        [http://dx.doi.org/10.1016/j.toxicon.2010.05.012] [PMID: 20677373]

[18]   Kisiel W, Hermodson MA, Davie EW. Factor X activating enzyme from Russell's viper venom: Isolation and characterization. Biochemistry 1976; 15(22): 4901-6.
        [http://dx.doi.org/10.1021/bi00667a023] [PMID: 990251]

[19]   Biggs R. Human Blood Coagulation, Haemostasis and Thrombosis. Oxford: Blackwell Scientific Publications 1972; pp. 587-601.

[20]   Roche MN. Studies on blood coagulation (Master's thesis, McGill University, Montreal, Canada) 1941. [cited 2018 May 11]. Available from: http://digitool.library.mcgill.ca/webclient/StreamGate?folder_id=0&dvs=1526884773042~562

[21]   European Pharmacopoeia 80, Directorate for the Quality of Medicines & Healthcare of the Council of Europe. Strasbourg: EDQM 2014; pp. 157.

[22]   Kenneth GM. In: Laszlo L, Ed. Prothrombin. Methods in Enzymology, Academic Press: New York 1976; 45: pp. 123-56.
        [https://doi.org/10.1016/S0076-6879(76)45016-4]

[23]   Palta S, Saroa R, Palta A. Overview of the coagulation system. Indian J Anaesth 2014; 58(5): 515-23.
        [http://dx.doi.org/10.4103/0019-5049.144643] [PMID: 25535411]

[24]   Takeda S, Igarashi T, Mori H. Crystal structure of RVV-X: An example of evolutionary gain of specificity by ADAM proteinases. FEBS Lett 2007; 581(30): 5859-64.
        [http://dx.doi.org/10.1016/j.febslet.2007.11.062] [PMID: 18060879]

[25]   Prohaska W, Kretschmer V. Simple method for preparation of cryoprecipitate (CP) and cryodepleted plasma (CDP). Infusionsther Klin Ernahr 1984; 11(6): 342-4.
        [PMID: 6441780]

[26]   Sparrow RL, Simpson RJ, Greening DW. In: Greening D, Simpson R, Eds. A Protocol for the Preparation of Cryoprecipitate and Cryo-depleted Plasma for Proteomic Studies. Serum/Plasma Proteomics. 1619. Methods and Protocols, Humana Press: New York 2017; 1619: pp. 23-30.

[27]   Furie BC, Furie B. Purification of proteins involved in Ca(II)-dependent protein-protein interactions (coagulant protein of Russell's viper venom). Methods Enzymol 1974; 34: 592-4.
        [http://dx.doi.org/10.1016/S0076-6879(74)34078-5] [PMID: 4449476]

[28]   Exner T, Rickard KA. The activation of prothrombin complex concentrates by calcium *in vitro*. Biomedicine (Paris) 1977; 27(2): 62-5.
        [PMID: 861352]

[29]   Vacca JP. In: Bristol JA, Ed. Thrombosis and Coagulation. Annual Reports in Medicinal Chemistry, Academic Press: San Diego 1998; 33: pp. 81-90. [https://doi.org/10.1016/S0065-7743(08)61074-X]

[30]   Jenny NS, Lundblad RL, Mann KG. In: Coleman RW, Marder VJ, Alexander WC, George JN, Goldhaber SZ. Eds. Thrombin. Hemostasis and Thrombosis: Basic Principles and Clinical Practice, 5th ed. Lippinott Williams & Wilkins: Philadelphia 2006; pp. 193-213.

[31]   Davie EW, Kulman JD. An overview of the structure and function of thrombin. Semin Thromb Hemost 2006; 32(Suppl. 1): 3-15. [http://dx.doi.org/10.1055/s-2006-939550] [PMID: 16673262]

11

Regulation of RNA Editing in Chloroplast

Wei Tang[*]

College of Horticulture and Gardening, Yangtze University, Jingzhou, Hubei 434025, China

Abstract: RNA editing is an important process involved in the modification of nucleotides in the transcripts of a large number of functional genes. RNA editing results in the restoration of conserved amino acid residues for protein function in plants. In this review, I only describe and discuss the identified RNA editing and the RNA editing associated regulation in chloroplast, including cytidine-to-uridine editing, adenosine-to-inosine editing, and regulation of RNA editing in model plants, crop plants, woody plants, and medical plants. Information described in this review could be valuable in future investigation of molecular mechanisms that determine the specificity of the RNA editing process.

Keywords: Adenosine-to-inosine editing, Chloroplasts, Cytidine-to-uridine editing, Pentatricopeptide repeat protein, RNA editing factor.

1. INTRODUCTION

Despite great progress has been made in technologies of analyzing RNA editing, critical editing factors remain to be identified in many of the plants [1]. RNA editing changes the sequence of transcripts of functional gens, the molecular mechanisms determining the specificity of the RNA editing process are not fully understood [2]. In plants, RNA editing factors and sequence-specific RNA maturation factors promote RNA editing and translatability [3]. Two types of RNA editing: cytidine-to-uridine editing in mRNAs and adenosine-to-inosine editing, have been identified in the model plant *Arabidopsis thaliana* [4]. RNA editing is a relatively rare molecular process that changes nucleotide sequences of a RNA molecule through the insertion, deletion, and base substitution of nucleotides within the edited RNA molecule under the function of RNA polymerase [5 - 8]. RNA editing has been reported in some tRNA, rRNA, and mRNA. However, RNA editing has also been discovered in non-coding RNA, such as microRNAs (miRNAs) molecules of eukaryotes. Micro RNAs are a large family of small RNAs with important functional roles in regulating gene expression in animals and plants [9 - 12]. In human and mouse, miRNA22 precursor molecules are subject to post-transcriptional modification by A-to-I RNA editing *in vivo*. A-to-I RNA editing of miRNA22 is predicted to have significant implications for the biogenesis and function of miRNA22. RNA editing of miRNA gene products could also take place in plants. In Arabidopsis, RNA editing in the precursor microRNAs of ath-miR854 family has a potential role in miRNA maturation [13 - 15].

Although RNA editing is an essential post-transcriptional modification in chloroplast gene expression, the factors mediating those processes are not fully identified in plants. It has been reported that the large Pentatricopeptide Repeat (PPR) protein family are required for RNA editing in chloroplasts [5].Chloroplasts contain their own genetic system. Most chloroplast genes are organized in clusters and are co-transcribed. Posttranscriptional RNA editing of chloroplast gene transcripts is an important step in the control of chloroplast gene expression and is required for gene function [6]. Considering RNA editing of nuclear gene transcripts has been widely reviewed, here I only review RNA editing and the RNA editing associated regulation in chloroplast (Table **1**).

* Address correspondence to this author at the College of Horticulture and Gardening, Yangtze University, Jingzhou, Hubei 434025, China; E-mail: wt10yu604@gmail.com

Table 1. RNA Editing in chloroplasts of different plant species.

Plant Species	Editing Type	Editing Factors	Edited Transcripts	Reference
Arabidopsis	C-to-U	CLB19	rpoA and clpP	[2]
	C-to-U	OTP80, OTP81, OTP85, and OTP86	CRR22, CRR28, CLB19, OTP82, and OTP84,	[15]
	A-to-I	tRNA adenosine deaminase arginine (TADA)	Chloroplast transcripts	[4]
	C-to-U	CRR22 and CRR28	Organelle transcripts	[16]
	C-to-U	Quintuple editing factor 1 (QED1) and RARE1	Chloroplast C targets	[23]
	C-to-U	PDM1/SEL1 MORF9, MORF2, and MORF8	Transcripts of trnK, ndhA, and accD-1	[24]
	C-to-U	PDM2, organellar RNA editing factor 2 (MORF2), and MORF9	Chloroplast transcripts	[26]
	C-to-U	ORRM6, RIP1/MORF8, RIP2/MORF2, and RIP9/MORF9	psbF-C77 and accD-C794	[25,26]
Tobacco	C-to-U	Pentatricopeptide repeat protein	rpoB, psbL, and rps14	[9,32]
	C-to-U	Pentatricopeptide repeat protein	Plastid ATPase alpha-subunit transcript	[31]
	C-to-U	Pentatricopeptide repeat protein	RNA-dependent RNA polymerase (RdRp)	[30]
	C-to-U	Pentatricopeptide repeat protein	psaC and ndhD	[28,29]
	C-to-U	Pentatricopeptide repeat protein	The asRNA and ndhB transcripts	[26,36]
Poplar	C-to-U	Pentatricopeptide repeat protein	The asRNA_ndhB transcripts	[11]
Spinach	C-to-U	Pentatricopeptide repeat protein	psbE, psbF, psbL and psbJ transcripts	[11]
Pea	C-to-U	Pentatricopeptide repeat protein	Acetyl coA carboxylase gene transcripts	[10]
Rice	C-to-U	CLB19, CRR4 and CRR21	Transcripts of rpoA and clpP	[2]
	C-to-U	CRR4, CRR21, and CLB19	Acetyl coA carboxylase gene transcripts	[10,33]
Potato	C-to-U	CLB19, RpoA, and ClpP1	Transcript of accD	[33,34]
Maize	C-to-U	Pentatricopeptide repeat protein	Transcript ofchloroplast ATP synthase gene	[5,34]
Sweet potato	C-to-U	Pentatricopeptide repeat protein	Chloroplast functional-gene transcript	[34]
Bell pepper	C-to-U	Pentatricopeptide repeat protein	Transcripts of the plastid psbL gene	[27,36]
Mustard	C-to-U	Pentatricopeptide repeat protein	psaA transcripts	[37]
Tomato	C-to-U	Pentatricopeptide repeat protein	accD transcripts	[38]
Larix decidua	C-to-U	Pentatricopeptide repeat protein	Protochlorophyllide oxidoreductase	[40]
Pinus sylvestris	C-to-U	Pentatricopeptide repeat protein	The chlB transcript	[39]
Picea abies	C-to-U	Pentatricopeptide repeat protein	Protochlorophyllide oxidoreductase gene transcripts	[40]
Dianthus superbus	C-to-U	Pentatricopeptide repeat protein	Transcripts of ribosomal protein subunit S19	[44]
Chlamydomonas reinhardtii	C-to-U	OPR, MCG1, and MBI1	petG transcript, PsbI transcripts	[45]
Rhazya stricta	C-to-U	Tetrapyrroles and pentatricopeptide repeat proteins	Dnaj6, UDP-glucosyl transferase 85a2 gene transcripts	[46]

2. RNA EDITING IN CHLOROPLAST

2.1. C-to-U Editing

RNA editing is an important process of gene regulation through nucleotide modification at post-transcriptional level. Many transcripts expressed in chloroplasts are modified by C-to-U RNA editing. C-to-U editing is the major type of RNA editing discovered in the chloroplasts of plants including Arabidopsis, tobacco, maize, rice, and potato. In maize chloroplasts, 27 C-to-U RNA editing sites have been identified to affect the expression of at least 15 different genes in 10 different maize tissues including chloroplasts, etioplasts, and amyloplasts. The role of editing in chloroplasts is to correct detrimental mutations rather than to produce protein diversity [7]. Although C-to-U editing modifies the chloroplast genes in maize, the mechanism for recognition of the targets of editing is not fully understood [8]. It has been reported that several nuclear-encoded proteins containing pentatricopeptide repeat motifs have been essential for chloroplast RNA editing through analysis of mutants affected in chloroplast biogenesis in Arabidopsis and rice. Many PPR proteins include a C-terminal DYW deaminase domain and the zinc binding motifs [9]. For example, the Arabidopsis nuclear-encoded gene RARE1 is required for editing of the chloroplast accD transcript [10]. The DYW deaminase domains of PPR proteins are involved in editing their cognate editing sites in chloroplasts [9]. C-to-U editing modification does not change the genome DNA sequence. RNA editing changes specific nucleotide sequences within a RNA molecule. RNA editing has generated by RNA polymerase [7, 8, 10]. RNA editing in mRNAs effectively alters the amino acid sequence of the encoded protein so that it differs from that predicted by the genomic DNA sequence. However, RNA editing including C-to-U editing modification does not change the genome DNA sequence. The events of RNA editing include the insertion, deletion, and base substitution of nucleotides within the edited RNA molecule. Therefore, RNA editing can serves as a safeguard because it can reverse harmful genomic mutations in corresponding RNA transcripts. Genomic studies in humans have uncovered a large number of RNA editing sites. Because cells employ RNA editing mechanisms to correct mistakes made during DNA replication, RNA editing can mean correcting detrimental mutations.

Computational prediction of RNA secondary structures in chloroplasts of Arabidopsis, *Nicotiana tabacum*, and poplar identified a long antisense RNA that is typical for the plastid-encoded RNA polymerase gene ndhB. The asRNA_ndhB transcripts accumulate in young leaves and are subject to C-to-U RNA editing [11]. In some cases, C-to-U RNA editing enzymes may participate in editing different sites in the chloroplast [9]. In the chloroplast DNA of moth orchid (*P. aphrodite* subsp. Formosana), 137 edits including 126 C-to-U and 11 U-to-C conversions were identified. 110 and 106 edits were present in leaf and floral tissues, respectively. RNA editing occurred in non-protein-coding transcripts such as tRNA, introns and some regulatory regions [12].

2.2. A-to-I Editing

Although cytidine-to-uridine editing has been reported to be the majority of RNA editing in the chloroplasts of plants, another type of RNA editing: adenosine-to-inosine editing in a plastid genome-encoded tRNA has been characterized in the chloroplasts of the model plant *Arabidopsis thaliana*. A-to-I editing was identified in the anticodon of the plastid tRNA-Arg(ACG). AtTadA gene expression is involved in A-to-I editing in the chloroplast [4].

3. RNA EDITING IN MODEL PLANTS

3.1. Arabidopsis

A large number of nucleus-encoded factors discovered *via* biochemical approaches in Arabidopsis may regulate plastid gene expression through RNA editing. The nucleus-encoded factors may modulate chloroplast gene expression in response to environmental cues *via* the assembly of the protein products into the photosynthetic apparatus [3]. The discovery of genes responsible for the specific RNA editing in the chloroplast *via* genetic approach is important in Arabidopsis [1]. The pentatricopeptide repeat proteins CLB19, CRR4, and CRR21 are required for editing of chloroplast transcripts rpoA and clpP that are important in chloroplast development and early seedling development [2]. Using the transgenomic suppression of point mutations, additional nuclear-encoded components have been found in chloroplast, indicating that several chloroplast-specific mechanisms evolved in the regulatory functions of the complexity of chloroplast gene expression in Arabidopsis [13]. Biochemical and mutant complementation studies showed that the Arabidopsis thaliana tRNA adenosine deaminase arginine (TADA) was required for deamination of chloroplast (cp)-tRNAArg(ACG) to cp-tRNAArg (ICG) and disruption of TADA reduced yields of chloroplast-encoded proteins and impaired photosynthetic function [14].

Analysis of mutants in chloroplast biogenesis resulted in the identification of several Pentatricopeptide Repeat (PPR) proteins that are essential for the editing of RNA transcripts in *Arabidopsis thaliana*. The editing factors OTP80, OTP81, OTP82, OTP84, OTP85, and OTP86 participate in more than 17 editing events and are involved in the editing of multiple sites CRR22, CRR28, CLB19, OTP82, and OTP84 in Arabidopsis chloroplasts [15]. A-to-I editing in the anticodon of the plastid tRNA-Arg(ACG) reduced chloroplast translational efficiency [4]. Different DYW family members have different functions in RNA editing. The DYW proteins CHLORORESPIRATORY REDUCTION22 and CRR28 are necessary for editing of multiple plastid transcripts [16]. Chloroplast ribonucleoproteins (cpRNPs) are involved in chloroplast RNA processing in chloroplasts and the cpRNP family member CP31A exhibits highly specific defects in chloroplast RNA metabolism, suggesting that these chloroplast proteins are functional equivalents of nucleocytosolic hnRNPs [17]. The AtECB2 gene codes a pentatricopeptide repeat protein with a C-terminal DYW domain that is required for chloroplast transcript accD RNA editing and early chloroplast biogenesis in *Arabidopsis thaliana* [18].

RNA editing sites found in chloroplasts of Arabidopsis could be involved in RNA maturation [11]. The editing of accD and ndhF transcripts is partially affected by the vanilla cream1 (vac1) albino mutant, suggesting that the VAC1 protein may be involved in plastid-to-nucleus retrograde signaling in addition to its role in chloroplast RNA editing and gene expression [19]. Functional significance of these processing events is an active area of current research [20]. SEL1 is a pentatricopeptide repeat gene that is involved in the regulation of plastid gene expression required for normal chloroplast development [21]. SVR7 gene that encodes a PPR-SMR protein in Arabidopsis thaliana is involved in translational activation of chloroplast ATP synthase in eudicotyledonous and monocotyledonous plants [5]. Many PPR proteins with a C-terminal DYW deaminase domain and zinc binding motifs are required in C-to-U editing. For example, OTP84 is required for editing three chloroplast sites in *Arabidopsis thaliana* [9].

CLB19 is essential for the editing and functionality of the subunit A of plastid-encoded RNA polymerase (RpoA) and the catalytic subunit of the Clp protease (ClpP1) [22]. The chloroplast PPR-DYW editing factor, Quintuple Editing factor 1 (QED1), was shown to affect five different plastid editing sites and is required for editing efficiency [23]. PDM1 interacts directly with MORF9, MORF2, and MORF8 to edit the transcripts of trnK and ndhA, as well as accD-1, suggesting that PDM1 is an important protein for post-transcriptional regulation in chloroplast [24]. PDM2 that encodes a pentatricopeptide repeat protein that interacts with Multiple Organellar RNA editing Factor 2 (MORF2) and MORF9 to affect plastid RNA editing efficiency [25]. ORRM6 (ORGANELLE RNA RECOGNITION MOTIF PROTEIN6) interacts with RIP1/MORF8, RIP2/MORF2, and RIP9/MORF9 to regulate RNA editing of psbF-C77 and accD-C794 in Arabidopsis [26].

Mutation stress, DNA duplication, ABA stress, abiotic stresses such as salt, drought, and low temperature stress can induce Pentatricopeptide Repeat (PPR) proteins [1, 2, 25]. Horizontal gene transfer has played a major role in the sporadic phylogenetic distribution of different PPR subclasses in both eukaryotes and prokaryotes. Pentatricopeptide repeat (PPR) proteins are a large family of RNA-binding proteins that regulate gene expression in organelles and the nucleus [1, 25]. PPRs have functions in RNA processing, splicing, editing, stability and translation. PPR proteins are classified into different subclasses based on their domain architecture and their function. Investigation into the structure and function of PPR-TGM proteins presents a novel opportunity for the exploitation of PPR proteins as drug targets to prevent disease [2, 25]. Functional and evolutionary studies provide insights into the different PPR subfamilies and their phylogenetic distributions will facilitate the understanding of the eukaryotic PPR protein families identified to date. However, mitochondrial and chloroplast Pentatricopeptide Repeat (PPR) proteins have been reported to regulate plant response to abiotic stresses [1, 2, 25].

3.2. Tobacco

RNA editing alters the transcript levels of tobacco chloroplast genes. Sequence analysis revealed 13 C-to-U editing sites in transcripts of 11 tobacco chloroplast genes. For example, the genes psbE, psbF, psbL and psbJ, encode two RNA editing sites. RNA editing process is differentially down-regulated in leucoplasts and proplastids and may function as a regulatory device in plastid gene expression [27]. In tobacco chloroplasts, RNA factors and RNA-binding proteins are involved in RNA editing and the regulation of gene expression in chloroplasts [6]. Transcripts of two tobacco chloroplast genes, the photosystem I component (psaC) and the NADH dehydrogenase subunit (ndhD), are modified through C-to-U editing [28]. Editing in tobacco rpoA mRNA restores the conserved leucine residue which is known to be important for transcriptional activation, suggesting that editing may be involved in the regulation of chloroplast-encoded RNA polymerase activity [29]. In the ndhD transcript of tobacco, editing of an ACG codon to a standard AUG initiator codon is a prerequisite for translation. The presence of an RNA-dependent RNA polymerase (RdRp) activity may have general implications in plastid gene expression [30].

The expression of tobacco chloroplast genes is modified by C-to-U RNA editing. It is not known whether complementary RNA is involved in chloroplast editing site recognition. Expressing RNA antisense to the sequences -20 to +6 surrounding the RpoB-2 C target of editing had shown that transcripts carrying sequences -31 to +60 surrounding the RpoB-2 sites were edited in chloroplast transgenic plants [8]. A defect in RNA editing of a tobacco-specific editing site in the plastid ATPase alpha-subunit transcript results in the albino phenotype of Nicotiana tabacum, suggesting that differences in RNA editing patterns contribute to the pigment deficiencies [31]. In tobacco chloroplast transcripts 34 nt are efficiently edited to U. Transgene transcripts carrying either the wild-type -31/+22 or -31/+60 sequence near NTrpoB C473, an editing site within tobacco rpoB transcripts, were highly edited *in vivo* and the transcripts of rpoB, psbL and rps14 are critical for efficient NTrpoB C473 editing [32]. Two nucleotide positions in the asRNA_ndhB transcripts that predominantly accumulate in young leaves are subject to C-to-U RNA editing and the accumulation of asRNA_ndhB and RNA editing appeared weak in a temperature shift experiment, suggesting that long asRNAs could be involved in RNA stability [11].

4. RNA EDITING IN CROP PLANTS

4.1. Rice

RNA editing changes the sequence of many transcripts in rice. The rice CLB19, which encodes a pentatricopeptide repeat protein, edits two chloroplast transcripts, rpoA and clpP [2]. Biochemical studies showed that the C-terminal domain is sufficient for tRNA deamination in plants [14]. Proteins containing pentatricopeptide repeat motifs have been identified to be trans-factors essential for chloroplast RNA editing in rice. Rice contains over 400 PPR genes including 80 C-terminal DYW domains. The rice editing factors CRR4, CRR21, and CLB19 are required for editing of the chloroplast accD transcript [10]. Gene expression in nongreen plastids is largely uncharacterized. Transcripts of photosynthesis-related genes showed a greater reduction in nongreen plastids compared with leaves. The transcripts of the fatty acid biosynthesis gene accD, displayed relatively high ribosome association [33]. Results of site-directed mutagenesis of the rpoA target demonstrated that the E domain of CLB19 interacts with the RNA-interacting protein MORF2/RIP2 for the editing and functionality of the subunit A of plastid-encoded RNA polymerase (RpoA) [22].

4.2. Maize

The expression of the maize plastid genome is dependent on many nucleus-encoded factors. The maize nucleus-encoded factors function in plastid gene expression and RNA editing [3]. In maize plastids, transcripts are modified at 27 C-to-U RNA editing sites and expression of 15 different genes is affected. The editing efficiency of different sites postulates individual trans-acting factors specific to each editing site [7]. RNA editing is an essential post-transcriptional step in chloroplast gene expression. The pentatricopeptide repeat proteins are required for chloroplast RNA editing. For example, maize ATP4 gene functions in maize chloroplast RNA editing [5]. The organization and structure of the chloroplast genome of maize were compared with other species and some gene gain-and-loss events were identified. In addition, the RNA editing events and differential expressions of the chloroplast functional-genes were detected [34].

4.3. Bell Pepper

The nucleotide sequence of the plastid psbL gene from bell pepper is an example of RNA editing. Sequencing of the psbL cDNA revealed that ACG codon is post-transcriptionally edited into an AUG initiation codon in leaves, indicating that the RNA editing machinery exists in bell pepper chloroplasts [35]. RNA editing in chloroplasts proceeds by the conversion of individual cytidine residues to uridine. The discovery of RNA editing in chloroplasts has provided researchers with a wealth of molecular puzzles and with extraordinarily high precision [36].

4.4. Mustard

The mustard chloroplast ycf3 gene was investigated for RNA editing. The ycf3 gene revealed two class-II introns that were removed, but no RNA editing seemed to be involved. Whereas transcripts of ycf3 and psaA initiated from PEP promoters, as well as transcripts of photosystem core-protein genes were edited [37].

4.5. Tomato

Transcriptomics and translatomics analysis of the tomato (*Solanum lycopersicum*) plastid genome demonstrated that RNA editing exists in plastid genes in fruits and leaves. Sequencing analysis during chromoplast development demonstrated that transcripts of the plastid-encoded gene, accD, which is involved in fatty acid biosynthesis, has the developmental patterns of RNA editing through specific developmental changes in RNA editing efficiency [38].

5. RNA EDITING IN WOODY PLANTS

5.1. *Pinus sylvestris*

RNA editing was less investigated in gymnosperms than in angiosperms. In pine species, RNA editing was first identified in the chlB transcript of *Pinus sylvestris*. ChlB is a chloroplast gene required for light-independent chlorophyll synthesis. Two C-to-U editing sites have been identified in the transcripts of ChlB. The editing of a CCG codon leads to an amino-acid substitution from proline to leucine and the editing of a CGG codon results in an arginine to tryptophan substitution [39].

5.2. *Larix decidua*

Light-independent Chlorophyll (Chl) biosynthesis is important for the photosynthetic pigment-protein complexes. Expression of CHl genes is regulated by RNA editing. In *Larix decidua* Mill seedlings, C-to-U RNA editing exists in the transcripts of the Chl biosynthesis genes chlL, chlN and chlB that encode subunits of the light-independent protochlorophyllide oxidoreductase. The accumulation of the ChlB subunit was developmentally regulated and the efficiency of chlB RNA-editing was reduced in the mature dark-grown larch seedlings. RNA editing was also identified in seedlings of *Picea abies* (L.) Karst [40]. Expression of plastid genes showed organelle-specific elaborations on a prokaryotic scaffold [41]. The nucleus provides essential factors that are involved in many processes inside the chloroplast including RNA editing that regulates the expression and assembly of the photosynthetic thylakoid membrane complexes [42]. Expression of most plastid genes involves specific post-transcriptional processing events, such as RNA editing. Expression of the psbB gene cluster and RNA editing factors stabilize many of the processed RNA transcripts in the chloroplast [43].

6. RNA EDITING IN MEDICAL PLANTS

6.1. *Dianthus superbus*

RNA editing has been identified in a few of medical plants. In *Dianthus superbus* var. longicalycinus, the ribosomal protein subunit S19 (rps19), the translation initiation factor IF-1 (infA), and ribosomal protein subunit L23 (rpl23) genes were contributed to the molecular biology and genetic engineering of RNA editing [44].

6.2. *Chlamydomonas reinhardtii*

RNA editing of transcripts of chloroplast genes has been identified in *Chlamydomonas reinhardtii*. MCG1 and MBI1, which code for subunit of photosystem II, are editing related factors that are essential for functional cytochrome b6f dimer and are involved in translation of plastid mRNAs [45].

6.3. *Rhazya stricta*

A large number of transcripts including Pentatricopeptide Repeat (PPR) protein genes have been identified in *Rhazya stricta* through RNA-Seq analysis. Chaperone protein Dnaj6 and protein transparent testa 12 are related to abiotic responses and participate Reactive Oxygen Species (ROS) production. They are targeted editing genes of the PPR gene family that were independent of the salt stress due to their RNA editing patterns were unchanged [46].

CONCLUSION

We have overviewed the identified RNA editing and the RNA editing associated regulation in chloroplast, including cytidine-to-uridine editing, adenosine-to-inosine editing, and regulation of RNA editing in model plants, crop plants, woody plants, and medical plants. RNA editing is conducted by an editing complex called editosome that includes the deaminase [29, 32, 36]. The mechanism of the editosome includes an endonucleolytic cut at the mismatch point between the guide RNA and the unedited transcript. After the endonucleolytic cut, a terminal U-transferase adds U from UTP at the 3' end of the mRNA and a U-specific exoribonuclease removes the unpaired base, followed by an RNA ligase to rejoin the ends of the edited mRNA transcript. The RNA editing complex can use only a single guide RNA at a time [29, 36]. Therefore, more than one guide RNA and editosome complex will be needed when a RNA transcript requires extensive editing. The RNA editing process has been observed in some tRNA, rRNA, mRNA, or miRNA molecules of eukaryotes [32, 36].

RNA editing occurs in the cell nucleus and cytosol, as well as within mitochondria and plastids. Editing of mRNA or other kinds of RNA is made to individual nucleotides by changing the base-pairing potential through the process of translation that involves base pairing between mRNA and tRNAs [29, 32, 36]. Editing of the mRNA occurs at the 3' end of the mRNA and may change the amino acid of the protein to increase the diversity of protein products that can be synthesized from the genome [29, 32]. Cells may also use different mechanisms that insert and delete nucleotides from RNAs in some eukaryotes. An example is in the mitochondria of trypanosomes [29, 36]. Uracil residues are added, or deleted from the mitochondrial mRNAs at many sites through the interaction of a large number of small guide RNAs. The best-characterized example of C-to-U RNA editing is the editing of the mRNA of apolipoprotein B that generates a stop codon and leads to down-regulation of protein expression [29, 32, 36].

RNA editing occurs during the process of transcription and the process of mRNA splicing in the nucleus, chloroplasts, or mitochondria [32, 36]. The process of RNA editing comes in a myriad of different biological and evolutional natures across the phylogenetic spectrum. RNA editing is a biochemical process that cells can make discrete changes to specific nucleotide sequences within a RNA molecule [29, 36]. RNA editing makes distinct changes to RNA after its synthesis and before the translation of the mRNA into a protein. In addition to specific editing of individual nucleotides in the nucleus, mitochondria, and plastids, RNA editing is also a dynamic mechanism that generates molecular and functional diversity [29, 32, 36]. Knowledge described in this review could be valuable in determining the specificity of the RNA editing process.

ACKNOWLEDGEMENTS

The authors are grateful to Wells Thompson for critical reading of the manuscript. This research was supported the National Natural Science Foundation of China (31270740).

REFERENCES

[1]   Kotera E, Tasaka M, Shikanai T. A pentatricopeptide repeat protein is essential for RNA editing in chloroplasts. Nature 2005; 433(7023): 326-30.
        [http://dx.doi.org/10.1038/nature03229] [PMID: 15662426]

[2]   Chateigner-Boutin AL, Ramos-Vega M, Guevara-García A, et al. CLB19, a pentatricopeptide repeat protein required for editing of rpoA and clpP chloroplast transcripts. Plant J 2008; 56(4): 590-602.
        [http://dx.doi.org/10.1111/j.1365-313X.2008.03634.x] [PMID: 18657233]

[3]   Barkan A, Goldschmidt-Clermont M. Participation of nuclear genes in chloroplast gene expression. Biochimie 2000; 82(6-7): 559-72.
        [http://dx.doi.org/10.1016/S0300-9084(00)00602-7] [PMID: 10946107]

[4]   Karcher D, Bock R. Identification of the chloroplast adenosine-to-inosine tRNA editing enzyme. RNA 2009; 15(7): 1251-7.
        [http://dx.doi.org/10.1261/rna.1600609] [PMID: 19460869]

[5]   Zoschke R, Qu Y, Zubo YO, Börner T, Schmitz-Linneweber C. Mutation of the pentatricopeptide repeat-SMR protein SVR7 impairs accumulation and translation of chloroplast ATP synthase subunits in Arabidopsis thaliana. J Plant Res 2013; 126(3): 403-14.
        [http://dx.doi.org/10.1007/s10265-012-0527-1] [PMID: 23076438]

[6]   Sugita M, Sugiura M. Regulation of gene expression in chloroplasts of higher plants. Plant Mol Biol 1996; 32(1-2): 315-26.
        [http://dx.doi.org/10.1007/BF00039388] [PMID: 8980485]

[7]   Peeters NM, Hanson MR. Transcript abundance supercedes editing efficiency as a factor in developmental variation of chloroplast gene expression. RNA 2002; 8(4): 497-511.
        [http://dx.doi.org/10.1017/S1355838202029424] [PMID: 11991643]

[8]   Hegeman CE, Halter CP, Owens TG, Hanson MR. Expression of complementary RNA from chloroplast transgenes affects editing efficiency of transgene and endogenous chloroplast transcripts. Nucleic Acids Res 2005; 33(5): 1454-64.
        [http://dx.doi.org/10.1093/nar/gki286] [PMID: 15755747]

[9]   Hayes ML, Dang KN, Diaz MF, Mulligan RM. A conserved glutamate residue in the C-terminal deaminase domain of pentatricopeptide repeat proteins is required for RNA editing activity. J Biol Chem 2015; 290(16): 10136-42.
        [http://dx.doi.org/10.1074/jbc.M114.631630] [PMID: 25739442]

[10]   Robbins JC, Heller WP, Hanson MR. A comparative genomics approach identifies a PPR-DYW protein that is essential for C-to-U editing of the Arabidopsis chloroplast accD transcript. RNA 2009; 15(6): 1142-53.
        [http://dx.doi.org/10.1261/rna.1533909] [PMID: 19395655]

[11]   Georg J, Honsel A, Voss B, Rennenberg H, Hess WR. A long antisense RNA in plant chloroplasts. New Phytol 2010; 186(3): 615-22.
        [http://dx.doi.org/10.1111/j.1469-8137.2010.03203.x] [PMID: 20202127]

[12]   Chen TC, Liu YC, Wang X, Wu CH, Huang CH, Chang CC. Whole plastid transcriptomes reveal abundant RNA editing sites and differential editing status in Phalaenopsis aphrodite subsp. formosana. Bot Stud (Taipei, Taiwan) 2017; 58(1): 38.
        [http://dx.doi.org/10.1186/s40529-017-0193-7] [PMID: 28916985]

[13]   Maier UG, Bozarth A, Funk HT, et al. Complex chloroplast RNA metabolism: just debugging the genetic programme? BMC Biol 2008; 6: 36.
        [http://dx.doi.org/10.1186/1741-7007-6-36] [PMID: 18755031]

[14]   Delannoy E, Le Ret M, Faivre-Nitschke E, et al. Arabidopsis tRNA adenosine deaminase arginine edits the wobble nucleotide of chloroplast tRNAArg(ACG) and is essential for efficient chloroplast translation. Plant Cell 2009; 21(7): 2058-71.

[http://dx.doi.org/10.1105/tpc.109.066654] [PMID: 19602623]

[15]   Hammani K, Okuda K, Tanz SK, Chateigner-Boutin AL, Shikanai T, Small I. A study of new Arabidopsis chloroplast RNA editing mutants reveals general features of editing factors and their target sites. Plant Cell 2009; 21(11): 3686-99.
[http://dx.doi.org/10.1105/tpc.109.071472] [PMID: 19934379]

[16]   Okuda K, Chateigner-Boutin AL, Nakamura T, et al. Pentatricopeptide repeat proteins with the DYW motif have distinct molecular functions in RNA editing and RNA cleavage in Arabidopsis chloroplasts. Plant Cell 2009; 21(1): 146-56.
[http://dx.doi.org/10.1105/tpc.108.064667] [PMID: 19182104]

[17]   Tillich M, Hardel SL, Kupsch C, et al. Chloroplast ribonucleoprotein CP31A is required for editing and stability of specific chloroplast mRNAs. Proc Natl Acad Sci USA 2009; 106(14): 6002-7.
[http://dx.doi.org/10.1073/pnas.0808529106] [PMID: 19297624]

[18]   Yu QB, Jiang Y, Chong K, Yang ZN. AtECB2, a pentatricopeptide repeat protein, is required for chloroplast transcript accD RNA editing and early chloroplast biogenesis in Arabidopsis thaliana. Plant J 2009; 59(6): 1011-23.
[http://dx.doi.org/10.1111/j.1365-313X.2009.03930.x] [PMID: 19500301]

[19]   Tseng CC, Sung TY, Li YC, Hsu SJ, Lin CL, Hsieh MH. Editing of accD and ndhF chloroplast transcripts is partially affected in the Arabidopsis vanilla cream1 mutant. Plant Mol Biol 2010; 73(3): 309-23.
[http://dx.doi.org/10.1007/s11103-010-9616-5] [PMID: 20143129]

[20]   Barkan A. Studying the structure and processing of chloroplast transcripts. Methods Mol Biol 2011; 774: 183-97.
[http://dx.doi.org/10.1007/978-1-61779-234-2_12] [PMID: 21822840]

[21]   Pyo YJ, Kwon KC, Kim A, Cho MH. Seedling Lethal1, a pentatricopeptide repeat protein lacking an E/E+ or DYW domain in Arabidopsis, is involved in plastid gene expression and early chloroplast development. Plant Physiol 2013; 163(4): 1844-58.
[http://dx.doi.org/10.1104/pp.113.227199] [PMID: 24144791]

[22]   Ramos-Vega M, Guevara-García A, Llamas E, et al. Functional analysis of the Arabidopsis thaliana CHLOROPLAST BIOGENESIS 19 pentatricopeptide repeat editing protein. New Phytol 2015; 208(2): 430-41.
[http://dx.doi.org/10.1111/nph.13468] [PMID: 25980341]

[23]   Wagoner JA, Sun T, Lin L, Hanson MR. Cytidine deaminase motifs within the DYW domain of two pentatricopeptide repeat-containing proteins are required for site-specific chloroplast RNA editing. J Biol Chem 2015; 290(5): 2957-68.
[http://dx.doi.org/10.1074/jbc.M114.622084] [PMID: 25512379]

[24]   Zhang HD, Cui YL, Huang C, et al. PPR protein PDM1/SEL1 is involved in RNA editing and splicing of plastid genes in Arabidopsis thaliana. Photosynth Res 2015; 126(2-3): 311-21.
[http://dx.doi.org/10.1007/s11120-015-0171-4] [PMID: 26123918]

[25]   Du L, Zhang J, Qu S, et al. The pentratricopeptide repeat protein pigment-defective mutant2 is involved in the regulation of chloroplast development and chloroplast gene expression in arabidopsisplant cell Physio 2017; 58(2-3): 747-59.
[http://dx.doi.org/10.1093/pcp/pcx004]

[26]   Hackett JB, Shi X, Kobylarz AT, et al. An organelle RNA recognition motif protein Is required for photosystem II subunit psbF transcript editing. Plant Physiol 2017; 173(4): 2278-93.
[PMID: 28213559]

[27]   Bock R, Hagemann R, Kössel H, Kudla J. Tissue- and stage-specific modulation of RNA editing of the psbF and psbL transcript from spinach plastids--a new regulatory mechanism? Mol Gen Genet 1993; 240(2): 238-44.
[http://dx.doi.org/10.1007/BF00277062] [PMID: 8355656]

[28]   Hirose T, Sugiura M. Both RNA editing and RNA cleavage are required for translation of tobacco chloroplast ndhD mRNA: a possible regulatory mechanism for the expression of a chloroplast operon consisting of functionally unrelated genes. EMBO J 1997; 16(22): 6804-11.
[http://dx.doi.org/10.1093/emboj/16.22.6804] [PMID: 9362494]

[29]   Hirose T, Kusumegi T, Tsudzuki T, Sugiura M. RNA editing sites in tobacco chloroplast transcripts: editing as a possible regulator of chloroplast RNA polymerase activity. Mol Gen Genet 1999; 262(3): 462-7.
[http://dx.doi.org/10.1007/s004380051106] [PMID: 10589833]

[30]   Zandueta-Criado A, Bock R. Surprising features of plastid ndhD transcripts: addition of non-encoded nucleotides and polysome association of mRNAs with an unedited start codon. Nucleic Acids Res 2004; 32(2): 542-50.
[http://dx.doi.org/10.1093/nar/gkh217] [PMID: 14744979]

[31]   Schmitz-Linneweber C, Kushnir S, Babiychuk E, Poltnigg P, Herrmann RG, Maier RM. Pigment deficiency in nightshade/tobacco cybrids is caused by the failure to edit the plastid ATPase alpha-subunit mRNA. Plant Cell 2005; 17(6): 1815-28.
[http://dx.doi.org/10.1105/tpc.105.032474] [PMID: 15894714]

[32]   Hayes ML, Reed ML, Hegeman CE, Hanson MR. Sequence elements critical for efficient RNA editing of a tobacco chloroplast transcript in vivo and in vitro. Nucleic Acids Res 2006; 34(13): 3742-54.
[http://dx.doi.org/10.1093/nar/gkl490] [PMID: 16893957]

[33]   Valkov VT, Scotti N, Kahlau S, et al. Genome-wide analysis of plastid gene expression in potato leaf chloroplasts and tuber amyloplasts: transcriptional and posttranscriptional control. Plant Physiol 2009; 150(4): 2030-44.
[http://dx.doi.org/10.1104/pp.109.140483] [PMID: 19493969]

[34]   Yan L, Lai X, Li X, Wei C, Tan X, Zhang Y. Analyses of the complete genome and gene expression of chloroplast of sweet potato [Ipomoea batata]. PLoS One 2015; 10(4): e0124083. [Ipomoea batata].
[http://dx.doi.org/10.1371/journal.pone.0124083] [PMID: 25874767]

[35]   Kuntz M, Camara B, Weil JH, Schantz R. The psbL gene from bell pepper (*Capsicum annuum*): plastid RNA editing also occurs in non-photosynthetic chromoplasts. Plant Mol Biol 1992; 20(6): 1185-8.
[http://dx.doi.org/10.1007/BF00028906] [PMID: 1463853]

[36]   Bock R. Sense from nonsense: how the genetic information of chloroplasts is altered by RNA editing. Biochimie 2000; 82(6-7): 549-57.
[http://dx.doi.org/10.1016/S0300-9084(00)00610-6] [PMID: 10946106]

[37]   Summer H, Pfannschmidt T, Link G. Transcripts and sequence elements suggest differential promoter usage within the ycf3-psaAB gene cluster on mustard (*Sinapis alba* L.) chloroplast DNA. Curr Genet 2000; 37(1): 45-52.
[http://dx.doi.org/10.1007/s002940050007] [PMID: 10672444]

[38]   Kahlau S, Bock R. Plastid transcriptomics and translatomics of tomato fruit development and chloroplast-to-chromoplast differentiation: chromoplast gene expression largely serves the production of a single protein. Plant Cell 2008; 20(4): 856-74.
[http://dx.doi.org/10.1105/tpc.107.055202] [PMID: 18441214]

[39]   Karpinska B, Karpinski S, Hällgren J. The chlB gene encoding a subunit of light-independent protochlorophyllide reductase is edited in chloroplasts of conifers. Curr Genet 1997; 31(4): 343-7.
[http://dx.doi.org/10.1007/s002940050214] [PMID: 9108142]

[40]   Demko V, Pavlovic A, Valková D, Slováková L, Grimm B, Hudák J. A novel insight into the regulation of light-independent chlorophyll biosynthesis in *Larix decidua* and *Picea abies* seedlings. Planta 2009; 230(1): 165-76.
[http://dx.doi.org/10.1007/s00425-009-0933-3] [PMID: 19404675]

[41]   Barkan A. Expression of plastid genes: organelle-specific elaborations on a prokaryotic scaffold. Plant Physiol 2011; 155(4): 1520-32.
[http://dx.doi.org/10.1104/pp.110.171231] [PMID: 21346173]

[42]   Lyska D, Meierhoff K, Westhoff P. How to build functional thylakoid membranes: from plastid transcription to protein complex assembly. Planta 2013; 237(2): 413-28.
[http://dx.doi.org/10.1007/s00425-012-1752-5] [PMID: 22976450]

[43]   Stoppel R, Meurer J. Complex RNA metabolism in the chloroplast: an update on the psbB operon. Planta 2013; 237(2): 441-9.
[http://dx.doi.org/10.1007/s00425-012-1782-z] [PMID: 23065055]

[44]   Raman G, Park S. Analysis of the complete chloroplast genome of a medicinal plant, *Dianthus superbus* var. *longicalyncinus*, from a comparative genomics perspective. PLoS One 2015; 10(10): e0141329.
[http://dx.doi.org/10.1371/journal.pone.0141329] [PMID: 26513163]

[45]   Wang F, Johnson X, Cavaiuolo M, Bohne AV, Nickelsen J, Vallon O. Two Chlamydomonas OPR proteins stabilize chloroplast mRNAs encoding small subunits of photosystem II and cytochrome b6 f. Plant J 2015; 82(5): 861-73.
[http://dx.doi.org/10.1111/tpj.12858] [PMID: 25898982]

[46]   Hajrah NH, Obaid AY, Atef A, *et al.* Transcriptomic analysis of salt stress responsive genes in *Rhazya stricta*. PLoS One 2017; 12(5): e0177589.
[http://dx.doi.org/10.1371/journal.pone.0177589] [PMID: 28520766]

12

Productivity and Biodiesel Quality of Fatty Acids Contents from *Scenedesmus obliquus* in Domestic Wastewater using Phototrophic and Mixotrophic Cultivation Systems

Alejandro Ruiz-Marin[1,*], Yunuen Canedo-López[1], Asteria Narvaez-García[1], Juan Carlos Robles-Heredia[1] and Jose del Carmen Zavala-Loria[2]

[1]*Faculty of Chemical Engineering, University Autónoma de Ciudad del Carmen, Concordia, 24180 Ciudad del Carmen, Campeche, México*
[2]*Department of Industrial Engineering, University International Hiberoamericana, IMI III, Mexico*

Abstract:

Background:

Microalgae remove nutrients from wastewater with the possibility of grow in mixotrophic and heterotrophic cultures. However, the effluent quality can modify the profile of fatty acids and biodiesel quality.

Methods:

Phototrophic and mixotrophic (light / dark; 12/12 h) cultures of *Scenedesmus obliquus* on domestic wastewater (WW) and Artificial Wastewater (AW) was carried out to evaluate the lipid accumulation and fatty acid methyl esters profile. The microalgae was first cultivated in an enriched medium (90 mg N-NH4 L^{-1}) and subsequently under nitrogen limitation (30, 20 and 10 mg N L^{-1}) using a two-stage process for both culture media.

Results:

A higher cell density in enriched AW medium was obtained in phototrophic and mixotrophic culture of 19 x 10^6 cell mL^{-1} and 20 x 10^6 cell mL^{-1}, respectively; than for WW (13 x 10^6 cell mL^{-1} and 14 x 10^6 cell mL^{-1}, respectively). The nitrogen limitation (from 90 to 20 mg N L^{-1}) for AW increased the lipid content by 5.0% and 17.28% under phototrophic and mixotrophic conditions, respectively and only 5% for WW in mixotrophic culture.

Conclusion:

The high Cetane Number (CN) show a positive correlation with high Saturated Fatty Acids (SFA) content and negative correlation with the Degree of Saturation (DU), suggesting a good ignition of fuel. The Cold Filter Plugging Point (CFPP) (-6.02 to -8.45 °C) and Oxidative Stability (OS) (3.53 - 6.6 h) propose to *Scenedesmus obliquus* as a candidate in the production of biodiesel and potential application for an integral urban wastewater treatment system.

Keywords: *Scenedesmus obliquus*, Autotrophic, Mixotrophic, Domestic wastewater, Lipids, Biodiesel quality.

1. INTRODUCTION

In recent years, environmental issues (climate change) in relation to the effects caused by the high consumption of fossil fuels and the progressive depletion of this resource in many countries have motivated new research aimed at the

* Address correspondence to this author at the Faculty of Chemical Engineering, University Autónoma de Ciudad del Carmen, Street 56 #4 x Av. Concordia, 24180 Ciudad del Carmen, Campeche, México; Tel: E-mail: aruiz@pampano.unacar.mx

exploration of new renewable energy sources such as the use of microalgal biomass to obtain lipids and synthesis of biodiesel [1]. Microalgae have shown important properties such as high photosynthetic efficiency, high rates of lipid accumulation and high biomass production. Some of the microalgae species reserve high lipid content (above 70%) under conditions of environmental stressors such as limitation of essential nutrients as nitrogen, phosphate and some metals.

Nitrogen limitation changes cellular carbon flux from protein synthesis to lipid synthesis, resulting in increases of 20% to 40% in microalgae [2, 3]. However, high lipids contents produced under nutrient limitation conditions are usually associated with low algal biomass productivity. A replenishing nitrogen source is generally necessary to maintain a high cell growth rate and achieve high cell density. Producing sufficient biomass with enhanced lipids contents can be done using a two-stage culture strategy [4]. In this strategy, an alga is first grown under nutrient-sufficient conditions to allow maximum cell density, and then deprived of certain nutrients to trigger lipids accumulation [4, 5].

Another factor that contributes to the increase in the content of lipids and algal biomass for purposes of biodiesel production is culture in the presence of carbon sources. The algae can consume both sources of organic carbon (sugars) and inorganic carbon (CO_2), therefore, it has been proposed to combine the algal culture in urban wastewater treatment systems, with the opportunity to produce a renewable energy (Biodiesel) and be able to reduce costs for both processes [6].

There are many applications that in recent years have been attributed to algal culture and in the search to minimize the costs of biomass production studies have turned their attention to the use of industrial and urban wastewater, where the goal is to reduce costs and utilization of the content of nitrogen and phosphorus present in the effluents, which suggests that the cultivation of algae in wastewater does not generate additional contamination when the biomass is harvested, making the treatment system more attractive than the activated sludge systems [7]. Recently the debate on the use of algae-wastewater system for production of raw material for biofuel production focuses on evaluating the algal biochemical composition as it is not completely satisfactory for the biodiesel synthesis. Although in experiments with algae strains a great vitality has been observed in wastewater, the variations in the biochemical composition are dependent on the algal species and environmental conditions [8].

Therefore, the objective of the present work was to evaluate the profile of fatty acids by *Scenedesmus obliquus*, estimating the quality of biodiesel produced in phototrophic and mixotrophic culture mode using domestic wastewater as a source of nutrients and organic carbon.

2. MATERIALS AND METHODS

2.1. Strain Selection and Culture Media

The microalgae *Scenedesmus obliquus* was obtained from the Biology Laboratory and microalgae culture collection of the Centro de Investigación Científica y de Educación Superior de Ensenada, Baja California (CICESE). The microalgae *Scenedesmus obliquus* was chosen due to its ability to grow in wastewater and its high N and P removal efficiencies. Basic growth culture medium composition was similar to that of primary treatment effluent from the urban wastewater treatment plants [7]: 7 mg NaCl; 4 mg $CaCl_2$; 2 mg $MgSO_4 \cdot 7H_2O$; 15 mg KH_2PO_4; 115.6 mg NH_4Cl; 100 mg glucose; all dissolved in 1 L drinking water. Trace metals and vitamins were added according to guidelines for f/2 medium [9]. Culture conditions were 28 ± 1 °C with a 100 μE m^{-2} s^{-1} light intensity. During the 1 month acclimatization period, the culture was transferred to fresh media every seven days.

2.2. Experimental Design

In all experiments, *Scenedesmus obliquus* was cultured in cylindrical bioreactors made of transparent Polyethylene Terephthalate (PETE) with a 3 L maximum capacity. Reactors were pre-washed with a chlorine solution (0.5%) to prevent bacterial contamination. All treatments were run in triplicate at a constant temperature of 28 ± 1 °C, an initial cell density of 2×10^6 cell mL^{-1} and aeration at 0.4 L L^{-1} min^{-1}. In the phototrophic treatments, continuous light (100 μE s^{-1} m^{-2}) was provided using cool white fluorescent lamps and for the mixotrophic treatments, the same light intensity was applied in 12/12 h light/dark cycles.

A two-stage nitrogen reduction process was used. In the first stage, *S. obliquus* was grown in 1 L of medium containing 90 mg-N L^{-1} (nitrogen sufficiency). When the exponential growth phase ended, 1 L of fresh medium was added to the dilution until reaching 2 L of media containing 10, 20 y 30 mg L^{-1} N-NH$_4$. All treatments were run in triplicate for the two culture media: artificial wastewater and urban wastewater (Table **1**). and the nitrogen concentration was determined according to the standard methods of water and wastewater [10].

Table 1. Experimental design and nitrogen reduction levels.

Culture Media	Photoperiod / Growth Mode	Nitrogen Reduction Levels
Artificial Wastewater (AW)	Photoautotrophic	90 mg L^{-1}
		From 90 to 30 mg L^{-1}
		From 90 to 20 mg L^{-1}
		From 90 to 10 mg L^{-1}
	Mixotrophic (12/12 h)	90 mg L^{-1}
		From 90 to 30 mg L^{-1}
		From 90 to 20 mg L^{-1}
		From 90 to 10 mg L^{-1}
Urban Wastewater (WW)	Photoautotrophic	90 mg L^{-1}
		From 90 to 30 mg L^{-1}
		From 90 to 20 mg L^{-1}
		From 90 to 10 mg L^{-1}
	Mixotrophic (12/12 h)	90 mg L^{-1}
		From 90 to 30 mg L^{-1}
		From 90 to 20 mg L^{-1}
		From 90 to 10 mg L^{-1}

The domestic effluents were collected at the discharge from the aeration tanks of three wastewater treatment plant in the municipality of Ciudad del Carmen, Campeche; Mexico. The characteristics of the samples for the three plants are listed in Table **2**. As the content of N-NH$_4$ of the urban effluents was considered with variations in months for cultivating microalgae, the NH$_4$ concentration was adjusted to 90 mg L^{-1} and 30, 20 y 10 mg L^{-1} in order to reach a similar value for all the treatments.

Table 2. Chemical parameters determined for the wastewater effluent samples (WPT). The results represent the means of 3 replicates and the Standard Deviations (SD) are indicated after the ± sign.

Parameters	Units	Wastewater Treatment Plants Effluents		
		WPT$_1$	WPT$_2$	WPT$_3$
Temperature	°C	26.3 ± 0.2	27.0 ± 0.1	27.6 ± 0.01
pH	-	7.96	7.47	7.10
Total phosphorus	mg L^{-1}	19.64 ± 0.1	3.75 ± 0.2	4.22 ± 0.02
Total nitrogen	mg L^{-1}	99.13 ± 0.02	18.64 ± 0.1	3.10 ± 0.1
N-NH$_4$	mg L^{-1}	90.75 ± 0.03	17.06 ± 0.02	2.84 ± 0.02
N-NO$_3$	mg L^{-1}	8.29 ± 0.01	1.81 ± 0.01	0.32 ± 0.03
N-NO$_2$	mg L^{-1}	0.08 ± 0.1	0.013 ± 0.2	0.002 ± 0.2
N/P ratio	-	5.04	4.97	0.73
Fats and oil	mg L^{-1}	20.26 ± 1.2	8.03 ± 2.2	10.75 ± 1.8
BOD	mg L^{-1}	212 ± 0.8	28 ± 1.2	43 ± 0.5
TSS	mg L^{-1}	117 ± 0.5	27 ± 0.2	40 ± 0.3

2.3. Fatty Acid Extraction and Composition Analysis

At the end of each treatment, approximately 1 L culture was collected from the bioreactors and centrifuged at 4500 rpm and 14 °C for 15 min. The recovered biomass was frozen at -4.0 °C for 48 h, and then lyophilized for 3-5 days; the resulting dry biomass was stored at 0 °C. Total lipids were extracted following the dry extraction procedure described by Zhu et al. [11] and Feng et al. [12], and then quantified using the method of Pande et al. [13] using a tripalmitin standard (99%, Sigma-Aldrich).

The transesterification of fatty acids was according to the methods by Sato and Murata [14]; and Canedo-Lopez et al. [5]. The FAMEs profiles were generated with a Gas Chromatographer (GC) (Agilent Technology 7890).

One microliter of the FAME-hexane solution was injected into the GC equipped with a Flame Ionization Detector (FID) and separation was done in a DB-23 column (60 m length, 0.32 mm ID, 0.25 µm thick) with helium as the vehicle. Injector and detector temperature was 250 °C. A five-step temperature program was run: 120 °C for 5 min; 10 °C min^{-1} increases until reaching 180 °C; 180 °C for 30 min; 10 °C min^{-1} increases until reaching 210 °C; and 210 °C for 21 min [5]. A calibration curve was prepared for all FAMEs by injecting known concentrations of an external standard mixture containing 37 FAMEs (Supelco, Bellefonte, PA, USA); the correlation coefficient was equal to or greater than 95% in all cases.

2.4. Productivity and Evaluation of Fatty acid Quality for Biodiesel

Lipid productivity P_L (grams of lipids per liter of culture per day) was calculated using equation (1) [15, 16].

$$P_L = \frac{w_1 X_1 - w_2 X_2}{t_2 - t_1} \tag{1}$$

Where w is the lipid content as a weight fraction (g lipid per g dry biomass); and X is biomass concentration (g dry biomass per mL culture). Thus, $\Delta(wX)$ represents the lipids accumulated from inoculation (time t_1) to harvest (time t_2), which occurs in time Δt.

The specific growth rate μ in the logarithmic growth phase was defined as (equation 2):

$$\mu = \frac{\ln(X_2) - \ln(X_1)}{t_2 - t_1} \tag{2}$$

Where t_1 and t_2 are cultivation times during the logarithmic phase.

Biodiesel quality (fuel characteristics) such as Saponification Value (SV), Cetane Number (CN), Iodine Value (IV), Long Chain Saturated Factor (LCSF) and Col Filter Plugging Point (CFPP) were determined based on the fatty acid composition of the microalgal using empirical equation described by Vidyashankar et al. [17] and Guldhe et al. [18].

$$SV = \sum \frac{560N}{M} \tag{3}$$

$$IV = \sum \frac{254DN}{M} \tag{4}$$

$$CN = 46.3 + \frac{5458}{SV} - 0.225\,IV \tag{5}$$

$$LCSF = [0.1\,C16(wt\%) + 0.5\,C18(wt\% + 1.\,C20(wt\%) + 1.5\,C22(wt\%) + 2.\,C24(wt\%)] \tag{6}$$

$$CFPP = (3.1417 * LCSF) - 16.477 \tag{7}$$

Where D, M and N denote the number of double bonds, molecular mass and % mass fraction of each fatty acid component, respectively.

$$DU = MUFA + (2\,x\,PUFA) \tag{8}$$

$$Oxidative\ stability\ (OS) = -(0.0384\,x\,DU) + 7.77 \tag{9}$$

Where, DU is the degree of unsaturated, calculated from the fatty acid profile [18, 19]. The properties of synthesized biodiesel were compared with the specifications given by ASTM 6751 and EN14214 standards.

2.5. Statistical Analysis

An analysis of covariance (ANCOVA; $P \leq 0.05$) was applied to measure the effects of the manipulated variables (nitrogen limitation, photoperiod) on the response variables (specific growth rate and lipids productivity). The Tukey test (HSD) was applied when results exhibited significant differences.

3. RESULTS AND DISCUSSION

3.1. Growth and Nutrients Removal

The growth kinetics obtained from the microalgae *S. obliquus* during the culture period showed favorable increases in cell density for both culture media: artificial and urban wastewater (Fig. **1**). The growth for each culture medium showed significant differences ($P \leq 0.015$); where the highest growth was observed mainly in culture with artificial wastewater with respect to the urban wastewater medium (Fig. **1**). The growth rate (μ) decreased in urban wastewater medium, similar to that reported by several investigations (Table **3**), which is attributed to competition for other microorganisms present in the effluents (bacteria and protozoa), as well as the different forms of bioavailable nutrients [7, 20].

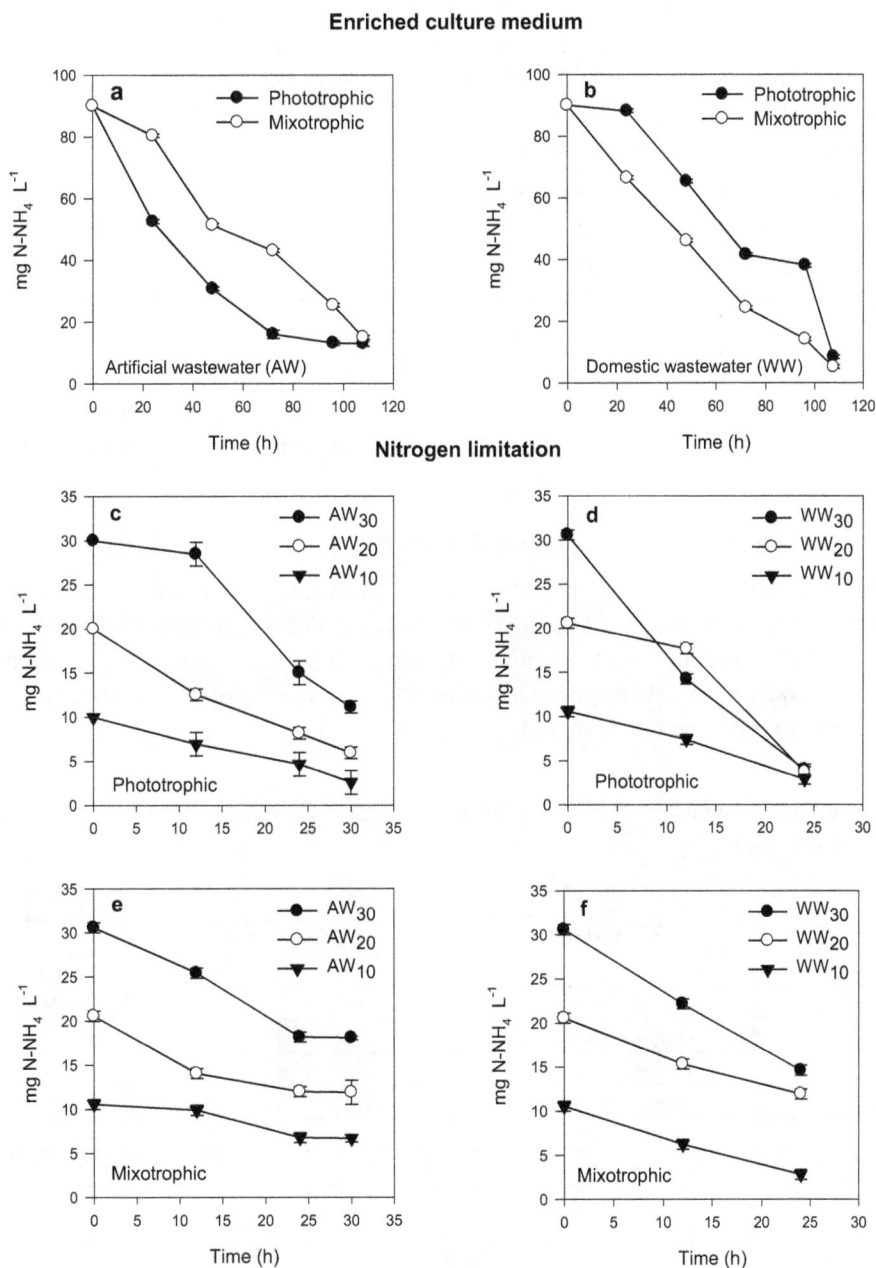

Fig. (1). Increase in cell density (cell mL^{-1}) in phototrophic and mixotrophic culture in Artificial Wastewater (AW) and urban wastewater (WW).

In particular, phototrophic and mixotrophic growth in similar culture medium (artificial and urban wastewater) did not present significant differences (P ≥ 0.121), suggesting that *S. obliquus* presented the ability to change its metabolism from phototrophic to mixotrophic (Figs. **1a**, **b**). The maximum cell density under phototrophic and mixotrophic culture obtained in artificial wastewater was 19 x 10^6 cell mL^{-1} and 20 x 10^6 cell mL^{-1}, respectively; while in urban wastewater culture, it was 13 x 10^6 cell mL^{-1} and 14 x 10^6 cell mL^{-1}, respectively. This shows the capacity of *S. obliquus* to adapt and grow in urban wastewater, as well as use inorganic and organic carbon sources in mixotrophic culture systems [5, 7, 21]. On the other hand, the treatments under nitrogen limitation (Figs. **1c**, **d**, **e**, **f**) growth showed significant differences during nitrogen variation (30, 20 and 10 mg L^{-1}) for both phototrophic and mixotrophic systems. The lowest cell density was for those mixotrophic cultures in urban wastewater (Fig. **1f**).

A probable explanation can be attributed to the fact that wastewater contains high levels of carbon, useful for some microalgae with the capacity to grow in a mixotrophic or heterotrophic environment, however, this organic carbon that has been previously degraded by bacteria under aerobic conditions, results inert and recalcitrant, becoming not available for consumption by microalgae. In the present study, the wastewater was not sterilized, which suggested that bacteria and the unavailable forms of carbon sources in the effluent were probably not favorable for the growth of *S. obliquus* under mixotrophic conditions [5, 22, 23]; suggesting that the rate of growth and removal of nutrients depends greatly on the organic loads present in the urban wastewater effluents.

A similar tendency was observed in the nitrogen removal for *S. obliquus* (Fig. **2**) in medium enriched for both artificial and urban wastewater, where the efficiency was 90% of nitrogen removed in both phototrophic and mixotrophic conditions (Figs. **2a**, **b**). Studies have reported that it is possible to achieve efficient nitrogen removal in wastewater treatments, similar to that reported in the present study [21, 24]. However, under nitrogen limitation at 30, 20 and 10 mg L^{-1}, it was observed that 80% of nitrogen was removed in the photoautotrophic system (figure c and d); while in mixotrophic culture, the nitrogen removal was not greater than 50% (Figs. **2e**, **f**).

Based on the results of growth and nitrogen removal under nitrogen limitation, it was possible selected the treatment with nitrogen reduction of 90 mg N-NH$_4$ L^{-1} at 20 mg L^{-1} with the objective of evaluating the fatty acid profile and quality of biodiesel.

3.2. Productivity and Fuel Quality of *S. obliquus*-Based Biodiesel

The analysis of productivity and fatty acids profile presented in this study was based on the Two-Stage method at a limiting concentration of 20 mg N-NH$_4$ L^{-1}. The lipids accumulation had an increase of 5.24% to 17.28% nitrogen limitation under phototrophic and mixotrophic conditions, respectively, using Artificial Wastewater (AW). While cultures with urban wastewater (WW) did not show a significant increase in lipids content for one autotrophic system, unlike a 5% increase in mixotrophic culture (Table **3**).

Table 3. Growth rate (μ), lipid production (g lipids / g biomass) and productivity (mg lipids L^{-1} d^{-1}) for *S. obliquus* grown under phototrophic and mixotrophic conditions.

Parameters	Culture Medium	Treatments			
		Phototrophic		Mixotrophic	
		WA	WW	WA	WW
μ (d^{-1})	Enriched culture	0.643	0.541	0.389	0.296
	Nitrogen limitation	0.425	0.179	0.354	0.225
g lipids/g biomass (% w/w)	Enriched culture	19.9	20.8	15.8	17.4
	Nitrogen limitation	21.0	19.0	19.1	18.3
Productivity (mg lipids L^{-1} d^{-1})	Enriched culture	26.49	23.41	19.24	16.31
	Nitrogen limitation	35.35	41.61	37.21	28.18

*(**AW**): Artificial Wastewater; (**WW**): Domestic Wastewater

Although the values of lipid production (% w / w) were low in WW compared to those obtained in AW, the results suggest that the microalga had the capacity to increase the lipid content under the mixotrophic condition and nitrogen limitation in urban wastewater. Mandal and Mallick [25] reported for *S. obliquus* a productivity (7. 14 mg lipid L^{-1} d^{-1}) under phototrophic condition lower than that reported in the present study (Table **3**). While Liang *et al.* [26] reported high productivity ranges of 11.6 - 58.6 mg lipid L^{-1} d^{-1} for *S. obliquus* under mixotrophic condition.

Enriched culture medium

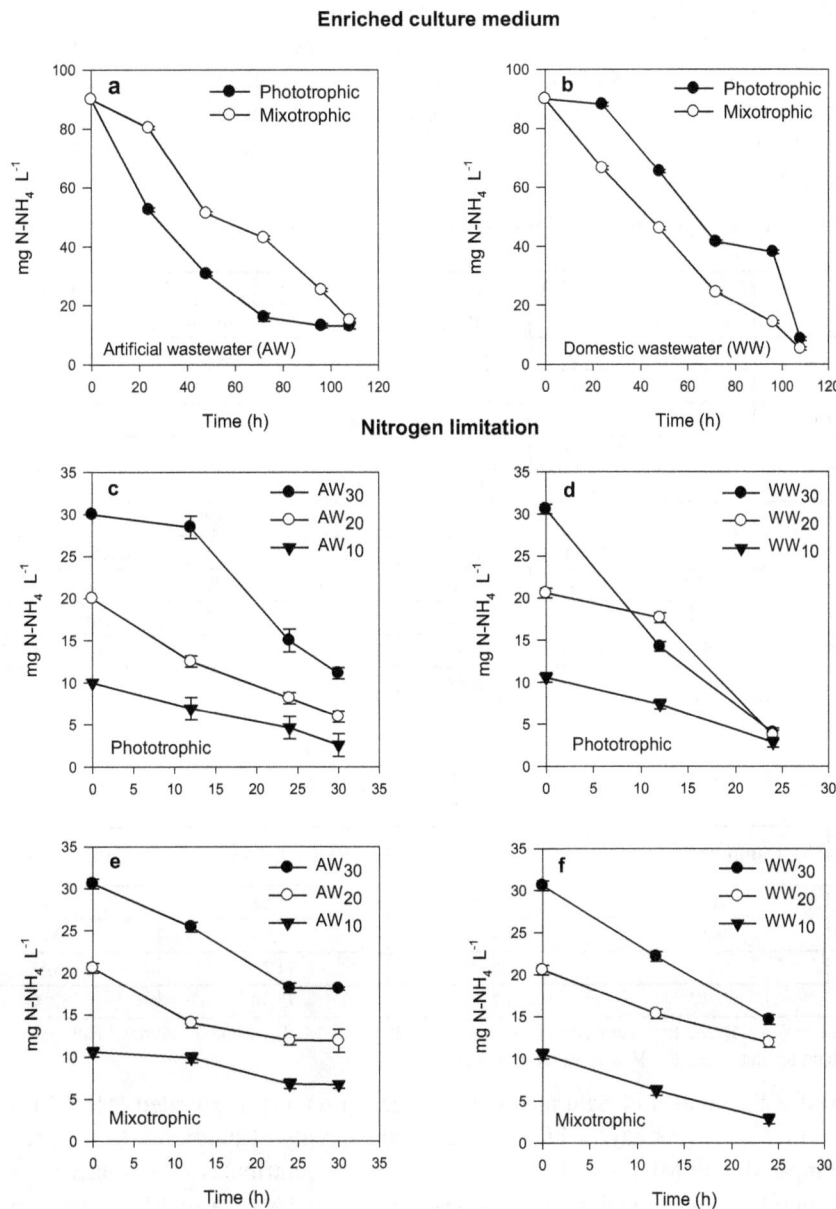

Fig. (2). Nitrogen removed in Artificial Wastewater (AW) and urban wastewater (AW). The values of 30, 20 and 10 mg N-NH$_4$ L^{-1} indicate the levels of nitrogen reduction in the culture medium.

This increase in productivity was also observed in the present studies when passing the cultures from a phototrophic to mixotrophic condition. Although studies have reported that productivity for *S. obliquus* may decrease (8 mg lipid L^{-1} d^{-1}) in cultures using secondary effluents [27], this was not observed in the present study, suggesting that wastewater can be proposed as a culture medium to substitute the conventional culture media, given that it is possible to achieve high productivity in phototrophic and mixotrophic conditions (Table **3**); offering an opportunity for microalgae technology in the removal of nutrients from wastewater treatment plants effluents, with the additional advantage of obtaining biomass with a favorable lipid content for the synthesis of biodiesel. However, some factors must be considered when using wastewater, such as organic load, organic and inorganic carbon content, nitrogen and phosphorus content, since the quality of the effluents varies according to the form of operation and technology used for the treatment of wastewater that could affect the biochemistry (carbohydrates, proteins and fatty acid profile) of the algal biomass.

The fatty acid methyl ester composition of biodiesel obtained by *S. obliquus* is shown in Table **4**. The Behenic acid (C22:0) and Myristic acid (C14:0) were the predominant Saturated Fatty Acids (SFA) contributing to phototropic culture to reach the 64.19% - 69.10% of SFA and for mixotrophic culture of 48.6% - 62.30% of SFA (Table **5**). While the Monounsaturated Fatty Acid (MUFA) predominant was the Eicosenoic acid (C20: 1) contributing to 25% -34.75% of the MUFA present in the algal biomass (Table **4**). It is a fact that the profile and composition of oil obtained by *S. obliquus* cultivated in domestic residual water, showed no differences to artificial wastewater cultures, suggesting that the fatty acid profile for *S. obliquus* was not affected; which proposes domestic wastewater as a good substitute growing medium for the production of algal biomass.

Table 4. Fatty acid methyl ester composition (%) of the biodiesel by *S. obliquus*.

FAME	Artificial Wastewater (AW)				Domestic Wastewater (WW)			
	Enriched Culture		Nitrogen Limitation		Enriched Culture		Nitrogen Limitation	
Phot	Mixotr	Phot	Mixotr	Phot	Mixotr	Phot	Mixotr	
Myristic acid (C14:0)	10.62	14.12	20.40	12.51	19.98	19.44	24.88	20.84
Palmitic acid (C16:0)	9.31	0.95	8.51	8.35	7.33	8.23	7.23	7.48
Arachidic acid (C20:0)	11.61	8.72	7.21	8.23	11.74	7.21	7.48	6.63
Behenic acid (C22:0)	17.14	15.69	14.71	19.49	11.44	13.20	13.55	13.03
Palmitoleic acid (C16:1)	8.73	8.55	6.71	7.68	7.88	7.06	6.52	6.81
Oleic acid (C18:1)	0.63	2.14	0.83	1.80	1.03	1.26	1.84	2.16
Eicosenoic acid (C20:1)	17.11	16.93	15.96	17.26	15.82	13.78	14.17	12.67
Erucic acid (C22:1)	9.13	11.47	9.34	8.00	9.60	11.30	8.53	10.28
Linoleic acid (C18:2)	9.64	12.01	9.83	8.43	9.13	10.28	8.95	9.56
Linolenic acid (C18:3)	6.07	9.39	6.48	8.19	6.01	8.20	6.82	10.51

*Phot: Phototrophic culture; Mixotr: Mixotrophic culture.

Table 5. Fatty acid content of *S. obliquus* in phototrophic and mixotrophic culture in artificial and urban wastewater.

Culture Mode	Culture Medium	SAF (% wt)	MUFA (% wt)	PUFA (% wt)	LCSF (% wt)
phototrophic	AW	69.10	25.83	5.00	2.98
	WW	64.19	25.00	10.80	3.33
Mixotrophic	AW	48.60	34.75	16.63	2.77
	WW	62.30	25.30	12.4	2.55

*(AW): Artificial wastewater; (WW): Domestic wastewater. SFA-Saturated Fatty Acid. MUFA-Monounsaturated Fatty Acid. PUFA-Polyunsaturated Fatty Acid. LCSF-Long Chain Saturation Factor. Values are average of triplicate.

A subject of interest is the content of Saturated Fatty Acids (SFA), monosaturated (MUFA) and polyunsaturated (PUFA) present in the oil since these have a high impact on the quality of biodiesel. In the present study, the high proportion of SFA compared with MUFA and PUFA in *S. obliquus* contributes to an increase in oxidative stability (Table **5**). An analysis of oil containing high concentrations of PUFA and MUFA on SFA tends to originate a biodiesel with high iodine value, a condition that determines the generation of a biodiesel that is oxidation prone; however, this was not observed in domestic residual water cultures (WW) and Artificial Wastewater (AW) (Table **5**). Therefore, the quality parameters evaluated as the cetane number and saponification value were related to the saturation of FAME in both cultures of *S. obliquus* in AW and WW.

The parameters determined in the present study that define the biodiesel quality were: Cetane Number (CN), iodine value (IV), Cold Filter Plugging Point (CFPP); Saponification Value (SV), Degree of Unsaturation (DU) and, Oxidative Stability (OS) for autotrophic and mixotrophic conditions in AW and WW medium (Table **6**).

Cetane Number (CN) is the main indicator of biodiesel quality related to ignition delay time and combustion quality. High values of CN mean better ignition and engine performance properties. The Cetane Number (CN) observed in all the treatments presented a low value to that recommended by the international standards ASTM 6751 and to the minimum by EN14214 (Table **6**). However, the high cetane number value and proportions of SFA obtained for all treatments can be associated with efficient combustion properties of biodiesel (Table **6**), similar to those reported by other authors [1, 19].

According to Katiyar *et al.* [28] a high amount of SFAs leads to a high cetane number, indicating complete combustion of biodiesel and smooth engine run, making biodiesel more suitable for vehicles. On the other hand, the formation of pollutants (low emission of NOx) is inversely proportional to the cetane number. The biodiesel generated from *S. obliquus* where the SFAs dominate the MUFAs has poor cold flow properties; therefore this biodiesel in cold climate regions may be limited, although this problem can be minimized by mixing oil from other species with opposite characteristics [19].

Studies by Wu and Miao [1] reported high CN values (59.30) for *S. obliquus* compared to the present work of 36.2 and 36.9 in the phototrophic and mixotrophic culture, respectively in urban wastewater (Table **6**). The CN values reported by the authors show a positive correlation with the high content of SFA and a negative correlation with DU, suggesting a good ignition of the fuel. Wu and Miao [1] suggest as criteria an SFA / PUFA ratio of 2.31 - 4.02 to obtain a good ignition of biodiesel from *S. obliquus*.

Comparatively, the SFA / PUFA ratio obtained in the present study was higher (5.02 - 5.94) in the cultures of *S. obliquus* in urban wastewater (WW), which suggests that the amount of lipids accumulated by microalgae should show a balance between the content of SFA and PUFA, since not always a high value of SFA with respect to PUFA may be the best condition for a good ignition of biodiesel. On the contrary, a high percentage of SFA affects the flow properties causing crystallization / solidification of the fuel in the filters of engine under colder climatic condition.

On the other hand, one of the problems that cause a high proportion of PUFAs is to affect the oxidative stability of the fuel, as can be verified with indicator IV (Iodine Value). This indicator is also a crude measurement of the total unsaturation of biodiesel, which has often been used in connection with issues such as oxidative stability, with the implication that biodiesel with high IV value is less stable to oxidation than that with low IV value [29].

In the present study, the cultures showed IV values lower than 120 g I_2 100 g^{-1} lipids; according to the values established by the European standard, which can be confirmed with the low values of the Degree of Unsaturated (DU) and with a positive correlation between IV and DU; suggesting for all treatments at minimum Oxidative Stability (OS) of 3 to 6 min (Table **5**). The IV, DU and OS values compared with other studies indicated for *S. obliquus* were similar, within the range established as quality criteria by the ASTM D6751 and EN 14214 standards (Table **6**), [1, 17 - 19, 31] indicating that biodiesel obtained from *S. obliquus* presents a good oxidative stability.

Table 6. Fatty acid content of *S. obliquus* in phototrophic and mixotrophic culture in artificial and urban wastewater.

Biodiesel Properties	Units	ASTM D6751	EN 14214	[1]	[17]	[18]	[19]	[31]	Phototrophic Culture		Mixotrophic Culture	
									AW	WW	AW	WW
CN	Min	47	51	57.93	57.13	51.74	63.63	59.98	40.2	36.9	31.5	36.2
IV	g I_2/100 g	-	120	72.81	77.91	98.86	35.38	77.36	27.2	41.7	67.42	45.03
CFPP	°C	-	≤5/≤-20	-5.68	-0.04	3.5	-11.87	-3.19	-7.1	-6.02	-7.7	-8.45
SV	mg KOH/g	-	-	194.8	192.4	-	216.0	164.6	315.3	325.9	245.6	313.5
DU	% wt	-	-	83.97	-	-	36.63	87.04	30.83	46.6	68.02	50.1
OS	h	3	6	7.31	-	-	-	3.53	6.6	5.9	5.2	5.84

* CN- Cetane Number; IV-Iodine value; CFPP-Cold Filter Plugging Point; SV- Saponification value; DU-Degree of unsaturation; OS-Oxidative stability. Values are average of triplicate. References [1 - 31].

Another quality criterion analyzed was the Cold Filter Plugging Point (CFPP), a parameter usually used to predict the flow performance of biodiesel at low temperature. Low temperature properties depend mostly on the saturated fatty acids content and the effect of unsaturated fatty acid composition can be considered negligible [30]. The value of CFPP (Value: -6.02 a - 8.45) obtained in the present study suggests that the quality of biodiesel satisfies international criteria with an acceptable limit of -5 °C to 13 °C (Table **6**) proposing *S. obliquus* as a candidate in the production of biodiesel and with potential application in urban wastewater treatment technologies.

The differences found in each of the quality parameters (CN, IV, CFPP, SV, DU and OS) with respect to other reported studies (Table **6**) is probably attributed to different factors such as: the composition of the culture medium, light intensity, culture systems (phototrophic, mixotrophic, heterotrophic), nitrogen limitation, lighting periods and the physiological potential of each microalgae strain. It is a fact that the quality of biodiesel depends strongly on the composition of the fatty acids, which can be modified according to the environmental conditions (culture medium) and the physiological potential of each algal strain. The difference in the fatty acids composition can be a reference since the degree of saturation / unsaturation represents an important discriminatory factor to estimate the quality of biodiesel. In accordance with other studies (Table **6**), [1, 17 - 19, 31] it can be concluded that microalgae rich in SFA will generate biodiesel with high CN and low IV and high oxidative stability, while species with a high PUFA content (high DU) are considered to generate biodiesel with low CN and high IV and are more prone to oxidation (Table **6**). The physico-chemical properties of *Scenedesmus obliquus* base biodiesel production compared with international standards (EN and ASTM).

CONCLUSION

The present study shows the potential of the use of a low-cost algal culture medium such as urban wastewater for the production of high-quality biomass and biodiesel compared to a conventional medium. In mixotrophic culture, *S. obliquus* showed the capacity to increase the lipids productivity over 5% in urban wastewater. The fatty acid profile for *S. obliquus* reveals a biodiesel used exclusively for urban vehicles since it has a low PUFA content and a high SFAs content. Similarly, high values of CN show a positive correlation with the high SFA content and negative correlation with DU, suggesting a good ignition of the fuel, which is favorable for the emission of pollutants. This proposes *S. obliquus* as a candidate in the biodiesel production and with potential application in urban wastewater treatment technologies with the possibility of use in phototrophic and mixotrophic cultures; However, the characteristics of wastewater (nutrients, organic carbon, *etc.*) can have a direct effect on the fatty acids profile and biodiesel quality, which should be considered in future research. In the present study, the effluents of domestic wastewater used in the cultures showed not to have a negative effect on the biodiesel quality.

HUMAN AND ANIMAL RIGHTS

No Animals/Humans were used for studies that are base of this research.

ACKNOWLEDGEMENTS

Authors acknowledgement to the members of CAEC in Environmental Engineering and Universidad Autónoma del Carmen (UNACAR) for encouragement and support.

REFERENCES

[1]   Wu H, Miao X. Biodiesel quality and biochemical changes of microalgae *Chlorella pyrenoidosa* and *Scenedesmus obliquus* in response to nitrate levels. Bioresour Technol 2014; 170: 421-7.
       [http://dx.doi.org/10.1016/j.biortech.2014.08.017] [PMID: 25164333]

[2]   Illman AM, Scragg AH, Shales SW. Increase in *Chlorella* strains calorific values when grown in low nitrogen medium. Enzyme Microb Technol 2000; 27(8): 631-5.
       [http://dx.doi.org/10.1016/S0141-0229(00)00266-0] [PMID: 11024528]

[3]   Liu ZY, Wang GC, Zhou BC. Effect of iron on growth and lipid accumulation in *Chlorella vulgaris*. Bioresour Technol 2008; 99(11): 4717-22.
       [http://dx.doi.org/10.1016/j.biortech.2007.09.073] [PMID: 17993270]

[4]   Mujtaba G, Choi W, Lee CG, Lee K. Lipid production by *Chlorella vulgaris* after a shift from nutrient-rich to nitrogen starvation conditions. Bioresour Technol 2012; 123: 279-83.
       [http://dx.doi.org/10.1016/j.biortech.2012.07.057] [PMID: 22940330]

[5]   Canedo-López Y, Ruiz-Marín A, Zavala-Loría JC. A two-stage culture process using *Chlorella vulgaris* for urban wastewater nutrient removal and enhanced algal lipid accumulation under photoautotrophic and mixotrophic conditions. J Renew Sustain Energy 2016; 8: 033102.
       [http://dx.doi.org/10.1063/1.4954078]

[6]   Rinna F, Buono S, Dominguez CIT, Nascimento AI, Sansone G, Assunta Barone CM. Wastewater treatment by microalgae can generate high quality biodiesel feedstock. J Water Proc Eng 2017; 18: 144-9.
       [http://dx.doi.org/10.1016/j.jwpe.2017.06.006]

[7]   Ruiz-Marín A, Mendoza-Espinosa LG, Stephenson T. Growth and nutrient removal in free and immobilized green algae in batch and semi-continuous cultures treating real wastewater. Bioresour Technol 2010; 101(1): 58-64.
        [http://dx.doi.org/10.1016/j.biortech.2009.02.076] [PMID: 19699635]

[8]   Li Q, Du W, Liu D. Perspectives of microbial oils for biodiesel production. Appl Microbiol Biotechnol 2008; 80(5): 749-56.
        [http://dx.doi.org/10.1007/s00253-008-1625-9] [PMID: 18690426]

[9]   Guillard RR, Ryther JH. Studies of marine planktonic diatoms. I. Cyclotella nana Hustedt, and Detonula confervacea (cleve) Gran. Can J Microbiol 1962; 8: 229-39.
        [http://dx.doi.org/10.1139/m62-029] [PMID: 13902807]

[10]   American Public Health Association-Standard Methods 1995.

[11]   Zhu M, Zhou PP, Yu LJ. Extraction of lipids from Mortierella alpina and enrichment of arachidonic acid from the fungal lipids. Bioresour Technol 2002; 84(1): 93-5.
        [http://dx.doi.org/10.1016/S0960-8524(02)00028-7] [PMID: 12137275]

[12]   Feng Y, Li C, Zhang D. Lipid production of Chlorella vulgaris cultured in artificial wastewater medium. Bioresour Technol 2011; 102(1): 101-5.
        [http://dx.doi.org/10.1016/j.biortech.2010.06.016] [PMID: 20620053]

[13]   Pande SV, Khan P, Venkitasubramanian TA. Microdetermination of lipids and serum total fatty acid. Anal Biochem 1963; 6: 415-23.
        [http://dx.doi.org/10.1016/0003-2697(63)90094-0] [PMID: 14077635]

[14]   Sato N, Murata N. Membrane lipids, in Methods in Enzimology. Kaplan (Academic Press) 1988; 167: pp. 251-9.

[15]   Khozin-Goldberg I, Cohen Z. The effect of phosphate starvation on the lipid and fatty acid composition of the fresh water eustigmatophyte Monodus subterraneus. Phytochemistry 2006; 67(7): 696-701.
        [http://dx.doi.org/10.1016/j.phytochem.2006.01.010] [PMID: 16497342]

[16]   Rodolfi L, Chini Zittelli G, Bassi N, et al. Microalgae for oil: Strain selection, induction of lipid synthesis and outdoor mass cultivation in a low-cost photobioreactor. Biotechnol Bioeng 2009; 102(1): 100-12.
        [http://dx.doi.org/10.1002/bit.22033] [PMID: 18683258]

[17]   Vidyashankar S, VenuGopal KS, Kavitha MD, et al. Characterization of fatty acids and hydrocarbons of chlorophycean microalgae towards their use as biofuel source. Biomass Bioenergy 2015; 77: 75-91.
        [http://dx.doi.org/10.1016/j.biombioe.2015.03.001]

[18]   Guldhe A, Singh P, Kumari S, Rawat I, Permaul K, Bux F. Biodiesel synthesis from microalgae using immobilized Aspergillus niger whole cell lipase biocatalyst. Renew Energy 2016; 85: 1002-10.
        [http://dx.doi.org/10.1016/j.renene.2015.07.059]

[19]   Nascimento IA, Izabel-Marques SS, Dominguez-Cabanelas IT, et al. Screening microalgae strains for biodiesel production: Lipid productivity an estimation of fuel quality base on fatty acids profiles as selective criteria. Bioenerg Res 2013; 6: 1-13.
        [http://dx.doi.org/10.1007/s12155-012-9222-2]

[20]   Lau PS, Tam NFY, Wong YS. Effect of algal density on nutrient removal from primary settled wastewater. Environ Pollut 1995; 89: 59-66.
        [http://dx.doi.org/10.1016/0269-7491(94)00044-E]

[21]   Martínez ME, Sánchez S, Jimenéz JM, Yousfi FE, Muñoz L. Nitrogen and phosphorus removal from urban wastewater by the microalga Scenedesmus obliquus. Bioresour Technol 2000; 73: 263-72.
        [http://dx.doi.org/10.1016/S0960-8524(99)00121-2]

[22]   Wang L, Li Y, Chen P, et al. Anaerobic digested dairy manure as a nutrient supplement for cultivation of oil-rich green microalgae Chlorella sp. Bioresour Technol 2010; 101(8): 2623-8.
        [http://dx.doi.org/10.1016/j.biortech.2009.10.062] [PMID: 19932957]

[23]   Zhou W, Min M, Li Y, et al. A hetero-photoautotrophic two-stage cultivation process to improve wastewater nutrient removal and enhance algal lipid accumulation. Bioresour Technol 2012; 110: 448-55.
        [http://dx.doi.org/10.1016/j.biortech.2012.01.063] [PMID: 22326332]

[24]   González LE, Cañizares RO, Baena S. Efficiency of ammonia and phosphorus removal from a Colombian agroindustrial wastewater by the microalgae Chlorella vulgaris and Scenedesmus dimorphus. Bioresour Technol 1997; 60: 259-62.
        [http://dx.doi.org/10.1016/S0960-8524(97)00029-1]

[25]   Mandal S, Mallick N. Waste utilization and biodiesel production by the green microalga Scenedesmus obliquus. Appl Environ Microbiol 2011; 77(1): 374-7.
        [http://dx.doi.org/10.1128/AEM.01205-10] [PMID: 21057012]

[26]   Liang Y, Sarkany N, Cui Y. Biomass and lipid productivities of Chlorella vulgaris under autotrophic, heterotrophic and mixotrophic growth conditions. Biotechnol Lett 2009; 31(7): 1043-9.
        [http://dx.doi.org/10.1007/s10529-009-9975-7] [PMID: 19322523]

[27]   Xin L, Hu HY, Ke G, Sun YX. Effects of different nitrogen and phosphorus concentrations on the growth, nutrient uptake, and lipid accumulation of a freshwater microalga Scenedesmus sp. Bioresour Technol 2010; 101(14): 5494-500.
        [http://dx.doi.org/10.1016/j.biortech.2010.02.016] [PMID: 20202827]

[28]   Katiyar R, Bharti KR, Gurjar BR, Kumar A, Biswas S, Pruthi V. Utilization of de-oiled algal biomass for enhancing vehicular quality biodiesel production from *Chlorella sp.* In mixotrophic cultivation systems. Renew Energy 2018; 122: 80-8.
[http://dx.doi.org/10.1016/j.renene.2018.01.037]

[29]   Knothe G. Improving biodiesel fuel properties by modifying fatty ester composition. Energy Environ Sci 2009; 2: 759-66.
[http://dx.doi.org/10.1039/b903941d]

[30]   Ramos MJ, Fernández CM, Casas A, Rodríguez L, Pérez A. Influence of fatty acid composition of raw materials on biodiesel properties. Bioresour Technol 2009; 100(1): 261-8.
[http://dx.doi.org/10.1016/j.biortech.2008.06.039] [PMID: 18693011]

[31]   Arora N, Patel A, Pruthi PA, Pruthi V. Boosting TAG accumulation with improved biodiesel production from novel oleaginous microalgae *Scenedesmus sp.* IITRIND2 utilizing waste Sugarcane Bagasse Aqueous Extract (SBAE). Appl Biochem Biotechnol 2016; 180(1): 109-21.
[http://dx.doi.org/10.1007/s12010-016-2086-8] [PMID: 27093970]

13

Production of Ligninolytic Enzymes from *Penicillium* Sp. and its Efficiency to Decolourise Textile Dyes

Sridevi Ayla[1], Narasimha Golla[2,*] and Suvarnalathadevi Pallipati[1]

[1]*Department of Applied Microbiology, Sri Padmavati Mahila University, Tirupati, A.P, India*
[2]*Applied Microbiology Laboratory, Department of Virology, Sri Venkateswara University, Tirupati -517502, A.P, India*

Abstract:

Background:

The present study discussed the bio decolourization of synthetic textile dyes using extracellular crude laccase from an Ascomycetes fungus *Penicillium* sp. Laccase based decolourization is found to be potentially advantageous to bioremediation technologies.

Methods:

In this study, the production of laccase was observed for 7 days of incubation under shaking conditions. Maximum laccase production was secreted by fungal strain on the 6th day of incubation under submerged fermentation. Incubation of fungal mycelium and culture filtrate as crude enzyme obtained from *Penicillium* sp. with textile dyes - Indigo, Reactive black-5, Acid blue -1 and Vat brown -5 on solid PDA medium and liquid PDA broth showed effective biological dye decolourisation.

Result:

Solid state dye decolourisation had shown 45%, 25%, 50% and 72% colour removal of dyes - Indigo, Reactive black-5, Acid blue -1 and Vat brown -5 whereas maximum decolourization of same dyes of 45%, 20%, 48%, and 75% was obtained in liquid state with crude enzyme within 3h.

Conclusion:

The results had shown the potential dye decolourisation capacity of the *Penicillium* sp. extracellular crude laccase and pave a way to apply this strain on an industrial scale.

Keywords: Laccase, Submerged fermentation, *Penicillium* sp., Synthetic dyes, Decolourization, Ligninolytic enzymes, .

---

1. INTRODUCTION

Laccase (EC 1.10.3.2, *p*-diphenol: dioxygen oxidoreductase), a major biocatalyst known for decades was first described in 1883 by Yoshida and he first reported it from the exudates of the Japanese lacquer tree, *Rhusvernicifera* [1]. They have been found in many taxa including plants, fungi, bacteria, and metazoa [2]. Laccases from fungi have been the focus of intense research due to their possible involvement in the transformation of a wide variety of phenolic compounds including the polymeric lignin and humic substances [3]. Laccases can be secreted extracellular or intracellular. Even though their function is different in the various organisms, all of them can catalyse polymerization or depolymerisation processes [4]. Fungi from Basidiomycota, Ascomycota and Deuteromycota phyla are the sources of fungal laccases, from various ecological niches. They have wide applications in industries such as textile, food, pharmaceuticals and synthetic chemistry as well as an eco-friendly approach with respect to the concept of "green

---

* Address correspondence to this author at Applied Microbiology Laboratory, Department of Virology, Sri Venkateswara University,Tirupati -517502, A.P, India; E-mail: gnsimha123@rediffmail.com

chemistry" [5 - 7]. Synthetic dyes are used extensively in some industries including textile, food, medicine. [8]. Among these industries, the textile industry generates huge effluent and discharge of this effluent is generating a serious problem [9 - 17]. Disposal of untreated dyeing effluent without treatment is a serious offence which causes both environmental and health hazards [18]. Even though recalcitrant synthetic dyes are generally removed by various chemical methods, biological removal was regarded as an inexpensive and effective way of treatment [19]. The extracellular ligninolytic enzymes produced by fungi in pure or crude form can decolourise various textile dyes. There is much research related to laccase and dyes degradation by various white rot fungi like *Funalia trogii* [20], *Phanerochaete chrysosporium* [21, 22], *Trametes versicolor* [23], *Trametes hirsute* [24], and *Lentinula edodes* [24, 25] but Studies on non-basidiomycete fungi that have the ability in production of lignolytic enzymes and degradation of dyes are scanty. Even though these fungi are also efficient for metabolizing a wide range of compounds, particularly by demethylation and oxidation [26], *Aspergillus* species [27], *Cunninghamella elegans* [28], *Penicillium geastrivorus* [29], *P.Ochrochloron* [27] *Penicillium simplicissimum* INCQS 40211 [30] and *Penicillium oxalicum* SAR-3 [31] are found to be successful for removing textile dyes from liquid media, evaluation of more fungal enzymes in remediation of toxic dye effluents is necessary. In this context, the present work was designed to evaluate laccase activity by *Penicillium* sp through submerged fermentation and decolourisation capability of obtained enzyme against synthetic industrial dyes (Indigo, Reactive black-5, Acid blue -1 and Vat brown -5). Thus, the results obtained had explained the ability of an Ascomycetes fungal bio decolourisation .

2. MATERIALS AND METHODS

2.1. Fungal Collection Site Description

Soil samples, decayed wood and tree barks with fruiting bodies were collected in and around Tirupati, Chittor dist., A.P, India. The samples collected were stored in sterile bags and stored in cool place for further isolation.

2.2. Isolation of Fungal Strain

The present fungus collected from decayed wood was isolated on Potato Dextrose Agar (PDA) medium by placing a small piece of wood and was incubated at 30 °C. The fungal colony grown on the plate was microscopically observed with lactophenol cotton blue stain and was recorded at 40X for preliminary morphological characterization. Molecular level identification of the fungus was done by 18s rDNA sequencing and sequence was searched with Blast. The fungus has been assigned the Accession no. and deposited in Genbank.

2.3. Lignolytic Enzymes Production

The isolated fungal strain was screened for lignolytic enzymes production by solid plate method and quantitatively by submerged fermentation. It was done with the inoculation of direct fungal mycelium and spore suspension ($2x10^6$ spores/ml) from a 7-day grown PDA slant. Initially, for the qualitative plate assay, a small piece of direct mycelium was placed on PDA plate supplemented with 0.01% guaiacol and incubated at 30°C for oxidation of guaiacol. Laccase efficiency was calculated by the formation of diameter of brown zone around the fungal colony. Further quantitative estimation of laccase, Lignin peroxidase and Manganese peroxidase activities was done by a 7-day grown fungus ($2x10^6$ spores/ml) of spore suspension in liquid PDA broth with 0.01% guaiacol at 150 rpm for 10 days. The culture filtrate was withdrawn at regular intervals after centrifugation at 5000 rpm for 5 min which was served as crude enzyme and was tested for activity of laccase, Lignin peroxidase and Manganese peroxidase enzymes.

2.4. Enzyme Assays

Assays of listed ligninolytic enzymes activities in the cultural filtrate were done by standard protocols. The reaction mixture containing 0.5ml of crude culture filtrate with 10 mM guaiacol in 10% (V/V) acetone containing 100 mM acetate buffer (pH 5.0) was incubated for 5 min and was assayed for Laccase activity. The change in absorbance of the reaction mixture was monitored at 470 nm ($\varepsilon = 6740$ M^{-1} cm^{-1}) and expressed as Laccase activity [32]. One international unit of activity corresponded to the amount of enzyme that oxidized one micromole of guaiacol per minute. Lignin peroxidase activity was determined within reaction mixture having 0.5 ml culture filtrate, 5 mM H_2O_2 and 0.25M tartaric acid (pH 2.5) and 2 mM veratryl alcohol. The enzyme activity was expressed in IU where one unit of Lignin Peroxidase corresponded to the amount of enzyme that oxidized one micromole of veratryl alcohol per min [33]. Manganese Peroxidase activity was quantified using a reaction mixture containing crude enzyme 1mM guaiacol, 10mM citrate phosphate buffer (pH 5.5), 1mM MnSO4, and 50 µM H_2O_2 in a total volume of 1 ml [34]. The reaction was monitored by measuring the increase in absorbance of the reaction product at 465 nm. One unit of Manganese Peroxidase was defined as the amount of enzyme that increased the absorbance at 465 nm by 1.0 per min.

2.5. Dye Decolourisation Potential (Solid and Liquid Studies)

Dye decolourisation potential of the isolated strain was carried out according to the method of Kiiskinen *et al.* [35]. Indigo, Reactive black-5, Acid blue -1 and Vat brown -5 (each dye at a concentration of 100mg/L) were added individually to 25 ml of 4% potato dextrose agar medium in Petri plates. The pH was adjusted to 5.5 before autoclaving at 121 °C for 15 min. Two millilitres of each dye from stock dyes were added to the media after autoclaving as sterile-filtered water solutions. The positive reaction on the plate was visualized after decolourisation within 3-5 days of incubation at 30 °C and percentage of decolourisation clearing zone was assessed.

The fungal strain was also examined for their ability to decolourise Indigo, Reactive black-5, Acid blue -1 and Vat brown -5 based on the modified method of Eichlerova *et al.* [36]. The 200 ml Erlenmeyer flask supplemented with above said concentration of dyes were inoculated with working culture filtrate(crude extract). To quantify dye decolourisation by UV visible spectrophotometer at 556nm, 640nm, 600nm, 455nm which are maximum absorbance wavelengths of dyes Indigo, Reactive black-5, Acid blue -1 and Vat brown -5. The samples were prepared in duplicate with 2 ml of each dye and 1ml of crude enzyme and a control was also made without the addition of culture filtrate. Percentage of dye decolourisation was calculated using the formula as shown below:

Percentage of dye decolourisation (%) = R % = [(Initial absorbance - final absorbance) / (initial absorbance)] x 100

Where the initial absorbance of the untreated dye and the final absorbance indicated absorbance of dye after treatment with enzyme source.

3. RESULTS AND DISCUSSION

3.1. Morphology, Growth Characters of the Fungus

After 5 days of incubation on PDA medium, the fungal culture was grown as greyish green colonies and designated as 'W' (Figs. **1a** and **b**). The microscopic examination revealed the development of septate mycelium with many branched conidiospores (2-3μm). Based on colony morphology and microscopic features, the fungus was identified as *Penicillium* (Fig. **1c**). Molecular characterisation was done by using ITS 4 and ITS 6 primers and was found that the sequence had a similarity with GU565119.1, GU565118.1 *etc* and was belonging to Ascomycetes family, thus, the strain 'W' based on morphological and phylogenetic analysis was identified as *Penicillium* sp. and was deposited in GenBank under accession no: KY930467.1 (Fig. **2**).

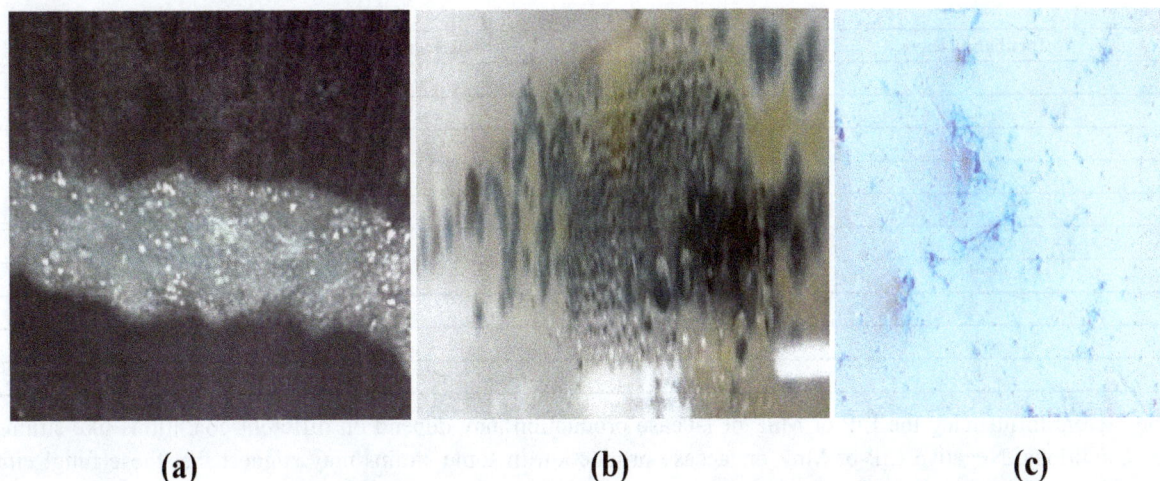

| (a) | (b) | (c) |

Fig. (1). (**a**) Sample collection site (**b**) Isolated fungal strain on PDA slant (**c**) Microscopic observation of isolate under 40 X.

Phylogram
Branch length: ● Cladogram ○ Real

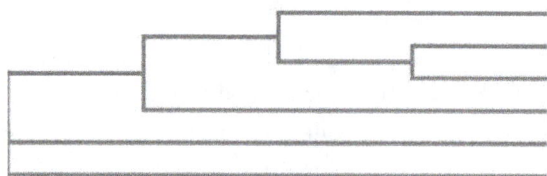

W 0.0122
gi|300684700|Penicillium_sp. 0.00086
gi|507847289|Penicillium_polonicum -0.00086
gi|608609301|Penicillium_cordubense 0
gi|626421393|Penicillium_viridicatum 0.00174
gi|484399651|Penicillium_thomii 0

Fig. (2). Phylogenetic relationship of *Penicillium* sp isolated from wood with other strains.

3.2. Ligninolytic Efficiency of *Penicillium* Sp.

To check for the ability of the present strain for the production of laccase, the strain was grown in PDA medium with guaiacol both on slants and plates. The grown fungal colonies on slant had shown oxidation of guaiacol on the first day itself (Fig. **3**) and produced reddish brown zone around and beneath the colony on PDA slant and plate due to oxidation of guaiacol. The positively screened plate photo was not placed and shown here due to a technical problem. The ligninolytic efficiency was further quantified in the same medium under shaking conditions with an rpm of 180 for the production of laccase, lignin peroxidase and manganese peroxidase enzymes, respectively. Production of laccase was maximum after 5 days of incubation and reached maximum on the 6th day. Later, the enzyme production was decreased from 7th day and was shown in Table **1**. Lignin peroxidase and manganese peroxidase enzymes production by this strain were observed negative.

Fig. (3). Fungal strain showing Laccase positive on PDA slant.

Table 1. Laccase production by *Penicillium* sp. during different days of incubation.

Incubation Days	Laccase Activity(U/Ml)
1	0.1
2	0.6
3	7.1
4	12.6
5	10.4
6	8.3
7	6.1
8	5.2
9	3.1
10	2.4

The factors influencing the LiP or MnP or laccase production may depend on different conditions like strain or culture conditions. Negative LiP or MnP or laccase production in some strains may suggest that these fungi either produce no significant levels of these enzymes or their production requires different conditions. These type of results were true and studied for several strains of *Trametes versicolor*, *Bjerkander aadusta* which are known as LiP producers but do not always produce lignin peroxidase depending on strain or culture conditions [37, 38]. In addition, previous results indicated the production of laccase reaches its maximum value with a delay of twenty-sixth day of incubation with a specific activity of 7.18 U/mg protein in a relatively gradual manner and then falls by *Penicillium martensii NRC 345* [39]. Though the degradation of lignocellulosics by ligninolytic enzymes of *Penicillium* sps. is different from that of the white-rot fungi [40], some reports had described the detection of the laccase, MnP and LiP enzymes by *P. pinophilum* TERIDB1 and *A. aisen* TERIDB6 was described by Pant *et al.* [41]. A similar type of work and secretion of ligninolytic enzymes was observed by Sack and Gunther, 1993 [42]. Recently Yang *et al.* reported the production of LiP by *Penicillium decumbens* [29]. In contrast, Zeng et al stated that the production of ligninolytic enzymes by most white-rot fungi was found to be a 'secondary metabolic' event [43]. Concerning the delay of optimum laccase formation which reached the 26[th]day of incubation and reported that *Botrytis cinerea* produced appreciable levels of laccase (9.8 U/ml) in a brief period (5-7 days) and they stated that, some fungal species required a longer production time (12-30 days) [44].

Crude extracellular culture filtrate of present strain *Penicillium* sp. was tested for the dye decolorizing ability on agar plate amended with four dyes used in this study at a concentration of 100 mg/L. *Penicillium* sp. was active to various levels in decolorizing the four dyes tested and started decolourization of Indigo and Vat brown -5 on the first day whereas Reactive black-5 and Acid blue -1 after 48hrs and with a range of 25-70% of decolourisation in 3 days and it does not proceed further. During decolourisation removal of dye from the plate was also may be due to adsorption of dye to the mycelium as colony growth was observed on plates. To confirm the dye decoloration was due to minimal adsorption and maximum by enzymatic mechanism, liquid decoloration experiments were carried out.

The crude extract after 7 days from fermented broth was evaluated for enzymatic decolourisation against Indigo, Reactive black-5, Acid blue -1 and Vat brown -5. About 75% decolourisation was observed within 3 hours with Vat brown -5 and 48% in case of Acid blue -1 whereas 45% and 20% in Indigo and Reactive black-5, respectively without any mediators (Figs. **4** and **5**, Table **2**).

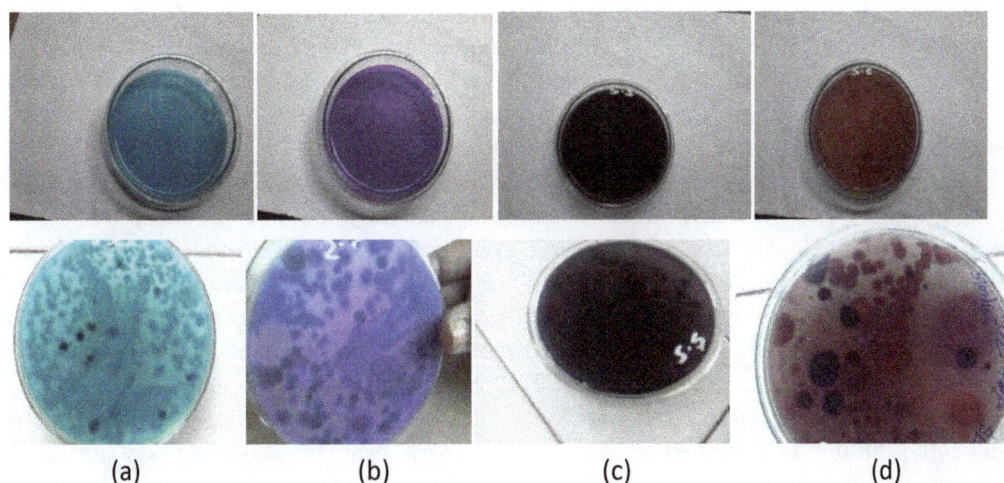

Fig. (4). Solid state decolourisation on plates along with dyes. (**a**) Indigo (**b**) Acid blue -1 (**c**) Reactive black-5 (**d**) Vat brown Top – controls with dyes Bottom - decolourised dyes with *Penicillium Sp.*

Table 2. % of dyes decolourisation by *Penicillium* sp in solid and liquid states.

Dyes	λ Max	% of Dye Decolourisation	
		Solid Decolourisation	Liquid Decolourisation
Indigo	455	45%	45%
Reactive black-5	600	25%	20%
Acid blue -1	640	52%	48%
Vat brown -5	556	70%	75%

Fig. (5). Liquid state decolourisation showing controls and with a crude enzyme derived from *Penicillium* sp.

4. DISCUSSION

Biotechnological methods like isolation of fungi from diverse environments and adopting their enzymes in the degradation of xenobiotics and various dyes are applying globally. The key enzymes that play a major role in bioremediation of environmental pollutants are lignolytic enzymes like manganese peroxidase, lignin peroxidise and laccase. In view of this importance, the present study was undertaken in orde to check the potentiality of the organism in secretion of major extracellular enzymes like laccase, manganese peroxidise and lignin peroxidise and their application in dye decolourisation. Screening different organisms and their enzymes for dye decolorization are beneficial as it may result in more efficient dye decolorizers than the commonly studied organisms. In this context, *Penicillium* sp. isolated from wood in this study was also found to be a laccase producer and production was increased in short incubation period. Results reported by Artexga *et al.* and Levin *et al.* that similar rmycelial results when performing dye degradation experiments using white rot fungi [45, 46]. It is to note in their results that the fungi *Peniophora* sp.hpF-04 *and Phellinus* sp.hpF-17 being predominant laccase producers and decolourisation are due to laccase production as well as media components. Several other authors had also discussed the role of enzymes in the decolourisation activity of ligninolytic fungi [47 - 49]. Number of researchers had worked on fungal decolourization studies to know the main and effective fungal enzymes. Various mechanisms have explained the role of lignin degrading enzymatic system of white-rot fungi and unknown enzymes in fungal dye decolourization. Among the ligninolytic enzymes, laccase is highly studied for its role in decolourisation and detoxification of various industrial and textile dyes. It was also reported that laccase is solely responsible for the decolourisation and degradation of dye [48, 50, 51]. Intracellular enzymes like vanillyl-alcohol oxidase and catalase-peroxidase of *P. simplicissimum* degraded a variety of compounds. The decolourisation of different dyes by many organisms and particularly various sps of *Penicillium* are discussed below. About 70% Reactive Brilliant Red X-3B was decolourised by a new strain *Penicillium* sps QQ [52]. Decolorization of direct blue by *Aspergillus flavus* and *Penicillium canescens* was after 6 days incubation in findings of Hefnawy *et al.* Husseiny [53] who found that the maximum degradation activity of *Penicillium* spp *of* reactive and direct dye was at 35 °C. The findings of Muthezhilan observed that *Penicillium citrinum* decolorized maximum number of dyes like methylene blue, gentian violet, crystal violet, cotton blue, Sudan black, malachite green, methyl red and corbol fushion to a great extent whereas *Penicillium oxalicum* expressed very low amount of color reduction [54]. Intracellular enzymes like vanillyl-alcohol oxidase and catalase-peroxidase of *P. simplicissimum* degraded a variety of compounds [55, 56]. The capacity of decolorization and detoxification of the textile dyes Reactive Red 198 (RR198), Reactive Blue 214 (RB214), Reactive Blue 21 (RB21) and the mixture of the three dyes (MXD) by *Penicillium simplicissimum* INCQS 40211. *Aspergillus flavus* and *Alternaria solani,* the decolorization and degradation of a triphenylmethane dye in 6 days. The study of Zheng *et al.* observed that *Penicillium sp.* removed and degraded the dyes Poly R-478 and Poly S-119 [57]. *Penicillium oxalicum* SAR-3 isolated and screened from dye contaminated soils decolourised and degraded Acid Red 183 (AR 183), Direct Blue 15 (DB 15) and Direct Red 75 (DR 75) at 100 mg L(-1) concentration with its manganese peroxidise enzyme [31]. High decolorization or degradation dyes also takes place either by laccase alone or with mediators [58, 59] and reduces toxicity also [58, 60]. The enzyme producing activity of White rot fungi makes them effective decolorizers and biode-graders [61 - 66]. Chung and Aust had reported that with laccase-mediator, there was high biodegradation than laccase alone [67]. Mediators such as phenolic compounds, ABTS, HBT, and 3HAA could enhance dye degradation [68]. Laccase of *Cyathus bulleri* decolorized and detoxified 80% the mixture of reactive and acidic dyes in presence of natural and synthetic mediators [69]. Complex mixtures of dyes were decolorized with an enzyme in the presence of HOBT [70]. Studies of Machado and Husain, 2007 obtained maximum decoloration in a complex mixture of dyes with the enzyme and 0.06 mM redox mediator 1hydroxybenzotriazole/violuric acid [71]. Whereas treatment of anthraquinone dyes with purified enzyme along with mediator degraded azo dyes [72]. *Momordica charantia* immobilised peroxidase was found to be effective in decolourization textile waste Water [73]. *Tricho santhesdioica* peroxidase- concanavalin A (PGP-Con A) complex, entrapped into calcium alginate pectin gel decolourised a mixture of two dyes [74]. Some other reports *e.g.* Rodriguez *et al.*, 2005, Wong and Yu, Gochev and Krastanov, Ismat Bibi are also explained the role of mediator along with laccase in dye decolourisation [75 - 78]. To remove color from dye effluents purified and/or immobilized enzym is used which is not economical. But the use of effective crude enzyme is economically viable and the present study explained very easy production and preparation of crude enzyme from this fungus to decolorize synthetic dyes without the role of mediators.

CONCLUSION

The Ascomycetes fungus *Penicillium* sp. had produced laccase under submerged fermentation. Four dyes were bio catalytically decolorized by extracellular laccase secreted by the present strain in solid and liquid cultivation. It was confirmed that the decolourisation was correlated with laccase and that *Penicillium sp.* was capable of decolorizing a range of industrially important textile dyes. The difference in decolourisation rates was according to dye types and due to various structures of the dyes. In conclusion, fungal decolourisation appears to be a promising alternative to replace currently available chemical treatment processes.

HUMAN AND ANIMAL RIGHTS

No animals/humans were used for studies that are the basis of this review.

ACKNOWLEDGEMENTS

First Author, grantee of UGC-PDF is very grateful to the UGC in India for financial support and also thankful to the Sri Padmavathi mahila University authorities for smooth conduct of research work.

REFERENCES

[1]   Thurston CF. The structure and function of fungal laccases. Microbiology 1994; 140: 19-26.
        [http://dx.doi.org/10.1099/13500872-140-1-19]

[2]   McMullan G, Meehan C, Conneely A, *et al*. Microbial decolourisation and degradation of textile dyes. Appl Microbiol Biotechnol 2001; 56(1-2): 81-7.
        [http://dx.doi.org/10.1007/s002530000587] [PMID: 11499950]

[3]   Baldrian P. Fungal laccases - occurrence and properties. FEMS Microbiol Rev 2006; 30(2): 215-42.
        [http://dx.doi.org/10.1111/j.1574-4976.2005.00010.x] [PMID: 16472305]

[4]   Riva S. Laccases: Blue enzymes for green chemistry. Trends Biotechnol 2006; 24(5): 219-26.
        [http://dx.doi.org/10.1016/j.tibtech.2006.03.006] [PMID: 16574262]

[5]   Rodgers CJ, Blanford CF, Giddens SR, Skamnioti P, Armstrong FA, Gurr SJ. Designer laccases: A vogue for high-potential fungal enzymes? Trends Biotechnol 2010; 28(2): 63-72.
        [http://dx.doi.org/10.1016/j.tibtech.2009.11.001] [PMID: 19963293]

[6]   Jeon JR, Baldrian P, Murugesan K, Chang YS. Laccase-catalysed oxidations of naturally occurring phenols: From *in vivo* biosynthetic pathways to green synthetic applications. Microb Biotechnol 2012; 5(3): 318-32.
        [http://dx.doi.org/10.1111/j.1751-7915.2011.00273.x] [PMID: 21791030]

[7]   Rivera-Hoyos CM, Morales-Alvarez ED, Poutou Pinales RA, Pedroza-Rodriguez AM, Rodriguez-Vazquez R, Delgado-Boada JM. Fungal laccases. Fungal Biol Rev 2013; 27: 67-82.
        [http://dx.doi.org/10.1016/j.fbr.2013.07.001]

[8]   Azmi W, Sani RK, Banerjee UC. Biodegradation of triphenylmethane dyes. Enzyme Microb Technol 1998; 22(3): 185-91.
        [http://dx.doi.org/10.1016/S0141-0229(97)00159-2] [PMID: 9463944]

[9]   Kalyani DC, Telke AA, Dhanve RS, Jadhav JP. Ecofriendly biodegradation and detoxification of Reactive Red 2 textile dye by newly isolated
        Pseudomonas sp. SUK1. J Hazard Mater 2009; 163(2-3): 735-42.
        [http://dx.doi.org/10.1016/j.jhazmat.2008.07.020] [PMID: 18718713]

[10]   Nigam P, Banat IM, Singh D, Marchant R. Microbial process for the decolorization of textile effluent containing azo, diazo and reactive dyes.
        Process Biochem 1996; 31: 435-42.
        [http://dx.doi.org/10.1016/0032-9592(95)00085-2]

[11]   Murugesan K, Kalaichelvan PT. Synthetic dye decolourization by white rot fungi. Indian J Exp Biol 2003; 41(9): 1076-87.
        [PMID: 15242299]

[12]   Khan IT, Jain V. Effects of Textile industry waste water on growth and some biochemical parameters of Triticum aestivum. J Environ Poll
        1995; 2: 50.

[13]   Schliephake K, Mainwaring DE, Lonergan GT, Jones IK, Baker WL. Transformation and degradation of the disazo dye Chicago Sky Blue by
        a purified laccase from Pycnoporus cinnabarinus. Enzyme Microb Technol 2000; 27(1-2): 100-7.
        [http://dx.doi.org/10.1016/S0141-0229(00)00181-2] [PMID: 10862908]

[14]   Chang JO-Shu, Chien Chou, Chen Shan-Yu. Decolorization of azo dyes with immobilized Pseudomonas luteola. Process Biochem 2001; 36:
        757-63.
        [http://dx.doi.org/10.1016/S0032-9592(00)00274-0]

[15]   Rafu MA, Salman Ashraf S. Removal of colored compounds from textile industry effluents. J Hazard Mater 2009; 166: 6-16.
        [PMID: 19128875]

[16]   Hu TL. Removal of reactive dyes from aqueous solution by different bacterial genera. Water Sci Technol 1992; 26: 357-66.
        [http://dx.doi.org/10.2166/wst.1992.0415]

[17]   Chen KC, Wu JY, Liou DJ, et al. Decolorization of the textile. bacterial strains. J Biotechnol 2003; 101: 57-68.
        [http://dx.doi.org/10.1016/S0168-1656(02)00303-6] [PMID: 12523970]

[18]   Shedbalkar U, Dhanve R, Jadhav J. Biodegradation of triphenylmethane dye cotton blue by Penicillium ochrochloron MTCC 517. J Hazard
        Mater 2008; 157(2-3): 472-9.
        [http://dx.doi.org/10.1016/j.jhazmat.2008.01.023] [PMID: 18282658]

[19]   Robinson T, McMullan G, Marchant R, Nigam P. Remediation of dyes in textile effluent: A critical review on current treatment technologies
        with a proposed alternative. Bioresour Technol 2001; 77(3): 247-55.
        [http://dx.doi.org/10.1016/S0960-8524(00)00080-8] [PMID: 11272011]

[20]   Yesilada O, Cing S, Asma D. Decolourisation of the textile dye Astrazon Red FBL by Funalia trogii pellets. Bioresour Technol 2002; 81(2):
        155-7.
        [http://dx.doi.org/10.1016/S0960-8524(01)00117-1] [PMID: 11762908]

[21]   Abadulla E, Tzanov T, Costa S, Robra KH, Cavaco-Paulo A, Gübitz GM. Decolorization and detoxification of textile dyes with a laccase
        from Trametes hirsuta. Appl Environ Microbiol 2000; 66(8): 3357-62.
        [http://dx.doi.org/10.1128/AEM.66.8.3357-3362.2000] [PMID: 10919791]

[22]   Martins MA, Queiroz MJ, Silvestre AJD, Lima N. Relationship of chemical structures of textile dyes on the pre-adaptation medium and the
        potentialities of their biodegradation by Phanerochaete chrysosporium. Res Microbiol 2002; 153(6): 361-8.
        [http://dx.doi.org/10.1016/S0923-2508(02)01332-3] [PMID: 12234010]

[23]   Ramsay JA, Nguyen T. Decoloration of textile dyes by Trametes versicolor and its effect on dye toxicity. Biotechnol Lett 2002; 24: 1757-61.
        [http://dx.doi.org/10.1023/A:1020644817514]

[24]   Boer CG, Obici L, de Souza CG, Peralta RM. Decolorization of synthetic dyes by solid state cultures of Lentinula (Lentinus) edodes
        producing manganese peroxidase as the main ligninolytic enzyme. Bioresour Technol 2004; 94(2): 107-12.
        [http://dx.doi.org/10.1016/j.biortech.2003.12.015] [PMID: 15158501]

[25]   Hatvani N, Mécs I. Effects of certain heavy metals on the growth, dye decolorization, and enzyme activity of Lentinula edodes. Ecotoxicol
        Environ Saf 2003; 55(2): 199-203.
        [http://dx.doi.org/10.1016/S0147-6513(02)00133-1] [PMID: 12742369]

[26]   Cha CJ, Doerge DR, Cerniglia CE. Biotransformation of malachite green by the fungus Cunninghamella elegans. Appl Environ Microbiol
        2001; 67(9): 4358-60.
        [http://dx.doi.org/10.1128/AEM.67.9.4358-4360.2001] [PMID: 11526047]

[27]   Abd El-Rahim WM, Moawad H. Enhancing bioremoval of textile dyes by eight fungal strains from media supplemented with gelatine wastes
        and sucrose. J Basic Microbiol 2003; 43(5): 367-75.
        [http://dx.doi.org/10.1002/jobm.200310267] [PMID: 12964179]

[28]   Ambrósio ST, Campos-Takaki GM. Decolorization of reactive azo dyes by Cunninghamella elegans UCP 542 under co-metabolic conditions.
        Bioresour Technol 2004; 91(1): 69-75.
        [http://dx.doi.org/10.1016/S0960-8524(03)00153-6] [PMID: 14585623]

[29]   Yang Q, Yang M, Pritsch K, Yediler A, Hagn A. Decolourisation of synthetic dyes and production of manganese-dependent peroxidase by
        new fungal isolates. Biotechnol 2003.

[30]   Fraaije MW, Mattevi A, van Berkel WJ. Mercuration of vanillyl-alcohol oxidase from *Penicillium* simplicissimum generates inactive dimers. FEBS Lett 1997; 402(1): 33-5.
[http://dx.doi.org/10.1016/S0014-5793(96)01494-9] [PMID: 9013853]

[31]   Saroj S, Kumar K, Pareek N, Prasad R, Singh RP. Biodegradation of azo dyes acid red 183, direct blue 15 and direct red 75 by the isolate *Penicillium oxalicum* SAR-3. Chemosphere 2014; 107: 240-8.
[http://dx.doi.org/10.1016/j.chemosphere.2013.12.049] [PMID: 24418068]

[32]   Das N, Sengupta S, Mukherjee M. Importance of laccase in vegetative growth of *Pleurotus florida*. Appl Environ Microbiol 1997; 63(10): 4120-2.
[PMID: 16535720]

[33]   Tien M, Kirk TK. Methods in enzymology – Biomass, part B, Lignin, Pectin and Chitin. San Diego, CA: Academic Press Inc, 1988;161: 238-49.

[34]   Bonnen AM, Anton LH, Orth AB. Lignin degrading enzymes of the commercial button mushroom, *Agaricus bisporus*. Appl Environ Microbiol 1994; 60(3): 960-5.
[PMID: 16349223]

[35]   Kiiskinen LL, Rättö M, Kruus K. Screening for novel laccase-producing microbes. J Appl Microbiol 2004; 97(3): 640-6.
[http://dx.doi.org/10.1111/j.1365-2672.2004.02348.x] [PMID: 15281946]

[36]   Eichlerova I, Homolka L, Nerud F. Decolourisation of high concentrations of synthetic dyes by the white rot fungus *Bjerkandera adusta* strain CCBAS 232. Dyes Pigments 2007; 75: 38-44.
[http://dx.doi.org/10.1016/j.dyepig.2006.05.008]

[37]   Kirk TK, Farrell RL. Enzymatic "combustion": The microbial degradation of lignin. Annu Rev Microbiol 1987; 41: 465-505.
[http://dx.doi.org/10.1146/annurev.mi.41.100187.002341] [PMID: 3318677]

[38]   Waldner R, Leisola MSA, Fiechter A. Comparison of ligninolytic activities of selected white-rot fungi. Appl Microbiol Biotechnol 1988; 29: 400-7.
[http://dx.doi.org/10.1007/BF00265826]

[39]   Yu HY, Zeng GM, Huang GH, Huang DL, Chen YN, Homolka L. Lignin degradation by *Penicillium simplicissimum*.. Huan Jing Ke Xue 2005; 26(2): 167-71.
[PMID: 16004322]

[41]   Pant D, Adholeya A. Biological approaches for treatment of distillery wastewater: A review. Bioresour Technol 2007; 98(12): 2321-34.
[http://dx.doi.org/10.1016/j.biortech.2006.09.027] [PMID: 17092705]

[42]   Sack and Gunther. Metabolism of PAH by fungi and correlation with extracellular enzymatic activities. J basic microbial 2007; 39(4): 269-77.

[43]   Zeng GM, Yu HY, Huang HL, Huang DL, Chen YN. Laccase activities of a soil fungus *Penicillium simplicissimum*in relation to lignin degradation. World J Microbiol Biotechnol 2006; 22: 317-24.
[http://dx.doi.org/10.1007/s11274-005-9025-0]

[44]   Fortina MG, Acquti A, Rossi P, Manachini PL, Gennaro CD. Production of laccase by *Botrytis ci-nerea* and fermentation studies with strain F226. J Ind Microbiol 1996; 17: 69-72.
[http://dx.doi.org/10.1007/BF01570044]

[45]   Aretxaga A, Romero S, Sarrà M, Vicent T. Adsorption step in the biological degradation of a textile dye. Biotechnol Prog 2001; 17(4): 664-8.
[http://dx.doi.org/10.1021/bp010056c] [PMID: 11485427]

[46]   Levin L, Papinutti L, Forchiassin F. Evaluation of Argentinean white rot fungi for their ability to produce lignin-modifying enzymes and decolorize industrial dyes. Bioresour Technol 2004; 94(2): 169-76.
[http://dx.doi.org/10.1016/j.biortech.2003.12.002] [PMID: 15158509]

[47]   Sathiyamoorthi P, Periyarselvam S, Sasikalaveni A, Murugesan K, Kalaichelvan PT. Decolourisation of textile dyes and their effluents using white rot fungi. Afr J Biotechnol 2007; 6: 424-9.

[48]   Liu W, Chao Y, Yang X, Bao H, Qian S. Biodecolorization of azo, anthraquinonic and triphenylmethane dyes by white-rot fungi and a laccase-secreting engineered strain. J Ind Microbiol Biotechnol 2004; 31(3): 127-32.
[http://dx.doi.org/10.1007/s10295-004-0123-z] [PMID: 15069603]

[49]   Verma P, Madamwar D. Decolorization of synthetic textile dyes by lignin peroxidase of *Phanerochaete chrysosporium*. Folia Microbiol (Praha) 2002; 47(3): 283-6.
[http://dx.doi.org/10.1007/BF02817653] [PMID: 12094739]

[50]   Rodríguez E, Pickard MA, Vazquez-Duhalt R. Industrial dye decolorization by laccases from ligninolytic fungi. Curr Microbiol 1999; 38(1): 27-32.
[http://dx.doi.org/10.1007/PL00006767] [PMID: 9841778]

[51]   Casieri L. Biodegradation and Biosorption of synthetic dyes by fungi. PhD thesis. University of Turin (Italy): Department of plant biology 2005.

[52]   Gou M, Qu Y, Zhou J, Ma F, Tan L. Azo dye decolorization by a new fungal isolate, *Penicillium sp.* QQ and fungal-bacterial cocultures. J Hazard Mater 2009; 170(1): 314-9.

[http://dx.doi.org/10.1016/j.jhazmat.2009.04.094] [PMID: 19473759]

[53]   Husseniy M. Biodegradation of reactive and direct dyes using Egyptian isolates. J Appl Sci Res 2008; 4: 599-606.

[54]   Muthezhilan R, Yogananth N, Vidhya S, Jayalakshmi S. Dye degrading mycoflora from industrial effluents. Resear J Microbiol 2008; 3: 204-8.
        [http://dx.doi.org/10.3923/jm.2008.204.208]

[55]   Fraaije MW, Mattevi A, van Berkel WJ. Mercuration of vanillyl-alcohol oxidase from *Penicillium simplicissimum* generates inactive dimers. FEBS Lett 1997; 402(1): 33-5.
        [http://dx.doi.org/10.1016/S0014-5793(96)01494-9] [PMID: 9013853]

[56]   Fraaije MW, Roubroeks HP, Hagen WR, Van Berkel WJH. Purification and characterization of an intracellular catalase-peroxidase from *Penicillium simplicissimum*. Eur J Biochem 1996; 235(1-2): 192-8.
        [http://dx.doi.org/10.1111/j.1432-1033.1996.00192.x] [PMID: 8631329]

[57]   Zheng ZX, Levin RE, Pinkham JL, Shetty K. Decolorization of polymeric dyes by a novel *Penicillium* isolate. Process Biochem 1999; 34: 31-7.
        [http://dx.doi.org/10.1016/S0032-9592(98)00061-2]

[58]   Kasturi K, Mohan V, Sridhar S, Pati B, Sarma P. Water Res 2009; 43: 3647-58.
        [http://dx.doi.org/10.1016/j.watres.2009.05.028] [PMID: 19540548]

[59]   Murugesan K, Yang I, Kim Y, Jeon J, Chang Y. Enhanced transformation of malachite green by laccase of *Ganoderma lucidum* in the presence of natural phenolic compounds. Appl Microbiol Biotechnol 2009; 82(2): 341-50.
        [http://dx.doi.org/10.1007/s00253-008-1819-1] [PMID: 19130052]

[60]   Abadulla E, Tzanov T, Costa S, Robra KH, Cavaco-Paulo A, Gübitz GM. Decolorization and detoxification of textile dyes with a laccase from *Trametes hirsuta*. Appl Environ Microbiol 2000; 66(8): 3357-62.
        [http://dx.doi.org/10.1128/AEM.66.8.3357-3362.2000] [PMID: 10919791]

[61]   Ponraj M, Gokila K, Zambare V. Bacterial decolorization of textile dye-orange 3R. Int J Adv Biotechnol Res 2011; 2(1): 168-77.

[62]   Ganappriya M, Logambal K, Ravikuma R. Investigation of direct red dye using *Aspergillus niger and Aspergillus flavus* under static and shacking conditions with modeling. Int J Sci Environ Technol 2012; 1(3): 144-5.

[63]   Radha K, Balu V. Kinetic study on decolorization of the dye acid orange the fungus *Phanerochate Chrysosporium*. J Med Appl Sci 2009; 3(7): 38-47.

[64]   Shertate RS, Thora PR. Biodecolorizarion and degradation of textile diazo dye reactive blue 171 by *Marinobactor* Sp. NB-6 -A bioremedial aspect. Int J Pharma Bio Sci 2013; 3(1): 330-442.

[65]   Tripathi ASK, Harsh NSK, Gupta N. Fungal treatment of industrial effluents: A mini-review. Life Sci J 2007; 4(2): 78-81.

[66]   McMullan G, Meehan C, Conneely A, *et al.* Microbial decolourisation and degradation of textile dyes. Appl Microbiol Biotechnol 2001; 56(1-2): 81-7.
        [http://dx.doi.org/10.1007/s002530000587] [PMID: 11499950]

[67]   Chung N, Aust SD. Veratryl alcohol-mediated indirect oxidation of phenol by lignin peroxidase. Arch Biochem Biophys 1995; 316(2): 733-7.
        [http://dx.doi.org/10.1006/abbi.1995.1097] [PMID: 7864628]

[68]   Andrea C, Carlo G, Patrizia G. Phenolic compounds as likely natural mediators of laccase: A mechanistic assessment. J Molecular Catalysis B Enzymes 2008; 51: 118-20.

[69]   Chhabra M, Mishra S, Sreekrishnan TR. Mediator-assisted decolorization and detoxification of textile dyes/dye mixture by Cyathus bulleri laccase. Appl Biochem Biotechnol 2008; 151(2-3): 587-98.
        [http://dx.doi.org/10.1007/s12010-008-8234-z] [PMID: 18506632]

[70]   Kulshrestha Y, Husain Q. Decolorization and degradation of acid dyes mediated by salt fractionated turnip (*Brassica rapa*) peroxidases. Toxicol Environ Chem 2007; (89): 255-67.
        [http://dx.doi.org/10.1080/02772240601081692]

[71]   Machado KM, Compart LC, Morais RO, Rosa LH, Santos MH. Biodegradation of reactive textile dyes by basidiomycetous fungi from Brazilian ecosystems. Braz J Microbiol 2006; (37): 481-7.
        [http://dx.doi.org/10.1590/S1517-83822006000400015]

[72]   Zeng X, Cai Y, Liao X, Zeng X. Anthraquinone dye assisted the decolorization of azo dyes by a novel *Trametestrogii* laccase. Process Biochem 2012; 47: 160-3.
        [http://dx.doi.org/10.1016/j.procbio.2011.10.019]

[73]   Akhtar S, Khan AA, Husain Q. Potential of immobilized bitter gourd (*Momordica charantia*) peroxidases in the decolorization and removal of textile dyes from polluted wastewater and dyeing effluent. Chemosphere 2005; 60(3): 291-301.
        [http://dx.doi.org/10.1016/j.chemosphere.2004.12.017] [PMID: 15924947]

[74]   Jamal F, Singh S, Khatoon S, Mehrotra S. Application of immobilized pointed gourd (*Trichosanthesdioica*) peroxidase-concanavalin A complex on calcium alginate pectin gel in decolorization of synthetic dyes using batchprocesses and continuous two reactor system. J Bioprocess Biotech 2013; (3): 1-5.

[75]   Rodríguez Couto S, Sanromán M, Gübitz GM. Influence of redox mediators and metal ions on synthetic acid dye decolourization by crude laccase from *Trametes hirsuta*. Chemosphere 2005; 58(4): 417-22.
[http://dx.doi.org/10.1016/j.chemosphere.2004.09.033] [PMID: 15620733]

[76]   Wong YX, Yu J. Laccase-catalyzed decolorizat ion of synthetic dyes. Water Res 199(33): 3512-20.

[77]   Gochev VK, Krastanov AI. Fungal and laccases. Bulg J Agric Sci 2007; 13: 75-83.

[78]   Bibi I, Bhatti HN. Biodecolorization of reactive black 5 by laccase-mediator system. Afr J Biotechnol 2012; 11(29): 7464-71.

Detection and Molecular Characterization of *Vibrio Parahaemolyticus* in Shrimp Samples

Daryoush Asgarpoor[1, 2], Fakhri Haghi[2] and Habib Zeighami[2,*]

[1]*Student Research Center, Zanjan University of Medical Sciences, Zanjan, Iran*
[2]*Department of Microbiology, Zanjan University of Medical Sciences, Zanjan, Iran*

Abstract:

Background:

Food safety has emerged as an important global issue with international trade and public health implications. Bacterial pathogens as *Vibrio parahaemolyticus* recognized as an important cause of foodborne diseases related to the consumption of raw, undercooked or mishandled seafood worldwide.

Methods:

A total of 70 individual wild shrimp samples were collected from shrimp retail outlets in Zanjan, Iran and investigated for the presence of potentially pathogenic strains of *V. parahaemolyticus.* The shrimp samples were immediately homogenized and cultured on TCBS agar and subjected to confirmatory biochemical tests. Polymerase Chain Reaction (PCR) was performed for detection of total and pathogenic *V. parahaemolyticus* by amplification of *vp–toxR, tdh* and *trh* genes.

Results:

The conventional method indicated that 16 (22.8%) of samples were positive for *V. parahaemolyticus*. However, PCR verified that only 12 (17.1%) shrimp samples were positive for *V. parahaemolyticus*. Of the 70 shrimp samples in our study, only 2 (2.8%) *tdh* and 1 (1.4%) *trh* positive strains were identified.

Conclusion:

Detection of *tdh* and/ or *trh* positive *V. parahaemolyticus* in shrimp marketed in Zanjan, Iran shows a probable risk for public health. Therefore, the reliable molecular methods for monitoring of potentially pathogenic *V. parahaemolyticus* are strongly recommended for the routine seafood examination.

Keywords: *Vibrio parahaemolyticus*, Shrimp, PCR, Molecular characterization, *vp–toxR*, *tdh*, *trh* genes.

1. INTRODUCTION

Vibrio parahaemolyticus is a halophilic marine bacterium and some strains can cause gastroenteritis in humans through the consumption of raw, undercooked or mishandled contaminated seafood [1, 2]. Although the gastroenteritis caused by *V. parahaemolyticus* is often self–limited and characterized by diarrhea, headache, vomiting, nausea, abdominal cramps and low fever, the infection may cause septicemia, a life-threatening infection, in immunocompromized patients [3 - 5]. The pathogenic strains of *V. parahaemolyticus* are characterized by the production of Thermo stable Direct Hemolysin (TDH) and/or TDH–Related Hemolys in·(TRH) that can lyse the red blood cells on Wagatsuma blood agar (referred to as the Kanagawa phenomenon) and encoded by *tdh* and *trh* genes,

* Address correspondence to this author at the Department of Microbiology, Zanjan University of Medical Sciences, Zanjan, Iran, E-mail: zeighami@zums.ac.ir

respectively [3, 6]. The ubiquitous nature of *Vibrio* spp. in marine and estuarine environments makes it impossible to obtain seafood completely free of these species. It has been implicated in several outbreaks of seafood poisoning worldwide [6, 7]. Previous epide miological studies showed that *V. parahaemolyticus* is an important cause of foodborne disease in Asia, South America and the United States. The *V. parahaemolyticus* is frequently isolated from shellfish including oysters, clams, mussels, lobsters, crabs, shrimps and cockles, which provide an excellent substrate for the growth of these micro organisms in the aquatic habitats [8 - 13]. Many studies have been carried out on shellfish and findings concerning the distribution of *V. parahaemolyticus* in oysters and mussels are well documented. However, few data are available forcrustaceans, despite the popularity of crabs and shrimps and their rising consumption worldwide [14]. Shrimp is one of the most important fishery products, and shrimp farming is an important economy characteristic of Iran [13].The frequency of pathogenic *V. parahaemolyticus* infrozen ready–to–eat shrimps for human consumption was recently studied, and 7 to 8% of samples tested positive for *tdh* or *trh* virulence genes in countries such as Malaysia. Therefore, these shrimps might have the potential to cause *V. parahaemolyticus*–associated infections if consumed without further processing [14].

In recent years, *V. parahaemolyticus* has been recognized as the causative agent of 50–70% of all cases of gastroenteritis associated with consumption of seafood [5].

There are different methods for detection of *V. parahaemolyticus* in seafood samples. The most–probable–number (MPN) method is used for enumeration of *V. parahaemolyticus* from food and water, but this method is cumbersome and the recovery of the organism is low [15].The culture-based approaches and PCR technique which is faster, easier and more sensitive can be used for identification of *V. parahaemolyticus* in seafood samples [10]. The *V. parahaemolyticus* strains possess a regulatory gene, *toxR*, which is present in all strains irrespective of their ability to produce *tdh* and/or *trh*.Therefore, the PCR targeted to the *toxR* gene can be used as a method for identification at the species level [16].

The objective of this study was to determine the frequency of pathogenic *V. parahaemolyticus* in wild shrimp samples using the culture and PCR methods based on detection of *tdh* and *trh* virulence genes in Zanan, Iran.

2. MATERIALS AND METHODS

2.1. Sampling

From March to June 2015, a total of 70 individual wild shrimp samples were collected from shrimp retail outlets in Zanjan, Iran. Shrimp samples were packed into a clean polyethylene bag then marked and transported to the laboratory of food microbiology in a cooler with ice packs for analysis within 1 h.

2.2. Reference Strain

The *V. parahaemolyticus* ATCC 17802 waskindly obtained from the Pasteur Institute of Iran and was grown on Thiosulphate Citrate Bile Salt Sucrose (TCBS) Agar (MERCK, Darmstadt, Germany).

2.3. Isolation and Identification of *V. Parahaemolyticus*

The isolation and biochemical identification of *V. parahaemolyticus* were carried out as recommended in the FDA's Bacteriological Analytical Manual. Twenty five gram of the samples were homogenized for 60 s in a stomacher (Heidolph, Schwabach, Germany) with 225 mL of alkaline peptone water (APW) containing 3% NaCl and then incubated for enrichment at 37°C for 24 h. After primary enrichment, a loopful (without shaking the flask) from each of the enriched homogenate was streaked onto the surface of Thiosulfate Citrate Bile Sucrose (TCBS) agar plates (MERCK, Darmstadt, Germany)and incubated at 37°C for 24 h.On TCBS plates, sucrose negative colonies (green or blue-greencolonies with 2–3 mm in diameter), were picked up and inoculated into tryptonesoya broth with 3% NaCl, incubated at 37°C for 24 h, then purified onto nutrient agar slants with 3% NaCl and subjected for confirmatory biochemical tests using different media contained 2.5% NaCl. Every single colony was screened for Gram staining, motility, oxidase and urease activity, NaCl requirement, citrate utilization, triple sugar iron agar, arginine dehydrolase, lysine and ornithine decarboxylase, O/129 sensitivity, vogese proskauer, indole and acid production from lactose, arabinose, cellobiose, mannitol and mannose.

2.4. Genomic DNA Extraction

A colony of *V. parahaemolyticus* (one colony per sample) was picked from nutrient agar and inoculated into 5 ml of

LB (Luria Bertani Broth, Merck) until the exponential phase with 2 McFarland turbidity with shaking at 120 rpm at 37°C. One ml from an overnight culture in LB was spun at 8,000 rpm for 5 min. The supernatant was discarded and the cell pellet was resuspended in 200μL sterile deionized water and boiled at 100°C for 10 min. The tube was immediately placed in ice for 5 min; then the cell lysate was centrifuged at 13000 r.p.m. for 3 min to pellet the cell debris and the clear supernatant was transferred to a new tube. A 5 μl aliquot of supernatant was used for PCR.

2.5. Molecular Confirmation of *V. Parahaemolyticus* Isolates

Confirmation of presumptive *V. parahaemolyticus* isolates was performed using PCR targeting the *vp–toxR* gene with the following primers and amplicon size 368bp: *toxR–F*:5'–GTCTTCTGACGCAATCGTTG–3' and *toxR–R*:5'–ATACGAGTGGTTGCTGTCATG–3' [2]. A specific primer pairs were also used for detection of *tdh* and *trh* virulence genes in *toxR* positive strains with the following sequences: *tdh–F*:5'–CCACTACCACTCTCATATGC–3' , *tdh–R*:5'–GGTACTAAATGGCTGACATC–3' with amplicon size 251 bp and *trh–F*:5'GGCTCAAAATGGTTAAGCG –3' and *trh–R*:5'–CATTTCCGCTCTCATATGC–3' with amplicon size 250 bp [10]. Single PCR was performed using Dream Taq PCR Master Mix (Thermo Fisher Scientific), which contains Taq polymerase, dNTPs, $MgCl_2$ and the appropriate buffer. Each PCR tube contained 25 μl reaction mixture composed of 12.5 μl of the master mix, 2.5 μl of each forward and reverse primer solution (in a final concentration of 200 nM), 5 μl of DNA and nuclease–free water to complete the final volume. PCR was performed using the Gene Atlas 322 system (ASTEC) with the same cycling conditions for *toxR, tdh* and *trh* genes. Amplification involved an initial Denaturation at 94°C, 5 min followed by 30 cycles of denaturation (94°C, 1 min), annealing (57°C, 1.5 min) and extension (72°C, 1.5 min), with a final extension step (72°C, 8 min). The amplified DNA was separated by submarine gel electrophoresis on 1.5% agarose, stained with ethidium bromide and visualized under UV trans illumination.

3. RESULTS AND DISCUSSION

The *V. parahaemolyticus* is an enteric human pathogen that occurs naturally in the marine and estuarine environments worldwide. Several outbreaks of seafood poisoning were caused by *V. parahaemolyticus* in many countries and regions of the world including USA, Japan, India and Taiwan [10]. In this study, a total of 70 individual wild shrimp samples were studied for the presence of pathogenic *V. parahaemolyticus*. A conventional cultural method based on the appearance of green or blue– green colonies on TCBS agar and microscopic examination was detected presumptive *V. parahaemolyticus* in 30 (42.8%) out of the 70 shrimp samples. However, the biochemical tests of the presumptive *V. parahaemolyticus* strains indicated that 22.8% (16/70) of shrimp samples were positive for *V. parahaemolyticus*. Variable incidences of *V. parahaemolyticus* in seafood had been demonstrated using conventional methods. The frequency of *V. parahaemolyticus* in our study was lower than some previous studies. According to Abd–Elghany & Sallam [10] and Quintoil *et al* [17], the frequency of *V. parahaemolyticus* in shellfish samples was 33.3%, and 36.8%, respectively. However, lower incidence of *V. parahaemolyticus* in seafood samples was reported from Italy and Netherlands with 6.2%, 24.3% and 8%, respectively [1, 18, 19].

This variation in *V. parahaemolyticus* frequency among seafood samples may be due to the difference of the geographical region, type of shellfish sample, watersalinity, seasons of sampling, post-harvest practices and hygienic standards applied during the handling, transport and storage of seafood products, as well as the methods used for isolation and identification of the organism [10].

Fast and accurate diagnosis of food–borne pathogens is very important for a positive outcome of eradication programs. PCR based methods which target the conserved region of *V. parahaemolyticus* such as *toxR* gene is more efficient, reliable and faster compare to the conventional techniques [20]. In our study, the biochemically identified isolates were further verified using PCR targeting the *vp–toxR* gene. It has been indicated that 12 (17.1%) samples out of a total 70 shrimp samples were positive for *vp–toxR* gene. In the present study, the frequency of *V. parahaemolyticus* positive samples based on *vp–toxR* gene, was approximately similar to those reported by Abd–Elghany & Sallam in Egypt [10], who found that 16.7% of shellfish samples were positive for *vp–toxR* gene and also by Hassan *et al.* [18] in the Netherlands, who detected that 19% (38/200) of retailed shellfish samples were positive for *toxR* gene. Only a few reports on the frequency of *V. parahaemolyticus* in seafood samples from Iran have been previously published. According to the previous reports from Iran, 9.3% and 11% of the shrimp samples [13, 21] and 21.4% of the fish samples [22] were positive for the presence of this pathogen.

As the presence of *V. Parahaemolyticus* strains carrying *tdh* and/or *trh* genes in seafood represents a public health risk, their detection would be of paramount importance. It is well known that only 1–2% of the environmental strains possess the *tdh* gene. Of the 70 shrimp samples in our study, only 2 (2.8%) *tdh* and 1 (1.4%) *trh* positive strains were identified. In the previous report from Iran, the prevalence of *tdh*–positive and *trh*–positive *V. parahaemolyticus* was 1.7% and 0.7%, respectively [21]. Similar to our results, in a study conducted in Malaysia, 5 (3.9%) and 1 (0.78%) strains isolated from live and frozen shrimp, respectively were positive for *tdh* gene, whilst 2 (1.56%) and 1 (0.78%) strains were positive for *trh* gene [23]. However, higher incidence of virulence *V. parahaemolyticus* was identified in several studies such as the study conducted in Turkey by Terzi *et al.* [24] who found that 24 (75%) out of the 32 strains isolated from mussel were potentially pathogenic depending on *tdh* and *trh* genes.

CONCLUSION

In conclusion, the detection of *tdh* and/or *trh* positive *V. parahaemolyticus* in shrimp marketed in Zanjan, Iran shows a probable risk for public health. Therefore, intensive and continuous monitoring of potentially pathogenic *V. parahaemolyticus* are strongly recommended in order to evaluate the human health risk arising from seafood consumption.

HUMAN AND ANIMAL RIGHTS

No Animals/Humans were used for studies that are base of this research.

ACKNOWLEDGEMENTS

This investigation was supported by Student Research Center, Zanjan University of Medical Sciences, Zanjan, Iran (A-12-392-10, ZUMS.REC.1393.265).

REFERENCES

[1]   Di Pinto A, Ciccarese G, De Corato R, *et al.* Detection of pathogenic *Vibrio parahaemolyticus* in southern Italian shellfish. Food Control 2008; 19: 1037-41.
        [http://dx.doi.org/10.1016/j.foodcont.2007.10.013]

[2]   Fujino T, Okuno Y, Nakada D, *et al.* On the bacteriological examination of shirasu food poisoning. Med J Osaka Univ 1953; 4: 299-304.

[3]   Drake SL, DePaola A, Jaykus LA. An overview of *Vibrio vulnificus* and *Vibrio parahaemolyticus*. Compr Rev Food Sci Food Saf 2007; 6: 120-44.
        [http://dx.doi.org/10.1111/j.1541-4337.2007.00022.x]

[4]   Gopal S, Otta SK, Kumar S, Karunasagar I, Nishibuchi M, Karunasagar I. The occurrence of Vibrio species in tropical shrimp culture environments; implications for food safety. Int J Food Microbiol 2005; 102(2): 151-9.
        [http://dx.doi.org/10.1016/j.ijfoodmicro.2004.12.011] [PMID: 15992615]

[5]   Wang JJ, Sun WS, Jin MT, *et al.* Fate of *Vibrio parahaemolyticus* on shrimp after acidic electrolyzed water treatment. Int J Food Microbiol 2014; 179: 50-6.
        [http://dx.doi.org/10.1016/j.ijfoodmicro.2014.03.016] [PMID: 24727382]

[6]   Chowdhury NR, Chakraborty S, Ramamurthy T, *et al.* Molecular evidence of clonal *Vibrio parahaemolyticus* pandemic strains. Emerg Infect Dis 2000; 6(6): 631-6.
        [http://dx.doi.org/10.3201/eid0606.000612] [PMID: 11076722]

[7]   Su Y-C, Liu C. *Vibrio parahaemolyticus*: A concern of seafood safety. Food Microbiol 2007; 24(6): 549-58.
        [http://dx.doi.org/10.1016/j.fm.2007.01.005] [PMID: 17418305]

[8]   Hosseini H, Cheraghali AM, Yalfani R, *et al.* Incidence of *Vibrio* Spp. in shrimp caught off the south coast of Iran. Food Control 2007; 15: 187-90.
        [http://dx.doi.org/10.1016/S0956-7135(03)00045-8]

[9]   Fuenzalida L, Armijo L, Zabala B, *et al. Vibrio parahaemolyticus* strains isolated during investigation of the summer 2006 seafood related diarrhea outbreaks in two regions of Chile. Int J Food Microbiol 2007; 117(3): 270-5.

[http://dx.doi.org/10.1016/j.ijfoodmicro.2007.03.011] [PMID: 17521760]

[10]   Abd Elghany SM, Sallam KI. Occurrence and molecular identification of *Vibrio parahaemolyticus* in retail shellfish in Mansoura, Egypt. Food Control 2013; 33: 399-405.
[http://dx.doi.org/10.1016/j.foodcont.2013.03.024]

[11]   Rahimi E, Tajbaksh E, Fadaeifard F, *et al.* Prevalence of *Vibrio* Spp. in marine shrimp (*Paeneus monodon*) caught off the persian gulf coast of Iran. Iranian J Food Sci Technol 2001; 21-6.

[12]   Suffredini E, Mioni R, Mazzette R, *et al.* Detection and quantification of *Vibrio parahaemolyticus* in shellfish from Italian production areas. Int J Food Microbiol 2014; 184: 14-20.
[http://dx.doi.org/10.1016/j.ijfoodmicro.2014.04.016] [PMID: 24810197]

[13]   Zarei M, Borujeni MP, Jamnejad A, *et al.* Seasonal prevalence of vibrio species in retail shrimps with an emphasis on *Vibrio Parahaemolyticus.* Food Cont 2012; 25: 107-9.
[http://dx.doi.org/10.1016/j.foodcont.2011.10.024]

[14]   Robert-Pillot A, Copin S, Gay M, Malle P, Quilici ML. Total and pathogenic *Vibrio parahaemolyticus* in shrimp: Fast and reliable quantification by real-time PCR. Int J Food Microbiol 2010; 143(3): 190-7.
[http://dx.doi.org/10.1016/j.ijfoodmicro.2010.08.016] [PMID: 20843573]

[15]   Raghunath P, Acharya S, Bhanumathi A, Karunasagar I, Karunasagar I. Detection and molecular characterization of *Vibrio parahaemolyticus* isolated from seafood harvested along the southwest coast of India. Food Microbiol 2008; 25(6): 824-30.
[http://dx.doi.org/10.1016/j.fm.2008.04.002] [PMID: 18620975]

[16]   Kim YB, Okuda J, Matsumoto C, Takahashi N, Hashimoto S, Nishibuchi M. Identification of *Vibrio parahaemolyticus* strains at the species level by PCR targeted to the toxR gene. J Clin Microbiol 1999; 37(4): 1173-7.
[PMID: 10074546]

[17]   Quintoil NM, Porteen K, Pramanik AK. Studies on occurrence of *Vibrio parahaemolyticus* in fin fishes and shellfishes from different ecosystem of West Bengal. Livest Res Rural Dev 2007; 19: 215-9.

[18]   Hassan ZH, Zwart kruis Nahuis JT, de Boer E. Occurrence of *Vibrio parahaemolyticus* in retailed seafood in The Netherlands. International Food Research Journal 2012. 19: 39e43

[19]   Ottaviani D, Santarelli S, Bacchiocchi S, *et al.* Presence of pathogenic *Vibrio parahaemolyticus* strains in mussels fromthe Adriatic Sea, Italy. Food Microbiol 2005. 22: 585e590.

[20]   Khan JA, Rathore RS, Khan S, Ahmad I. *In vitro* detection of pathogenic *Listeria monocytogenes* from food sources by conventional, molecular and cell culture method. Braz J Microbiol 2014; 44(3): 751-8.
[http://dx.doi.org/10.1590/S1517-83822013000300013] [PMID: 24516442]

[21]   Rahimi E, Ameri M, Doosti A, Gholampour AR. Occurrence of toxigenic *Vibrio parahaemolyticus* strains in shrimp in Iran. Foodborne Pathog Dis 2010; 7(9): 1107-11.
[http://dx.doi.org/10.1089/fpd.2010.0554] [PMID: 20528175]

[22]   Basti AA, Misaghi A, Salehi TZ, *et al.* Bacterial pathogens in fresh, smoked and salted Iranian fish. Food Control 2006; 17: 183-8.
[http://dx.doi.org/10.1016/j.foodcont.2004.10.001]

[23]   Sujeewa AK, Norrakiah AS, Laina M. Prevalence of toxic genes of *Vibrio parahaemolyticus* in shrimps (*Penaeus monodon*) and culture environment. Int Food Res J 2009. 16: 89e95.

[24]   Terzi G, Büyüktanir O, Yurdusev N. Detection of the tdh and trh genes in *Vibrio parahaemolyticus* isolates in fish and mussels from middle black seacoast of Turkey. Let Appl Microbiol 2009; 49: 757e -63.

15

Comparison of Antibacterial Activity of ZnO Nanoparticles Fabricated by Two Different Methods and Coated on Tetron Fabric

Ebrahim Zohourvahid Karimi[*] and Mohammad Ansari

Department of Metallurgy and Ceramic, Mashhad Branch, Islamic Azad University, Mashhad, Iran

Abstract:

Background:

Zinc Oxide Nanoparticles (ZnO NPs) have wide applications in various industries, especially they have been known for their antibacterial effects in polymers and textile fibers. ZnO NPs were produced by two different solutions and milling methods. Different techniques were used in order to select the most effective methods for coating the fabric with ZnO NPs. The microstructures and the composition of the ZnO NPs were investigated using Field Emission Scanning Electron Microscopy (FE-SEM) coupled with Energy Dispersive X-ray Spectroscopy (EDS) and X-ray diffraction analysis (XRD). Additionally, the antibacterial activity of the treated fabric against *Staphylococcus* aureus and *Escherichia coli* bacteria was investigated. The overall experimental findings show that the highest inhibitory effect against *Staphylococcus* aureus in the sample of fabric which covered with ZnO NPs synthesized by the solution method.

Methods:

In the solution method, ZnO NPs were synthesized by dissolving zinc chloride in 1, 2 Ethanediol and mixing with aqueous solution of sodium hydroxide. In milling method, firstly, zinc sulfide nanoparticles were prepared through reaction between zinc acetate and Thioacetamide and then by milling and oxidation the zinc sulfide nanoparticles, ZnO NPs were synthesized. In order to deposition ZnO NPs on the Tetron fabric, it was fully drawn and fixed on a frame. After that, acrylic copolymer resin was added into distilled water and ZnO NPs were added in another beaker to ethanol. The two beakers were then placed in the ultrasonic bath for a certain time. Finally, the fabric was dipped into the beaker containing resin for some moment and then immersed into the beaker containing ZnO NPs. During these processes, both beakers were in the ultrasonic bath. After drawing out the fabric from second beaker, it was dried in air. This procedure was performed for both types of ZnO NPs fabricated by two mentioned methods. Antibacterial activity of ZnO NPs coated on the fabric against two types of bacteria was studied by agar diffusion method.

Results:

XRD patterns of synthesized powders from both methods were identified as ZnO NPs. Sharp diffraction peaks indicate good crystallinity of ZnO NPs. The morphology of the ZnO NPs fabricated by both methods which was analyzed by field emission SEM shows that the ZnO particles synthesized by milling and solution methods are in nano scale at the range of 26 - 29 nm and 9 - 11 nm, respectively. The highest inhibitory effect against Staphylococcus aureus was shown for the fabric which coated by ZnO NPs produced by the solution method. It was seen, the antibacterial activity of ZnO NPs fabricated by solution method was higher than that of milling method.

Conclusion:

ZnO NPs were synthesized by two different methods and the antibacterial activity of Tetron fabric coated with ZnO NPs was studied. Distribution and stability of ZnO NPs on the fabric depend on fabrication method and particle size which means that the smaller particles have more stability and better distribution than larger particles. The particle size and deposited concentration of ZnO NPs were effective on antibacterial activity, so that the smaller particles tend less agglomeration and have more surface area and because

---

* Address correspondence to this author at the Department of Metallurgy and Ceramic, Mashhad Branch, Islamic Azad University, Mashhad, Iran; E-mail: ebrahimzk88@gmail.com

of that better antibacterial activity. Overall the results demonstrated a good antibacterial activity against *Staphylococcus aureus* than *Escherichia coli*.in the sample of fabric which covered with ZnO NPs synthesized by the solution method.

Keywords: ZnO nanoparticles, Antibacterial, Fabric, ZnO synthesis, Coating, Tetron fabric.

1. INTRODUCTION

Inorganic nanoparticles recently have been considered as non-viral vectors [1]. Some of the inorganic materials and metal oxides are known as a safe material for humans and animals and they include Zinc Oxide (ZnO), Magnesium Oxide (MgO), Copper Oxide (CuO), Calcium Oxide (CaO), Titanium Oxide (TiO$_2$) and Silver (Ag) [2 - 5]. Zinc oxide and silver nanoparticles are used for infectious diseases because of their antimicrobial properties and also zinc oxide has been used in the formulation of personal care products [3, 6, 7]. Zinc oxide has more usage in different areas such as semiconductor materials (UV light emitting diodes, laser diodes, solar cells and acoustic devices) [8 - 10], photo catalysts (due to their high activity, used for degradation of pollutants in water) [11], protecting oil paintings on paper (against dirt, fungal attack, and UV aging) [12]. Moreover, ZnO NPs stability under harsh processing conditions and relatively low toxicity together with the strong antimicrobial properties fortify their application as antimicrobials [2]. Also, many researchers have shown that some nanoparticles, such as ZnO NPs, have selective toxicity to bacteria and only exhibit minimum effect on human cells [13]. Therefore, ZnO NPs can be used in textile industry for adding the additional properties like antibacterial for using in hospital such as healing, hygienic and medical applications [14, 15]. Many attempts to investigate the antimicrobial effects of zinc oxide nanoparticles on the fabric have been performed. Abramova *et al.* [16] coated cotton fabric by ultrasonic method with ZnO NPs and evaluated its antibacterial effect against Escherichia coli. The result showed the agglomerated particles on cotton fabric with good resistance against *Escherichia coli* microorganism. Perelshtein *et al.* [17] covered cotton bandages with ZnO NPs under ultrasound irradiation and evaluated its antibacterial effect against *Escherichia coli* (Gram negative) and *Staphylococcus aureus* (Gram positive). Their results showed that the cotton bandages coated with 30nm ZnO NPs exhibited antibacterial effect against both microorganisms. C. Balakumar *et al.* [18] prepared ZnO NPs by wet chemical method and directly applied on to the 100% cotton woven fabric using pad-dry-cure method. They showed that the finished fabric demonstrated significant antibacterial activity against *Staphylococcus aureus* in both qualitative and quantitative tests. In this research, ZnO NPs were prepared by two different methods (solution and milling) and were deposited under ultrasonic waves onto the tetron fabric and then their antibacterial properties against two types of bacteria were studied.

2. EXPERIMENTAL PROCEDURE

2.1. Materials

Zinc acetate (purity >99.5%) and Thioacetamide (purity >99.0%) were purchased from Merck company. Also, Zinc chloride (>90%), 1, 2 Ethanediol (>90%), NaOH (>95%) and acrylic copolymer resin (DM5 from Azar chemical company) were prepared for the experiment.

2.2. Synthesis of ZnO NPs by Milling Method

The method includes two steps. In this process, zinc sulfide nanoparticles were produced thereafter zinc oxide nanoparticles synthesized. In the first step, zinc sulfide nanoparticles were prepared through reaction between Zinc Acetate (Zn (CH$_3$COO)$_2$.2H$_2$O) and Thioacetamide (CH$_3$CSNH$_2$). At first, zinc acetate was dried in an oven at 150°C for 30 min. Then, zinc acetate and thioacetamide were milled individually using a planetary ball mill, under the rate of 280 rpm for 20 min. Ball to powder ratio was 20:1. Afterwards, zinc acetate and thioacetamide with the reaction stoichiometric ratio were mixed together (15.64 grams zinc acetate and 5.53 grams thioacetamide) and then milled for 30 min. At the final, the milled powder mixture was heated in an oven at 150°C for 1 hour to achieve zinc sulfide nanoparticles. In the second step, ZnO NPs were synthesized through oxidation of ZnS NPs. For this purpose, ZnS NPs were heated at 600°C in an oven for 4 hours [19].

2.3. Synthesis of ZnO NPs by Solution Method

All chemicals used were purchased from Merck Company (Germany) and were of analytical grade. 11 grams zinc chloride were dissolved in 400 ml of 1, 2 Ethanediol in a beaker. The solution was stirred using a hot plate with magnetic stirrer until zinc chloride totally dissolved in 1, 2 Ethanediol. The temperature of the beaker was kept constant at 150°C. In the meantime, 40 grams sodium hydroxide (NaOH) was dissolved in 200 ml distilled water in a separate container. 32 ml of prepared sodium hydroxide solution was added in the form of drops to the 1, 2 Ethanediol solution under constant stirring. The resulting solution changed into a white colloid without any obvious precipitation. After adding the whole sodium hydroxide, it was given 30 min to do reaction. The solution was allowed to sediment and the supernatant solution was removed by washing 5 times with distilled water. Finally, the sediment was collected by filtering and dried in the air and then changed into powder form [20].

2.4. Deposition of ZnO NPs on the Tetron Fabric

Both types of ZnO NPs which prepared with previous methods were used in this section. The tetron fabric which is made of 65% polyester/35% rayon blend was cut into 8×8 cm^2 and it was washed by 2g Sodium Lauryl Sulfate (SLS) in 150 ml distilled water in a beaker to remove the surface pollution. Then, the fabric was fully drawn and fixed on a frame. After that, 5g of acrylic copolymer resin was added into 550 ml distilled water and the same time 0.1g ZnO nanoparticles was added in another beaker containing 500 ml ethanol 96%. The two beakers were then placed in the ultrasonic bath for 10 min. After this time, the fabric was dipped into the beaker containing resin for some moment. Finally, the fabric was immersed into the beaker containing ZnO NPs for 30 min. During these processes both beakers were in the ultrasonic bath. After drawing out the fabric from second beaker it was dried in air. This procedure was done for both types of ZnO NPs fabricated by two mentioned methods.

3. EVALUATION OF ANTIBACTERIAL ACTIVITY

3.1. Preparation of Microbial Strains

Bacterial strains included *Staphylococcus aureus* ATCC 25923 (PTCC 1431) and *Escherichia coli* ATCC 25922 (PTCC 1399) which were purchased from the collection of bacteria of Iran.

3.2. Fabric Discs Sterilization

Samples containing antibacterial compounds (ZnO NPs) as well as control sample were cut into discs with a diameter of 10 mm and put them into the glass plates. The plates were then sterilized for 15 minutes at 120° C autoclave.

3.3. Study of Antibacterial Effects

Diffusion method was used to investigate the antibacterial effects of ZnO NPs on fabric samples [21]. At first, one loop of standard strain culture media was cultured on the surface of plates. Then, sterile fabrics discs with a diameter of 10 mm were fixed on the surface of culture medium. After that, the plates were incubated for 24 hours at 37° C. In the disk diffusion method 1.5×108 CFU /ml (equivalent to 0.5 McFarland standards) of standard culture of each strain was cultured on agar surface at the first step, then it was spread on the surface of agar by sterile glass spreader. Antibacterial activity was observed as inhibition zone on Petri plates. Size of the inhibition zone was measured in millimeters using a metric ruler [22, 23].

4. CHARACTERIZATION

Field emission scanning electron microscopy was carried out using a Mira 3-XMU to evaluate structure and morphology of ZnO NPs. X-ray diffraction analysis was carried out using Philips X'pert diffractometer with Cu-kα (λ=1.54 Å) radiation operated at 40 kv and 30 mA to detect impurities and phases. The crystallite sizes of samples were estimated using the scherrer method. The diffraction angle was varied between 10° and 80°. Ultrasonic bath with 28 KHz frequency was used to disperse the ZnO NPs in the solution.

5. RESULTS AND DISCUSSION

5.1. X-ray Diffraction Analysis

Fig. (1a) shows the X-ray diffraction pattern of the ZnS particles precursor fabricated by milling and heating method. The peaks at $2\theta = 28.53, 47.45, 56.30$ are appointed to the (111), (220) and (311) reflection lines of Cubic ZnS particles, respectively. Fig. (1b) shows the X-ray diffraction pattern of the synthesized ZnO NPs by milling method. The reflections were indexed according to the diffraction pattern of hexagonal wurtzite-type ZnO. This pattern is in good agreement with reports [19]. The peaks at $2\theta = 31.83, 34.48, 36.32, 47.60, 56.66, 62.97, 68.01, 69.09, 72.49, 77.17$ are appointed to the (100), (002), (101), (102), (110), (103), and (112) reflection lines of hexagonal ZnO particles, respectively. Sharp diffraction peaks shown in Fig. (**1b**) indicate good crystallinity of ZnO NPs. The peak with very low height at $2\theta = 28.71$ is attributed to ZnS phase which indicates that the calcination is not done completely. Grain size of products was calculated using Debye-Scherrer equation:

$$D = 0.9\, \lambda / B\, cos\, \theta \qquad (1)$$

Where D is the average grain size, λ is the wavelength of X-ray which is equal to 1.54Å, θ is the Bragg diffraction angle, and B is the line broadening at half the maximum intensity (FWHM), in radians. Due to the Debye-Scherrer formula, the grain size of nanoparticles was 8 nm.

Fig. (**1c**) shows the X-ray diffraction pattern of the particles synthesized by the solution method. XRD patterns of powder were identified as ZnO NPs indicating that the solution method also leads to zinc oxide particles. Based on the Debye-Scherrer equation, the grains size is about 2 nm. Fig. (**2**) shows the different thermal heating regimes for milled samples. It shows that the peaks of ZnS as an impurity decrease with increasing time and temperature. Therefore, the optimum time and temperature were chosen from the heating regimes.

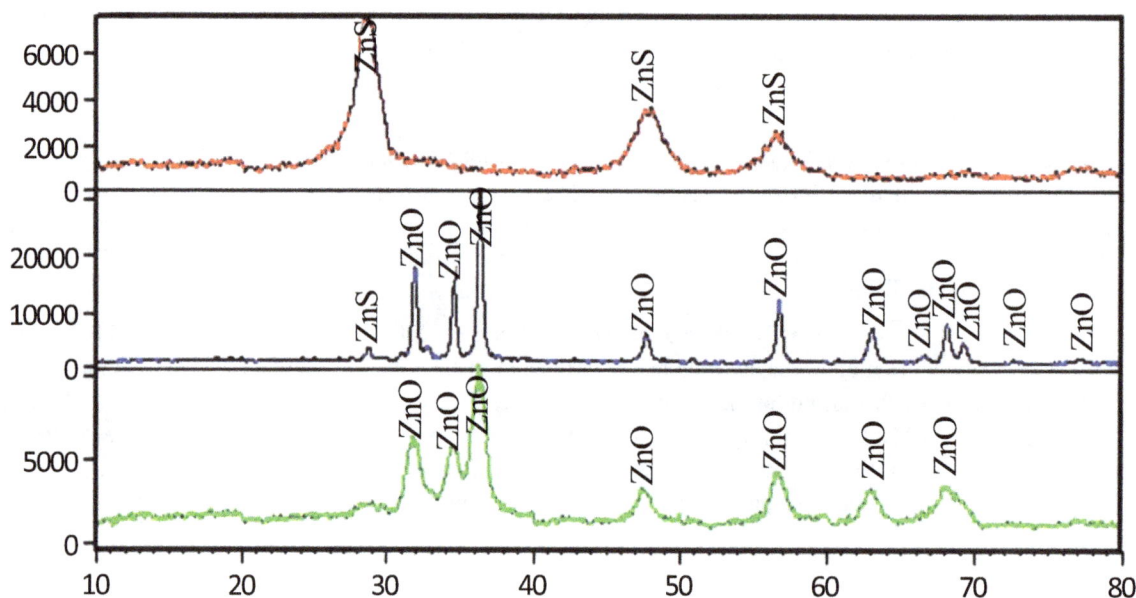

Fig. (1). XRD pattern of the (**a**) ZnS particles synthesized by milling method (**b**) ZnO particles synthesized by milling method (**c**) ZnO particles synthesized by solution method.

5.2. Field Emission SEM

The morphology of the ZnO NPs fabricated by both milling and solution methods which was analyzed by field emission SEM is presented in Fig. (**3**). Fig. (**3a, b**) shows the ZnO particles synthesized by milling method which are in nano scale at the range of 26 – 29 nm. Fig. (**3c, d**) shows the ZnO particles produced by solution method with the particle size of 9 to 11 nm.

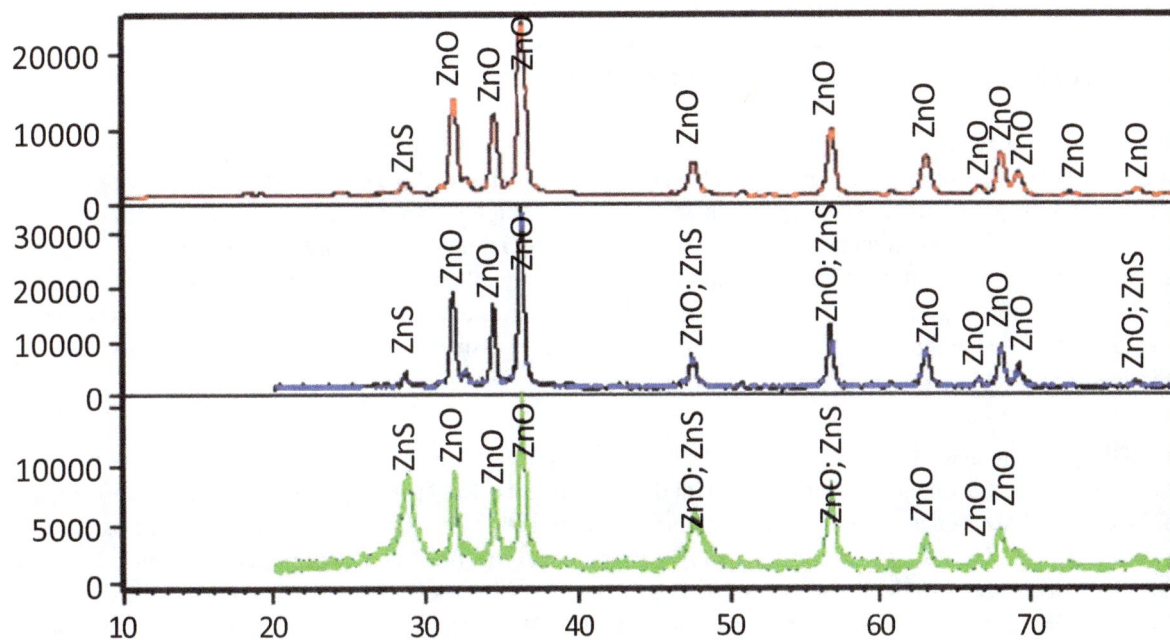

Fig. (2). Thermal comparison of Zinc oxide nanoparticles synthesized by milling method.

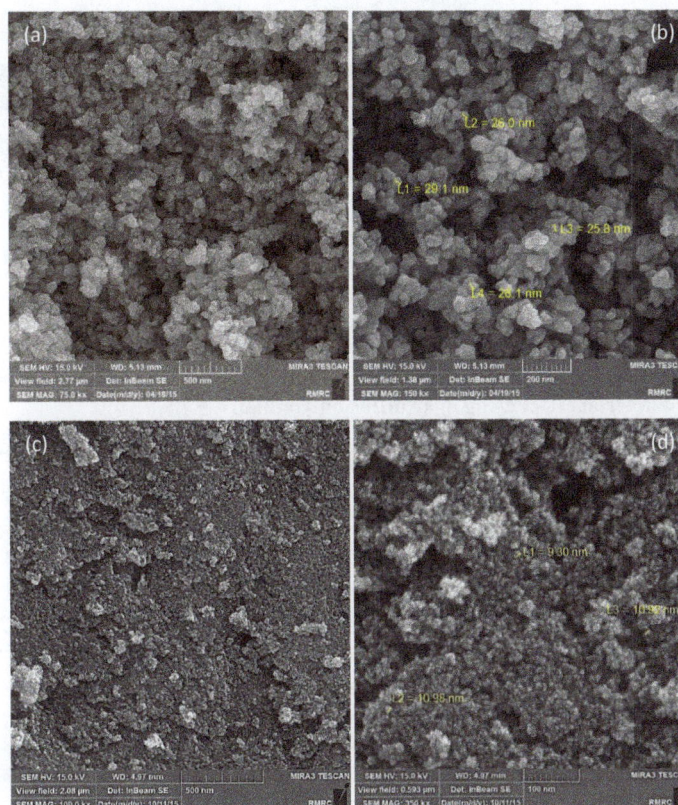

Fig. (3). FE-SEM images of ZnO NPs synthesized by (**a, b**) milling method (**c, d**) solution method.

After coating the surface of Tetron fabric with ZnO NPs, the morphology and distribution of the powders were analyzed using a field emission scanning electron microscope. SEM images of the sample coated with ZnO NPs produced by milling and solution methods are presented in Figs. (**4** and **5**), respectively. The SEM images of the treated fabrics presented Zinc oxide NPs deposited and embedded onto the surface of fabric fibers. Figs. (**4** and **5**), the top left images, show several fabric fibers coated by ZnO NPs in lower magnification. The scale bar is indicated 20 micrometers which is relatively equal to fibers width. It can also be seen from Figs. (**4a** to **d**) and (**5a** to **d**), that the distribution of ZnO NPs fabricated by solution method on the surface of the fibers is much better and the agglomeration is much less than that of milling method. From the SEM images the higher aggregation density of ZnO nanoparticles is observed on the fabric fibers in solution method and it causes to achieve better antibacterial activity and stability because of the smaller size and distribution of ZnO NPs in solution method. The large agglomerated particles of ZnO seen in SEM micrographs are probably due to sintering of ZnS powders during calcination treatment. The composition of the nanoparticles on the fabric was proven by quantitative energy dispersive X-ray elemental microanalysis (EDS) shown in Fig. (**6**).

The EDS spectrums show the peaks of Zn, C and O. The presence of C is probably due to the fabric and the peak in 2.7 Kev is appointed to S which is in agreement with the XRD results. The agar diffusion method is a relatively quick and easily executed semi-quantitative test to assess antibacterial activity of diffusible antibacterial agents on treated textile material [21]. The use of zinc oxide nanoparticles due to the reactive oxygen species formed from these nanoparticles, which results in the production of hydrogen peroxide from the surface of zinc oxide can be effectual for the inhibition of bacterial growth and antibacterial activity [24]. Therefore, antibacterial activity of ZnO NPs against two types of bacteria was studied by agar diffusion method (Fig. **7**). According to this method, the antibacterial activity of fabrics was shown by the diameter of the zone of inhibition in comparison to the control fabric. The experiment was performed six times and the mean value was taken. Fabric discs (10 mm) with or without ZnO NPs coating were tested and the results are presented in Table **1**. As shown in Table **1**, the highest inhibitory effect against *Staphylococcus aureus* is related to the sample 1 which coated by ZnO NPs produced by solution method. The average inhibition zone was 14 mm.

The average inhibition zone diameter of the fabric sample 2 coated by ZnO NPs fabricated by milling method was about 12.2 mm. As a consequence, the antibacterial activity of ZnO NPs fabricated by solution method was higher than that of milling method as well as the small size and high surface-to-volume ratio of nanoparticles allow for better interaction with bacteria [4] On the other hand, the concentration of H_2O_2 increases with decreasing particle size [25]. This is most likely because of the ZnO particles size which is smaller in sample 1 than that of sample 2 [26, 27]. Although antibacterial activity is observed against *Staphylococcus aureus*, the fabrics coated with ZnO NPs do not show any inhibition against *Escherichia coli*. This could be related to the concentration of ZnO NPs on fabric which was not enough to prevent bacterial growth. It should be noted that Gram-negative bacteria such as *E. coli* are more resistant to antibacterial compounds than Gram-positive bacteria [28].

Table 1. Evaluation of the antibacterial effect using agar diffusion method.

Fabric Samples	Microorganisms	Inhibition Zone Diameter (mm)						Average	Standard. Deviation
		1	2	3	4	5	6		
Control	*Staphylococcus aureus* (PTCC 1431)	0	0	0	0	0	0	0	0
	Escherichia coli (PTCC 1399)	0	0	0	0	0	0	0	0
Sample 1 (solution)	*Staphylococcus aureus* (PTCC 1431)	15	15	13	13	14	14	14	0.89
	Escherichia coli (PTCC 1399)	0	0	0	0	0	0	0	0
Sample 2 (milling)	*Staphylococcus aureus* (PTCC 1431)	15	12	12	11	11	12	12.2	1.47
	Escherichia coli (PTCC 1399)	0	0	0	0	0	0	0	0

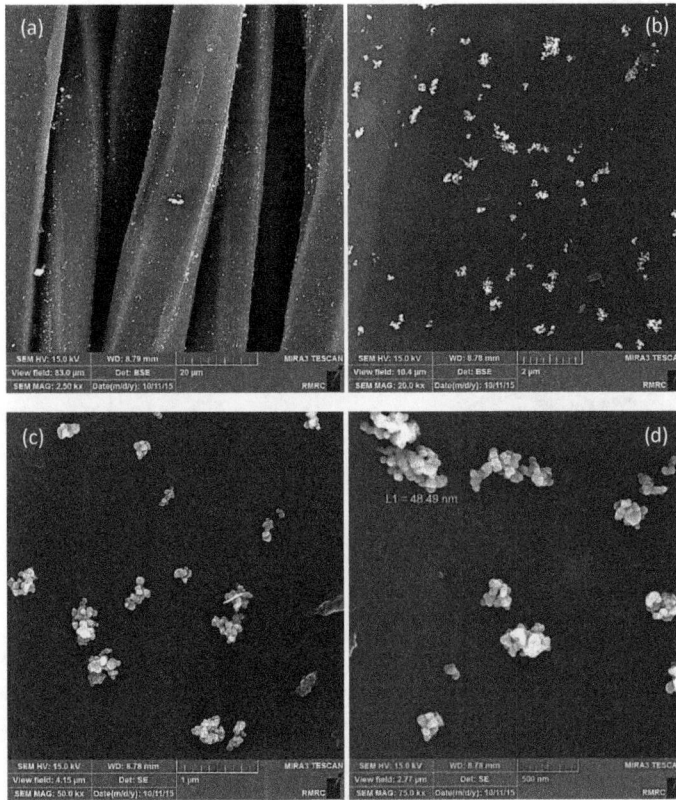

Fig. (4). ZnO NPs produced by milling method and deposited on fabric by immersion.

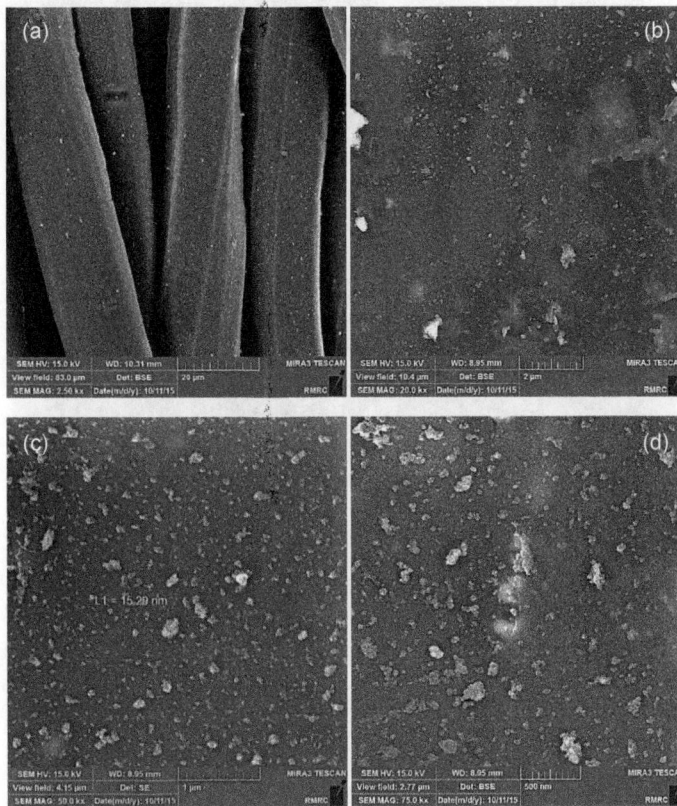

Fig. (5). FE-SEM images of ZnO NPs produced by solution method and deposited on fabric by immersion.

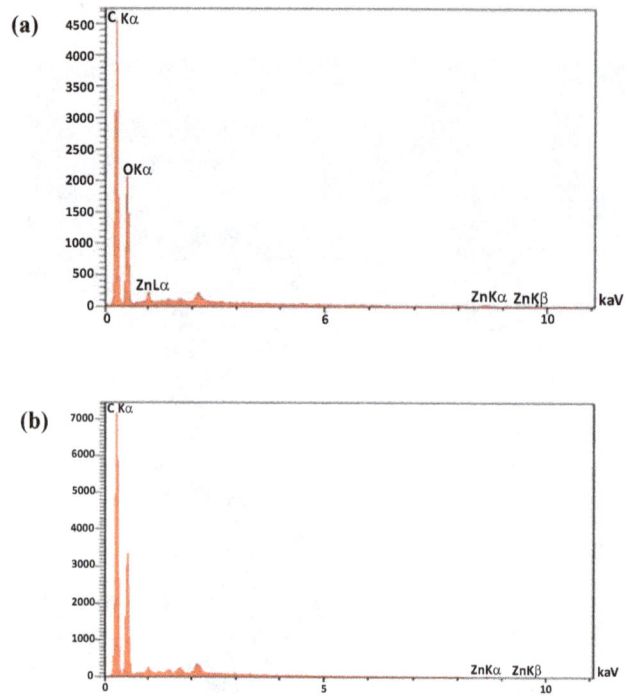

Fig. (6). EDS analysis of ZnO NPs produced by (**a**) milling method (**b**) solution method and deposited on fabric by immersion.

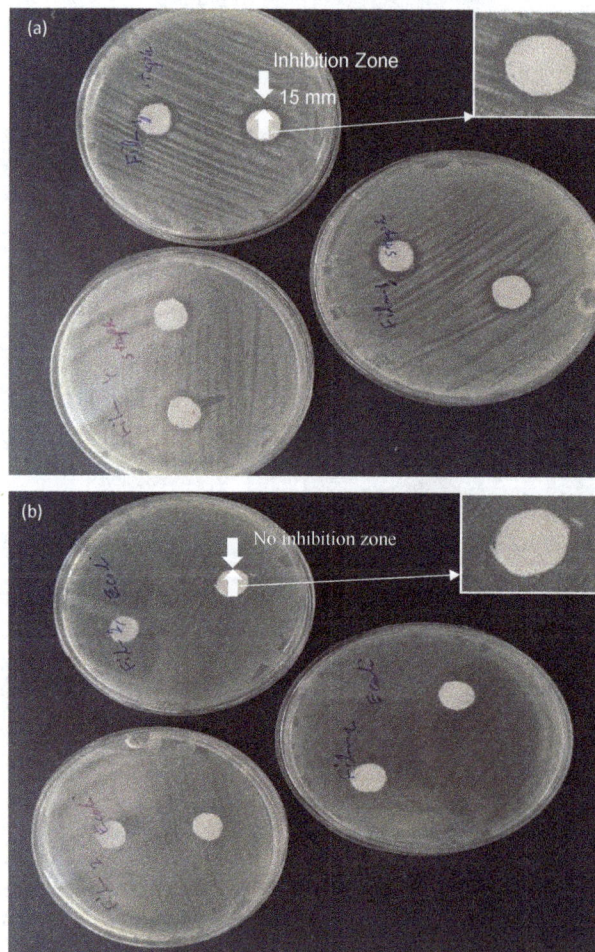

Fig. (7). Antibacterial effect of ZnO NPs against (**a**) *Staphylococcus aureus* (**b**) *Escherichia coli*.

CONCLUSION

ZnO NPs were synthesized by two different methods and the antibacterial activity of tetron fabric coated with ZnO NPs was studied. Distribution and stability of ZnO NPs on the fabric depend on fabrication method and particle size which means that the smaller particles have more stability and better distribution than larger particles. The particle size and deposited concentration of ZnO NPs were effective on antibacterial activity, so that the smaller particles tend less agglomeration and smaller particles have more surface area and because of that better antibacterial activity. Overall, the results demonstrated a good antibacterial activity against *Staphylococcus aureus* than *Escherichia coli*.

HUMAN AND ANIMAL RIGHTS

No animals/humans were used for studies that are the basis of this research.

ACKNOWLEDGEMENTS

Declared none.

REFERENCES

[1]   Xu ZP, Zeng QH, Lu GQ, Yu AB. Inorganic nanoparticles as carriers for efficient cellular delivery. Chem Eng Sci 2006; 61(3): 1027-40.
        [http://dx.doi.org/10.1016/j.ces.2005.06.019]

[2]   Stoimenov PK, Klinger RL, Marchin GL, Klabunde KJ. Metal oxide nanoparticles as bactericidal agents. Langmuir 2002; 18(17): 6679-86.
        [http://dx.doi.org/10.1021/la0202374]

[3]   Axtell H, Hartley S, Sallavanti R. United States patent US 5026778. 2005.

[4]   Reddy KM, Feris K, Bell J, Wingett DG, Hanley C, Punnoose A. Selective toxicity of zinc oxide nanoparticles to prokaryotic and eukaryotic systems. Appl Phys Lett 2007; 90(213902): 2139021-3.
        [PMID: 18160973]

[5]   Fu L, Liu Z, Liu Y, Han B, Hu P, Cao L, *et al.* Beaded cobalt oxide nanoparticles along carbon nanotubes: Towards more highly integrated electronic devices. Adv Mater 2005; 17(2): 217-21.
        [http://dx.doi.org/10.1002/adma.200400833]

[6]   Dickson RM, Lyon LA. Unidirectional plasmon propagation in metallic nanowires. J Phys Chem B 2000; 104(26): 6095-8.
        [http://dx.doi.org/10.1021/jp001435b]

[7]   Yamada H, Suzuki K, Koizumi S. Gene expression profile in human cells exposed to zinc. J Toxicol Sci 2007; 32(2): 193-6.
        [http://dx.doi.org/10.2131/jts.32.193] [PMID: 17538243]

[8]   Makino T, Segawa Y, Kawasaki M, Koinuma H. Optical properties of excitons in ZnO-based quantum well heterostructures. Sci Technol 2005; 20(4): S78.

[9]   Nakada T, Hirabayashi Y, Tokado T, Ohmori D, Mise T. Novel device structure for Cu (In, Ga) Se 2 thin film solar cells using transparent conducting oxide back and front contacts. Sol Energy 2004; 77(6): 739-47.
        [http://dx.doi.org/10.1016/j.solener.2004.08.010]

[10]   Pearton S, Heo W, Ivill M, Norton D, Steiner T. Dilute magnetic semiconducting oxides. Sci Technol 2004; 19(10): R59.

[11]   Zhai J, Tao X, Pu Y, Zeng X-F, Chen J-F. Core/shell structured ZnO/SiO 2 nanoparticles: preparation, characterization and photocatalytic property. Appl Surf Sci 2010; 257(2): 393-7.
        [http://dx.doi.org/10.1016/j.apsusc.2010.06.091]

[12]   El-Feky OM, Hassan EA, Fadel SM, Hassan ML. Use of ZnO nanoparticles for protecting oil paintings on paper support against dirt, fungal attack, and UV aging. J Cult Herit 2014; 15(2): 165-72.
        [http://dx.doi.org/10.1016/j.culher.2013.01.012]

[13]   Ravishankar Rai V, Jamuna Bai A. Nanoparticles and their potential application as antimicrobials. Formatex Research Center 2011; pp. 197-209.

[14]   Gouda M. Enhancing flame-resistance and antibacterial properties of cotton fabric. J Ind Text 2006; 36(2): 167-77.
        [http://dx.doi.org/10.1177/1528083706068677]

[15]   Seshadri DT, Bhat NV. Synthesis and properties of cotton fabrics modified with polypyrrole. Seni Gakkaishi 2005; 61(4): 103-8.
        [http://dx.doi.org/10.2115/fiber.61.103]

[16]   Abramova AV, Abramov VO, Gedanken A, Perelshtein I, Bayazitov VM. An ultrasonic technology for production of antibacterial

nanomaterials and their coating on textiles. Beilstein J Nanotechnol 2014; 5: 532-6.
[http://dx.doi.org/10.3762/bjnano.5.62] [PMID: 24991488]

[17]   Perelshtein I, Applerot G, Perkas N, *et al.* Antibacterial properties of an *in situ* generated and simultaneously deposited nanocrystalline ZnO on fabrics. ACS Appl Mater Interfaces 2009; 1(2): 361-6.
[http://dx.doi.org/10.1021/am8000743] [PMID: 20353224]

[18]   Rajendra R, Balakumar C, Ahammed HAM, Jayakumar S, Vaideki K, Rajesh E. Use of zinc oxide nano particles for production of antimicrobial textiles. IJEST 2010; 2(1): 202-8.

[19]   Lu H-Y, Chu S-Y, Tan S-S. The characteristics of low-temperature-synthesized ZnS and ZnO nanoparticles. J Cryst Growth 2004; 269(2): 385-91.
[http://dx.doi.org/10.1016/j.jcrysgro.2004.05.050]

[20]   Parthasarathi V, Thilagavathi G. Synthesis and characterization of zinc oxide nanoparticle and its application on fabrics for microbe resistant defence clothing. Int J Pharm Pharm Sci 2011; 3(4): 392-8.

[21]   Singh G, Joyce EM, Beddow J, Mason TJ. Evaluation of antibacterial activity of ZnO nanoparticles coated sonochemically onto textile fabrics. J microb biotech food sci 2012; 2(1): 106.

[22]   Bauer AW, Kirby WM, Sherris JC, Turck M. Antibiotic susceptibility testing by a standardized single disk method. Am J Clin Pathol 1966; 45(4): 493-6.
[http://dx.doi.org/10.1093/ajcp/45.4_ts.493] [PMID: 5325707]

[23]   Awoyinka O, Balogun I, Ogunnowo A. Phytochemical screening and *in vitro* bioactivity of Cnidoscolus aconitifolius (Euphorbiaceae). J Med Plants Res 2007; 1(3): 63-5.

[24]   Yamamoto O, Sawai J, Sasamoto T. Change in antibacterial characteristics with doping amount of ZnO in MgO–ZnO solid solution. Int J Inorg Mater 2000; 2(5): 451-4.
[http://dx.doi.org/10.1016/S1466-6049(00)00045-3]

[25]   Yamamoto O, Hotta M, Sawai J, Firstenberg-Eden R, Eden G. Impedance microbiology. J Am Ceram Soc 1998; 1007-11.
[http://dx.doi.org/10.2109/jcersj.106.1007]

[26]   Meruvu H, Vangalapati M, Chippada SC, Bammidi SR. Synthesis and characterization of zinc oxide nanoparticles and its antimicrobial activity against Bacillus subtilis and Escherichia coli. Chem 2011; 4(1): 217-22.

[27]   Yamamoto O. Influence of particle size on the antibacterial activity of zinc oxide. Int J Inorg Mater 2001; 3(7): 643-6.
[http://dx.doi.org/10.1016/S1466-6049(01)00197-0]

[28]   Azam A, Ahmed AS, Oves M, Khan MS, Habib SS, Memic A. Antimicrobial activity of metal oxide nanoparticles against gram-positive and gram-negative bacteria: A comparative study. Int J Nanomedicine 2012; 7: 6003-9.
[http://dx.doi.org/10.2147/IJN.S35347] [PMID: 23233805]

16

Microbial Diversity of *Mer* Operon Genes and their Potential Rules in Mercury Bioremediation and Resistance

Martha M. Naguib[1], Ahmed O. El-Gendy[2] and Ahmed S. Khairalla[2,*]

[1]*Department of Biotechnology and Life Sciences, Faculty of Post Graduate Studies for Advanced Sciences, Beni-Suef University, Beni-Suef 62511, Egypt*

[2]*Department of Microbiology and Immunology, Faculty of Pharmacy, Beni-Suef University, Beni-Suef 62511, Egypt*

Abstract:

Background:

Mercury is a toxic metal that is present in small amounts in the environment, but its level is rising steadily, due to different human activities, such as industrialization. It can reach humans through the food chain, amalgam fillings, and other sources, causing different neurological disorders, memory loss, vision impairment, and may even lead to death; making its detoxification an urgent task.

Methods:

Various physical and chemical mercury remediation techniques are available, which generally aim at: (i) reducing its mobility or solubility; (ii) causing its vaporization or condensation; (iii) its separation from contaminated soils. Biological remediation techniques, commonly known as bioremediation, are also another possible alternative, which is considered as cheaper than the conventional means and can be accomplished using either (i) organisms harboring the *mer* operon genes (*merB*, *merA*, *merR*, *merP*, *merT*, *merD*, *merF*, *merC*, *merE*, *merH* and *merG*), or (ii) plants expressing metal-binding proteins. Recently, different *mer* determinants have been genetically engineered into several organisms, including bacteria and plants, to aid in detoxification of both ionic and organic forms of mercury.

Results:

Bacteria that are resistant to mercury compounds have at least a mercuric reductase enzyme (MerA) that reduces Hg^{+2} to volatile Hg, a membrane-bound protein (MerT) for Hg^{+2} uptake and an additional enzyme, MerB, that degrades organomercurials by protonolysis. Presence of both *mer*A and *mer*B genes confer broad-spectrum mercury resistance. However, *mer*A alone confers narrow spectrum inorganic mercury resistance.

Conclusion:

To conclude, this review discusses the importance of mercury-resistance genes in mercury bioremediation. Functional analysis of *mer* operon genes and the recent advances in genetic engineering techniques could provide the most environmental friendly, safe, effective and fantastic solution to overcome mercuric toxicity.

Keywords: Mercury, Mercury toxicity, Biogeochemical cycle, Mercury remediation, Resistance mechanisms, *Mer operon*.

1. INTRODUCTION

For a long time, the term "heavy metals" had been widely used for metals associated with contamination and eco

* Address correspondence to this author at the Department of Microbiology and immunology, Faculty of Pharmacy, Beni-Suef University, Beni-Suef 62511, Egypt, E-mails: ahmed.elgendy@pharm.bsu.edu.eg, ahmedkhairalla@pharm.bsu.edu.eg

-toxicity. The International Union of Pure and Applied Chemistry (IUPAC) recommended using the term "toxic metal" as an alternative to "heavy metal" [1]. Toxic metals are stable and persistent environmental contaminants [2]. Many metals such as mercury, cadmium, chromium, zinc, lead, copper, arsenic *etc.*, used in different industries, are releasing its toxic ions and introducing it into the ecosystem leading to toxic effects, affecting humans, animals, plants, and microbial communities [3]. Also, some toxic metals naturally exist in very low concentrations in the ecosystem and are required in trace amounts as nutrients by microbial communities but in relatively higher concentration, they form toxic complexes on the biological cell [3, 4].

Mercury is the 16th rarest element on the earth [5] and considered as one of the mobile and toxic metals that exist naturally in low concentrations in the environment, and can be changed between different forms. It is the only metal to be liquid at room temperature. It can also exist as gas due to its high vapor pressure [6 - 8].

It is a major environmental pollutant especially methylmercury (MeHg) form in the aquatic regions. It can accumulate in biota so can reach and affect both wildlife and human seriously. It is one of the environmentally stable and persistent toxins for long periods, also it can be accumulated in different biological tissues [9, 10].

This review aims to cover all aspects related to the environmental biogeochemical cycle of mercury, the roles of *mer* genes in microbial adaptation to mercury, and potential bacterial remediation strategies of this toxic metal.

2. MERCURY BIOGEOCHEMICAL CYCLE

Environmental mercury cycle, as illustrated in Fig. (**1**), is usually facilitated biotically and abiotically between soils, water and atmosphere [11]. Mercury exists in the atmosphere in gaseous, particulates, and aqueous soluble forms [12] but gaseous form represents about 95% of atmospheric mercury [13] and it remains in the atmosphere for long periods. So, it can reach the far distance that should be considered as a huge environmental concern [12, 14, 15].

2.1. Oxidation Processes

In the atmosphere, Hg abiotically oxidized to Hg^{+2} through photo-oxidation reactions, mediated by O_2 through its interaction with hydrogen peroxide, ozone, sulfhydryl compounds, free radicals as Br, and by UV-B in presence of Cl_2 and photoreactive compounds as benzoquinone in presence of water droplets [11, 12, 16].

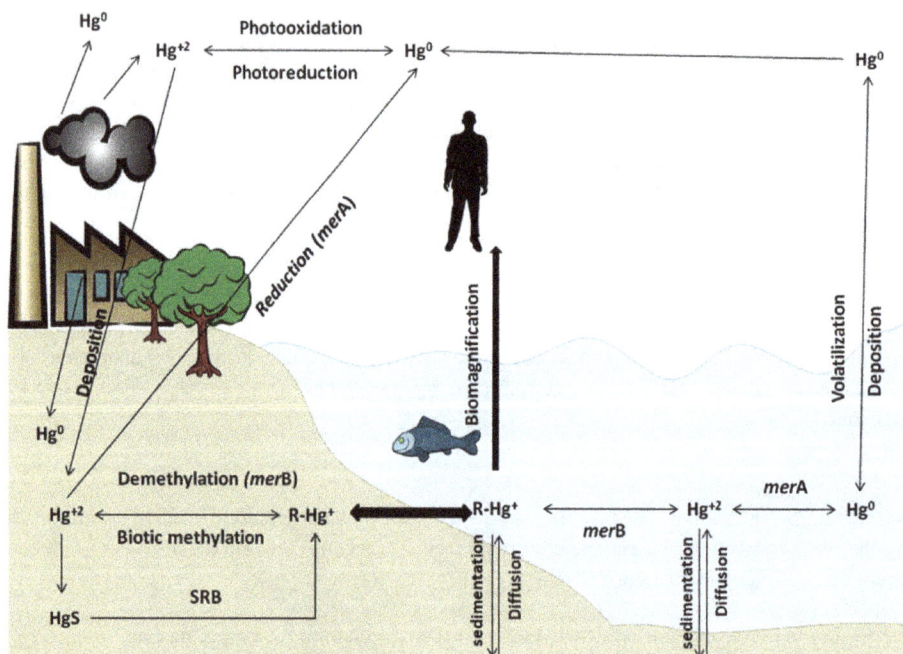

Fig. (1). Environmental mercury biogeochemical cycle [10].

Biotic Hg oxidation (bio-oxidation) in aerobic and phototrophic microorganisms is catalyzed by hydroperoxidases, *kat*G and *kat*E [13] or other oxidases [17]. Algae, plants and animals catalase and peroxidases are also capable of oxidizing Hg [17 - 19].

Anaerobic Hg oxidation mechanism by *Desulfovibrio desulfuricans* ND132 bacteria is unknown as obligate anaerobes do not carry such catalase or peroxidase genes but Colombo, Ha *et al.* 2013 suggested an alternative oxidation pathway influenced by reactive functional thiol groups of different anaerobic bacteria [17, 19] consistent with previous studies in which thiol moiety of organic compounds such as glutathione [20] and reduced humic acid [21] can bind and oxidize Hg.

2.2. Reduction Processes

Biotic Hg^{+2} reduction to Hg occurs by bacterial mercuric reductase enzyme encoded by *mer*A gene on *mer* operon (described later) [8, 11, 22], anaerobic *Geobacter sulfurreducens* PCA, *Geothrix fermentans* H5, *Cupriavidus metallidurans* AE104, *Shenwella oneidensis* MR-1 and *Geobacter metallireducens* GS-15 was found to reduce Hg^{+2} to Hg independent on the *mer* system but its need an electron acceptors and electron donors to suggest activity of respiratory electron transport chains. Its activity is effective at too low Hg^{+2} concentrations compared to amounts required for *mer* operon induction. Anaerobic bacteria showed dual role in the Hg redox cycle by both oxidizing Hg and reducing Hg^{+2} by unidentified reduction system other than *mer* system [19, 22].

The abiotic Hg^{+2} reduction is done not only by photochemical reactions but also *via* dark reactions [23] using organic matter free radicals as fulvic [24] and humic acid-associated free radicals [18].

2.3. Methylation Process

Biotic Hg^{+2} methylation is a natural bacterial process mainly occurred in seawater and coastal environment sediments [25], invertebrate digestive tracts, thawing permafrost soils, and extreme environments [26] through anaerobic methylators as Sulfate-Reducing Bacteria (SRB) [27 - 30], Iron Reducing Bacteria (IRB) and methanogens [27, 31, 32]. Two-gene cluster, *hgc*A and *hgc*B encode a putative corrinoid protein facilitating methyl transfer and a ferredoxin carrying out corrinoid reduction, were reported to be involved in mercury methylation process [27]. According to Podar *et al.* 2015, *hgc*A and *hgc*B genes were found in nearly all anaerobic environments but not in aerobic and not in human and mammalian microbiomes, reducing the expected risk of Hg methylation by microbiomes [26].

Another unrecognized oxidation/methylation pathway in the mercury cycle by anaerobic bacteria as in *D. desulfuricans* ND132 and *G. sulphurreducens* PCA [17, 33] produces MeHg using dissolved Hg as their sole Hg source. Further investigation is required for detecting the reactions involved in and the connection between Hg methylation, oxidation and produced toxic MeHg exportation out of cells as *D. desulfuricans* lacks the *mer* operon system that detoxifies organic/inorganic mercury [17].

Moreover, abiotic Hg^{+2} methylation carried out chemically [34] with the help of humic and fulvic acids, carboxylic acids, and compounds as fungicides or antifouling agents [11, 35]. It was found that sunlight played a minor and slower role in methylation reaction [36].

2.4. Demethylation Process

Biotic demethylation occurs simultaneously in the methylation sites as a reverse process [34]. It is catalyzed through reduction [37] or oxidation processes [38]. Aerobic reductive demethylation of CH_3Hg^+ occurs through *mer* operon functions (*mer*A and *mer*B) forming CH_4 and Hg [37]. Formed Hg will be evaporated into the air, and the cycle is repeated [39, 40]. Anaerobic reductive demethylation, *Geobacter bemidjiensis* Bem is an iron-reducing bacterium capable of simultaneously both methylating Hg^{+2} and degrading MeHg, due to the the presence of homologues of an organomercurial lyase and a mercuric reductase [41]. Oxidative demethylation occurs in more anaerobic conditions yielding CO_2 and small amounts of Hg^{+2} with an unknown mechanism [38, 39]. Produced Hg^{+2} may be available for re-methylation process or reduction to its vapor form [11]. Both anaerobic methylation and demethylation processes are affected by environmental dissolved organic matter, iron-sulfate biogeochemistry, and Hg^{+2} concentration [17, 27, 42].

Abiotic demethylation of MeHg affected by photo-degradation, especially UV-A and UV-B, at a wavelength range of 200-400 nm [11, 43, 44]. A study on Hg cycling found that dissolved MeHg concentration was decreased in daylight and increased in non-daylight suggesting that photo-degradation in water has a major role in methylation/demethylation processes aquatic systems [45].

All mercury forms (Hg, Hg^{+2}, and CH_3Hg^+) are interconvertible and can be introduced into the aquatic system [8]. Their concentration in the aquatic system depends mainly on reduction, methylation, and demethylation ratio which depend on the microbial community [6].

CH_3Hg^+ had a different pathway as bio-accumulator in the food chain. Its concentration is higher in top food chain organisms due to CH_3Hg^+ biomagnifications [46]. As CH_3Hg^+ concentration is higher in predatory animals such as beluga and polar bears than marine mammals which have higher concentrations of Hg than freshwater fish and overland mammals, due to their higher position in marine food webs [47]. Marine and freshwater fishes are still the main sources of dietary Hg exposure for humans [6, 47]. CH_3Hg^+ is the main specie absorbed in the gut; it enters the bloodstream and distributes to body tissues and organs [47]. In fish, CH_3Hg^+ first accumulates in the viscera (kidney, spleen, and liver) and later redistributed to other tissues as muscles and brain tissues. Hence, mammals and birds can demethylate CH_3Hg^+. So, a large amount of accumulated Hg in liver and kidney are found in an inorganic form [47].

3. MERCURY BIOREMEDIATION

It is a remediation technique using a wide range of living organisms (algae, fungi, yeasts, plants and bacteria) or their enzymes. Bioremediation is preferred as environmental friendly, promising technique. Microbial remediation using bacteria is widely used as they can be easily cultivated, grow faster and can accumulate metals in different conditions [48 - 51]. Moreover, different Gram-negative and positive bacterial isolates can resist, accumulate, adsorb and transform toxic mercury forms to less toxic forms by different mechanisms. These bacteria are named Mercury Resistant Bacteria (MRB) [8, 12, 48]. Bioremediation using bacterial strains showed promising results, reached 76.4% compared to other remediation techniques in removing mercury pollutant leached from spent fluorescent lamps [52].

3.1. Phytoremediation

Green plants or its associated microorganisms are used to remove or destroy contaminants from the soil. The insertion of bacterial mercury resistance genes as (merP, merC, merF and merT) encodes for different transporter proteins, (merA) encodes for mercury reductase or (merB) encodes for organomercurial lyase into plant cells after their sequences modification according to preferred plant codons. Genes inserted as merB remove organic mercury by protonolysis of C-Hg to Hg^{+2} while merA helps in the reduction of Hg^{+2} by the formation of volatile elemental mercury Hg which is then volatilized out of plant cells [53 - 57]. Different plant types and species were engineered as *Arabidopsis thaliana* as described in Table 1, yellow poplar [58], tobacco [59, 60], peanut [61], salt marsh cordgrass [62], rice [63] and eastern cottonwood [64, 65] that showed successful high resistance levels than their wild types.

Hence, genetically engineered plants can get rid of ionic and organic mercury by phytovolatilization. However, phytovolatilization major concern is the release of mercury vapors back to the environment, but this can be reduced by increasing efficacy of phytoextraction/phytosequestration. This could be achieved by transforming plants with certain bacterial genes as merP, merC, merF and merT [53]. For example, the increase of Hg^{+2} bioaccumulation in transgenic tobacco by expressions of the merP gene producing bacterial Polyphosphate Kinase (PPK) [54, 66].

An MRB Enterobacter strain exhibited a novel property of Hg immobilization by synthesis of nanoparticle Hg. The strain could intracellularly synthesize Hg nano-particles sized 2–5-nm [67].

Table 1. Effect of different *mer* genes, inserted and expressed in different regions of *A. thaliana* plant, on mercury phytoremediation.

Expression	Inserted gene	Effect on Resistance	References
Cytoplasm	merB	Increase resistance of phenylmercuric acetate PMA to 2 μM	[157].
Cytoplasm	merA	Increase resistance of Hg^{+2} to 100 Mm	[56].
Cytoplasm	merA/B	Increase resistance of organic and inorganic Hg to 5–10 μM	[158].
Cell membrane	merP (ppk)	Increase resistance of Hg^{+2} to 10 μM	[159].
Cell membrane	merC	Increase resistance of Hg^{+2} to 10 μM	[160].
Cytoplasm	merE	Increase of inorganic and organic-Hg accumulation	[55].

4. MERCURY RESISTANCE MECHANISMS

To adapt to toxic metals in the environment, bacteria and other organisms have developed different resistance mechanisms as a defense systems against these toxic materials. This defense systems that help bacteria to eliminate toxic materials from their growth medium includes:

1. Mercury bioaccumulation whether by simultaneous synthesis of mercury as nanoparticles [67] or by Hg^{+2} binding to carboxyl phosphates, hydroxyl, thio, or pyridine functional groups located on some bacterial cell wall [49] so, mercury is trapped and can't be vaporized back into the environment [67],

2. Sequestration and chelation of mercury using intracellular binding Metallothionein protein [68], a cysteine-rich protein able to bind mercury ions to form Mercury-cysteine complexion or extracellular polysaccharides in the cell wall [69, 70] as mercury compounds have high ability to bind with thiols of bacterial cysteine and reduce mercury toxicities [11],

3. Blocking mercury entry into cells through permeability barriers [11, 71],

4. Efflux and volatilization to convert toxic ionic mercury, Hg^{+2}, to much less toxic, elemental mercury, Hg through genetic manner [70], reductase enzymes as those expressed by the help of mercury resistance (mer) operon that will be described later [3, 70, 72] or through cytochrome c oxidase enzymes [73].

For these valuable mechanisms, different bacterial strains and other biological systems were engineered to be used for remediation and monitoring of environmental hazards such as increasing the bioaccumulation of Hg^{+2} by expression of the bacterial Polyphosphate Kinase (PPK) in transgenic tobacco [66].

5. COMPONENTS AND FUNCTIONS OF *MER* OPERON IN MERCURY RESISTANT BACTERIA (MRB)

All *mer* determinants (*mer* operon) are widely distributed by both Horizontal (HGT) and Vertical Gene Transfer (VGT) which explain their presence in different bacterial populations [70]. It was found located on chromosomal DNA [74], mobile elements as plasmids [40, 75], transposons as components of the Tn21 [22, 76] and Tn501 [46], or on integrons [77]. These *mer* determinants were identified in a wide range of previously isolated gram-negative [78] as seen in Fig. (2) and Gram-positive [79] bacterial strains from clinical [80, 81] and environmental [82, 83] samples. Genes in *mer* operon express different enzymes that can transform toxic to less toxic mercury forms as organomercuriallyase and mercuric reductase as illustrated in Table 2. In addition to mercury detoxification, some MRB can also detoxify other metals [84].

These *mer* determinants are classified into two types; narrow-spectrum that detoxifies only inorganic mercury through the main *mer*A gene or broad-spectrum that detoxifies both organic and inorganic mercury through *mer*A and *mer*B genes [8, 85, 86]. The *mer* operon is composed of the operator, promoter, regulator genes, and functional genes such as *mer*R, *mer*P, *mer*T, *mer*D, *mer*A, *mer*F, *mer*C, *mer*E, *mer*H, *mer*G and *mer*B. All these genes code for different proteins that participate in the detection, transportation and reduction or methylation of mercury ions [7, 75, 87, 88].

Table 2. All *mer* operon genes and their expressed proteins.

Gene	Gene Location in Operon	Protein Encoded	Protein Location	Function	References
*mer*A	Immediately after transport genes	Mercuric reductase	Cytoplasm	Reduction of Hg^{+2} to Hg	[8, 22].
*mer*B	Immediately after transport genes	Organomercuriallyase	Cytoplasm	Lysis of C-Hg bond	[8, 99].
*mer*F	May be in plasmid between *mer*P and *mer*A	Transporter protein	Inner membrane	Narrow and broad spectrum system transporters of mercuric.	[90, 92].
*mer*E	Downstream of *mer*A and *mer*B	Transporter protein	Inner protein	Uptake of inorganic and organo-mercurials (MeHg) into cytoplasm	[55, 89, 69].
*mer*G	Between *mer*A and *mer*B	Organomercurial compounds resistance protein	Periplasm	Resistance to phenyl-mercury by efflux mechanism (reduce their cellular permeability)	[11, 89, 86]
*mer*P	Upstream *mer*A	Periplasmic Hg^{+2} binding protein	Periplasm	Transfer of Hg^{+2}	[8, 11, 76].
*mer*T	Upstream *mer*A	Transporter protein	Inner membrane	Transporters of mercury in both narrow (Hg^{+2}) and broad (phenyl mercury) spectrum systems.	[89, 90].
*mer*D	Downstream *mer*A	Regulator protein	Cytoplasm	Negatively regulates the *mer* operon	[109, 110,.
*mer*C	Upstream *mer*A	Transporter protein	Inner membrane	Narrow and broad spectrum system transporters of mercuric.	[90],
*mer*R	Before transport genes	Regulator protein	Cytoplasm	Positively regulates the *mer* operon	[11, 22, 76, 89, 106].

(Table 2) contd.....

Gene	Gene Location in Operon	Protein Encoded	Protein Location	Function	References
*mer*H	Immediately adjacent to *mer*A	Mercury trafficking protein (metallochaperone)	Cytoplasm	Trafficking of mercury	[95, 161].
*mer*I	Located 3' to *mer*A	Unknown		Have no role in mercury resistance or regulation	[95, 161].

5.1. Hg^{+2} Binding and Transportation Genes

*mer*T, *mer*P, *mer*C, *mer*E, *mer*F, and newly discovered *mer*H express different proteins that have different roles in mercury transportation as shown in Fig. (2) [89].

*mer*T, expresses an inner membrane (cytoplasmic) proteins [22], helps in the uptake of organic phenylmercury [90] and inorganic mercury transport into the cytoplasm [91]. MerT protein has three transmembrane regions with cysteine pair located in its first transmembrane region Cys-Cys [90].

*mer*P, expresses a periplasmic protein [22] that has two cysteine residues to help in replacing the nucleophiles (Cl^-) linked to Hg^{+2} so, Hg^{+2} could bind to merP cysteine residues then transferred to other two cysteine residues on merT protein located on cytoplasmic membrane [8, 11, 76] then passed to other pair of merT's cytosolic cysteines residues. Hg^{+2} enters the cytoplasmic membrane where cytosolic thiols (cysteines or glutathions) compete with merT's cytosolic cysteines to bind with Hg^{+2} to be ready for merA activity [11].

*mer*T and *mer*P deletion from transposon Tn501 lead to Hg^{+2} sensitivity while the expression of Tn501 *mer*T and *mer*P in the absence of mercuric reductase causes Hg^{+2} supersensitivity [22]. Moreover, mutations in both *mer*T and *mer*P increase Hg^{+2} concentration required for induction of merA-lacZ transcriptional fusions. So, *mer*P is important for Hg^{+2} resistance [22]. However, Sone and Nakamura *et al.* showed that absence of *mer*P in presence of other alternative transporters *mer*C, *mer*E, *mer*F and *mer*T would increase both inorganic and organic mercury transportation. While the presence of *mer*P with all other transporters did not cause any difference in organic mercury transportation but increased the inorganic transportation [90]. MerE, MerT, MerC and MerF are broad-spectrum mercury transporter proteins that can transport organic phenylmercury and inorganic Hg^+2 into cells [11, 90 - 92].

*Mer*C, expresses an inner membrane (cytoplasmic) protein [22], that is involved in both inorganic and organic phenylmercury [90] uptake across the cytoplasmic membrane till it reaches active site reductase [70]. *Mer*C protein was found to have roles in mercury accumulation in Arabidopsis thaliana [93].

Absence or mutation of *mer*C could show no effects on mercury transportation and resistance in case of presence of *mer*T and *mer*P as shown in Tn501 and Tn5053. However, in some bacterial strains when both *mer*T and *mer*P do not exist, *mer*C can act alone as the main Hg^{+2} transporter [22, 76]. Other new findings showed that *mer*C is more preferred in Hg^{+2} transportation than *mer*E, *mer*F and *mer*T. Moreover, *mer*C was more efficient for designing successful mercurial bioremediation system [90].

*mer*F helps in both organic phenylmercury [90] and inorganic Hg^{+2} transport across the cytoplasm [11, 92].

*mer*H expresses a membrane protein and according to Schué, it was able to transport Hg^{+2} across the inner membrane using its two cysteine residue. Replacing *mer*T gene in an *E-coli* strain by *mer*H resulted in Hg^{+2} MIC reduction, although the strain is still resistant when compared with a control strain which has no transporter proteins [94]. As described by Schelert, a metallochaperone, homolog to C-terminal domain, called TRASH has a role in metal sensing and trafficking that explains *mer*H role in trafficking of Hg^{+2} to the MerR transcription factor. By non-sense or in-frame gene mutation of *mer*H, the new mutant strain was highly sensitive to Hg^{+2} and analysis of *mer*H deleted operon by mass spectroscopic. They found increased retention of Hg^{+2} intracellular that explains the low rate of *mer* operon induction and the need for *mer*H in metal trafficking [95].

*mer*E is involved in the transport system of inorganic [89] and organic phenylmercury compounds selectively across the membrane [96]. Organomercurial compounds are lipid-soluble and can pass through the cell membrane by simple diffusion [11, 97] or are transported inside by *mer*E or *mer*G [98]. Their transport system is poorly understood [96, 97].

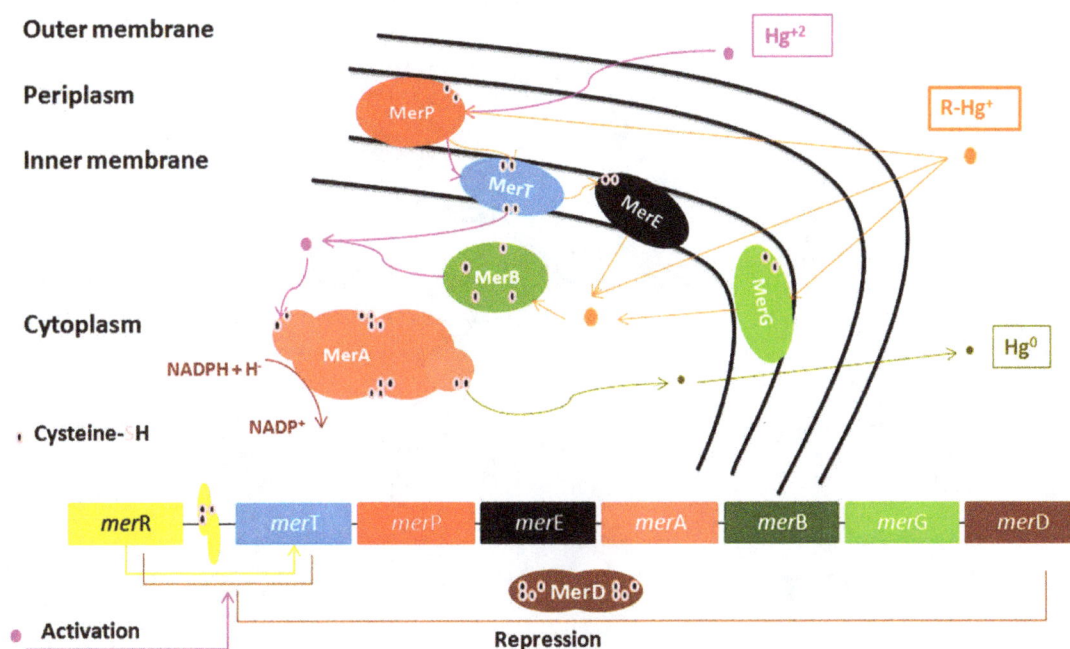

Fig. (2). Model *of bacterial mer* operan determinants and expressed genes. Organic methylmercury can enter inside cell through merP, T, and E transporter while, phenylmercury transported by merG protein also both may passively diffused inside the cell. Inorganic mercury transported by merP and T to be reduced by merA(10).

5.2. *mer* Detoxification Genes

*mer*A is the main gene in *mer* operon encodes for mercuric ion reductase enzyme, which is a flavoprotein catalyzes reduction of Hg^{+2} into volatile Hg by using NADPH located in the cytoplasm as source of electrons for the reduction of Flavin Adenine Dinucleotide (FAD) [76, 99, 100]. Hg is released out of the bacteria through cell membrane due to its lipid solubility without efflux system [11, 22, 75, 76, 101]. The released Hg back to the environment causes repetition of mercury cycle. To overcome this problem, a new study used an engineered bacteria to express polyphosphate, a chelator for divalent metal, in addition to *mer* operon determinants to overcome metal volatilization [102], help reducing environmental re-pollution. *mer*A has an amino-terminal domain (NmerA) Fig. (**3**), which is homologous to *mer*P and contains a pair of cysteines that directly remove Hg^{+2} from transport proteins cytosolic cysteines to be ready for reduction [11]. Mutations by insertion or deletion in *mer*A cause Hg^{+2} hypersensitive strains [22].

*mer*B, codes for organomercurial lyase catalyzes demethylation of organic mercury compounds (alkyl and aryl) *via* cleavage of its C-Hg bond releasing a protonated organic moiety (as methane (CH_4) in case of MeHg) and Hg^{+2} which is then reduced to Hg by merA using same NADPH-dependent mechanism [11, 78, 99, 103, 104].

5.3. Mechanism of Organomercurials Protonolysis by *Mer*b

Organomercuriallyase acts as a monomer containing four cysteine residues Cys117, Cys 96, Cys 159, and Cys160. Asp99, Cys96, and Cys159 residues of the enzyme and water molecule all are involved in protonolysis reaction [11, 103]. MeHg forms bond with Cys96, while the Cys159 site is fully reduced then the organic moiety released through two mechanisms, Fig. (**4**). In mechanism 1; proton from Cys159 transferred to methyl carbon and Cys159 forms covalent bond with Hg^{+2}. In mechanism 2, Cys159 first transfer proton to Asp99 then forms a covalent bond with Hg^{+2} in presence of methyl group. Asp99 then transfer a proton to the methyl group and the protonated methyl moiety is released. Hg^{+2} is attached to the enzyme by two sulfurs of Cys96 and Cys159, and oxygen from a water molecule [103]. Organomercuriallyase remains attached to Hg^{+2} until two solvent thiols separately bind two Hg^{+2} and removes it [11].

*mer*G helps in cellular permeability of organomercurial compounds (Phenylmercury) through effluxing [11, 96, 98, 105]. Mutation by *mer*G deletion, phenylmercury resistance was affected however, Hg^{+2} volatilization activity wasn't affected. So, *mer*G have no role in inorganic mercury compounds detoxification [96].

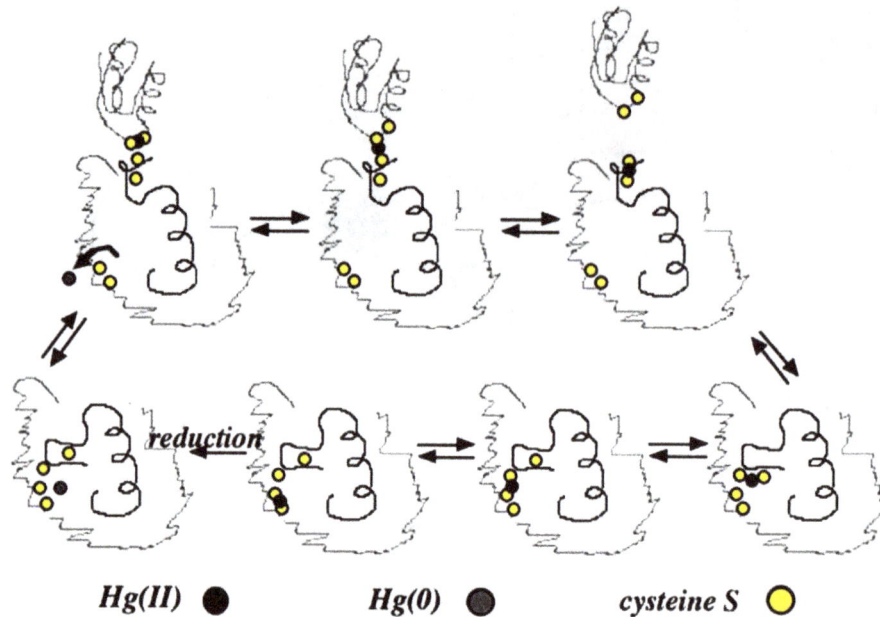

Fig. (3). Illustration for the role of NmerA in Hg^{+2} removal from transport proteins cytosolic cysteine to be reduced by mercuric reductase into Hg (11).

5.4. *mer* Operon Regulatory Genes

Regulatory genes are *mer*R and *mer*D [89]. *mer*R, a regulatory gene encodes a metalloregulatory protein. It is Hg^{+2} dependent transcriptional repressor-activator that can sense metal concentration and control the expression of other functional *mer* operon genes [11, 106, 107]. It binds with the promoter-operator region (*mer*OP) and positively and negatively regulate *mer* operon genes expression in presence or absence of Hg^{+2}, respectively [8, 22, 75, 76, 87, 106]. Hg-free-merR (apo-merR) undergoes conformational structure changes upon Hg^{+2} binding and converted from repression to activation state Hg-merR [108].

*mer*D, a gene encodes for a secondary regulatory protein, present downstream *mer*A [109]. MerD protein is expressed in very small amounts and downregulates the *mer* operon [110]. Its absence causes an increase in *mer* operon expression. It binds with the same operator-promoter region (*mer*OP) as MerR [8, 22, 76]. It acts as activation antagonist for MerR function. *Invitro* experimental showed that, after Hg^{+2} volatilization by *mer*A, the MerR does not give up its bound Hg^{+2} quickly causing *mer*A to be active. *mer*A expression should be stopped when Hg^{+2} is exhausted because *mer*A has an oxidase activity and produces toxic hydrogen peroxide in the absence of Hg^{+2} [11].

6. *MER* OPERON EVOLUTION, MOBILITY AND DIVERSITY

mer operon is an ancient system [97] may be located on the chromosome [74] and transferred vertically from parents to offspring or on Mobile Genetic Elements (MGE). DNA parts encode proteins that help in its mobility within bacterial genomes or between bacterial cells, facilitating Horizontal Gene Transfer (HGT) independent on reproduction [46, 77, 106, 111 - 114] so, this flexibility in genome [115] can help bacterial adaptation, social interactions and evolution by transferring good genes for the host as the *mer* genes [76, 114, 116 - 119]. MGE can help in the rapid spread of rare, spontaneous resistance mutants to a new bacterial population [114]. As plasmid-borne resistance genes can be originated as point mutations in sensitive bacteria and then transferred when they are flanked by short transposons, picked up by Tn*3* family transposons or as mobile cassettes by integrons [76, 114, 120]. *mer* operon as part of different types of group II transposons [121, 122] as Tn2, Tn501, pKLH2, pMERPH and Tn5053 in Gram-negative bacteria [11, 97, 121] and Tn5085, Tn5083 [122, 123], and newly identified Tn6294 [123] in Gram-positive bacteria, All of them encode genes for transposition (tnpA) and resolution functions (tnpR) [11, 121 - 124].

Successive exposure to mercury for a long time caused *mer* operon persistence and transfered between microbes through HGT and to be evolved rapidly and became more complex as *mer*A and species evolved. Evolution happened to adapt with environment, increasing mercury toxicity due to increased industrialization [89] that can explain the global distribution of *mer* determinants and their associated mobile elements between different bacterial strains [11, 97]. So, *mer*A was considered as a biomarker for measuring diversity in Hg detoxification and *mer* operon evolution [89, 125].

A recent study on different *mer* operons of different *Bacillus species* (as tndMER3, tn6294, and others) isolated from thirteen different countries is a good example for *mer* diversity, mobility, evolution and horizontal distribution between different *Bacillus* species worldwide [123].

Genes that express transporter proteins *mer*P and *mer*T were more common to occur in earliest evolved operons which were less complex while, other alternative transporter genes as *mer*C, *mer*F, *mer*E, and *mer*H commonly occurred in more complex operons evolved recently. In organomercurial detoxification system, *mer*G gene occurred in more recently evolved operons [89]. Proteins as merE, merC, merP, merD, and merT were also found to be more likely encoded on plasmids than others so, prefer to be transferred horizontally [89].

A recent research depended on sequencing of four different pQBR mercury resistance plasmids showed that mercury resistance transposons Tn5042 were totally similar in the four plasmids except for one base pair of *mer*R in pQBR103 and pQBR44 differed from pQBR55 and pQBR57, indicating that Tn5042 had been transferred between different pQBR plasmids by recombination. They also found that occurrence of pQBR55, pQBR57 and pQBR103 separately in *Pseudomonas fluorescens* SBW25 host resulted in different response to some environmental factors as Hg(II) concentration, although similarity in their *mer* operon suggested the effect of other plasmid-encoded genes [8, 126].

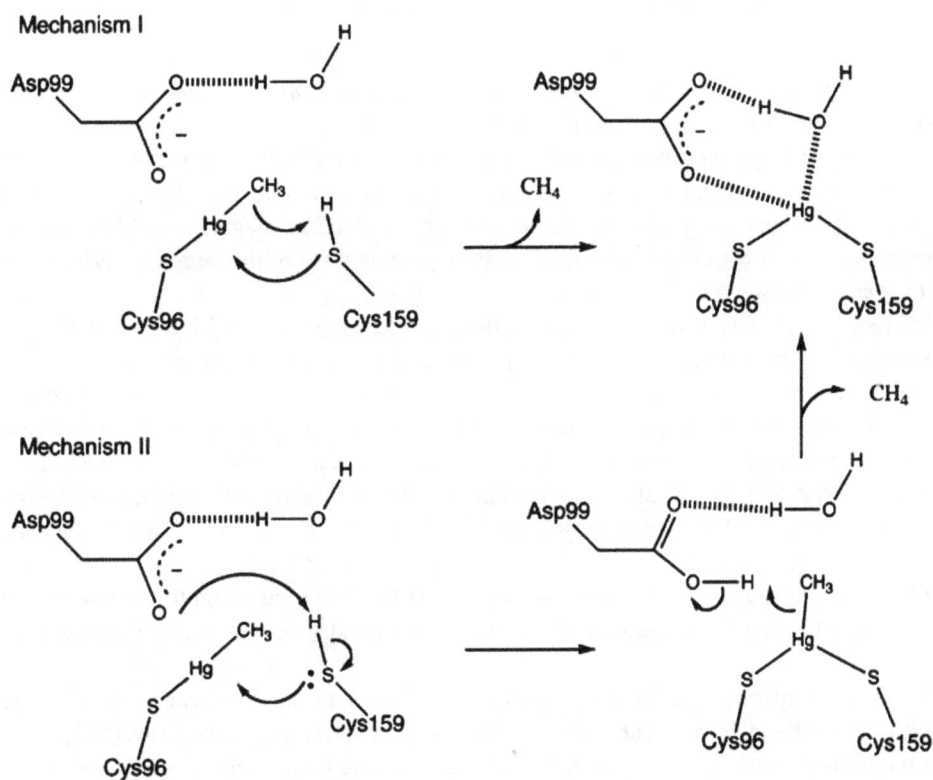

Fig. (4). Mechanisms of MerB catalysis for the Hg-C protonolysis reaction(103). Mechanism 1; Cys159 protonates the methyl group then reacts with Hg^{+2}. Mechanism 2; hydrogen was attached to the sulfur from Cys159 transferred to Asp99 then Asp99 utilized for methyl group protonation.

6.1. Diversity of *mer* Operon

Several variations in structure and organization of *mer* operons genes are known between both Gram-negative and Gram-positive bacterial strains as in Figs. (**5** and **6**), for examples:

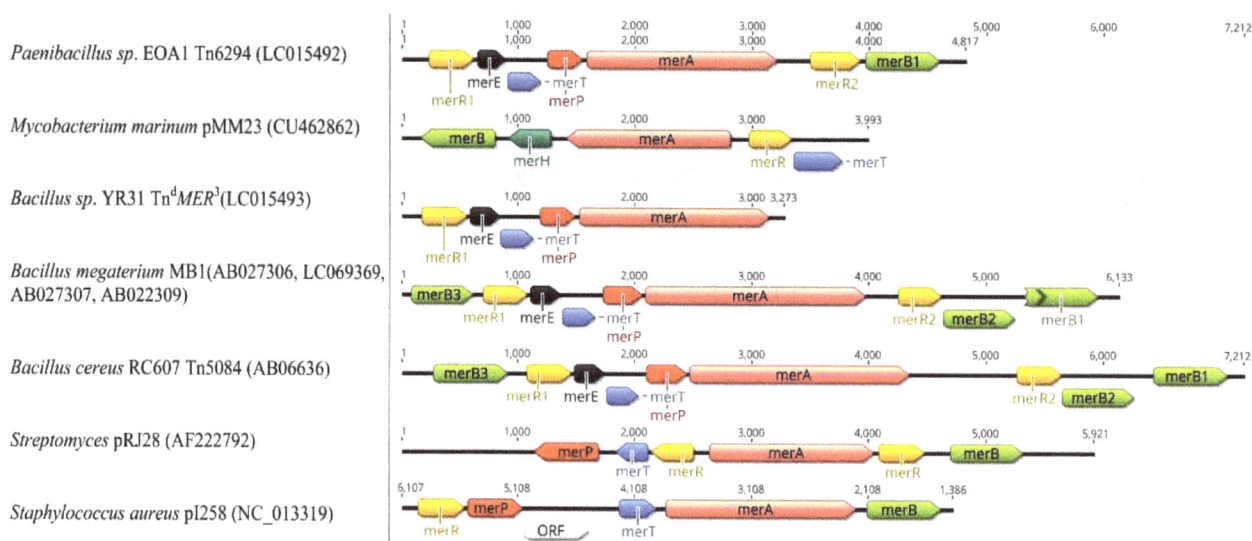

Fig. (5). Diversity *mer* operons in Gram-positive bacteria. Arrows indicate the gene product translation direction.

1. *mer*B is more common in Gram-positive *mer* operons [123, 127, 128], while *mer*D and *mer*C or *mer*F are more common in Gram-negative *mer* operons [11, 76, 97, 129, 130].
2. Narrow-spectrum *mer* operon genes are highly divergent compared to broad-spectrum operon genes. According to Narita, *et al.* in certain broad-spectrum mercury resistant *Bacillus species*, Polymerase Chain Reaction (PCR) product sizes *of mer* operon genes are identical to that of *Bacillus megaterium* MB1. However, in narrow-spectrum *mer* operon of certain *Bacillus species,* PCR product sizes of the targeted *mer*P and *mer*A regions are smaller than *mer*P and *mer*A of the *B. megaterium* MB1 [131].
3. Three different *mer*B genes were identified in different *Bacillus species* [127] as in Tn*MER*11-like transposons [123] resistant to different organomercurial compounds and also have multiple *mer*R.
4. Transcription direction of *mer*R is same as other *mer* operon functional genes in Gram-positive *bacteria mer* operons while, *mer*R transcription is divergent from the structural genes in the high-GC Gram-positive *mer* operons and Gram-negative operons except *Pseudoalteromonas haloplanktis* [11, 129, 132, 133].
5. Transposons Tn5084 and Tn5085 are identical in the genetic orientation and contain *mer*B3, *mer*R, *mer*E, *mer*T, *mer*P and *mer*A while, Tn5083 lacks *mer*R2, *mer*B2 and *mer*B1, compared to Tn5084 and Tn5085 [122, 123, 132].
6. Newly identified transposon Tn6294 has one *mer*B. It is the first transposon to carry *mer*A with only one N-terminal mercury binding domain, unlike all other reported broad spectrum *Bacillus species* that carry duplicate N-domain [123].
7. Tn^{d}MER3, a newly identified deleted transposon, was similar to the fragment in *Bacillus species* that carry *mer*RETPA with>90% identity to Tn6294 but with no *mer*B and transposonase genes [123].
8. *mer* operon of *Shewanella putrefaciens* pMERPH does not have both *mer*D or *mer*R, compared to other Gram-negative bacteria suggesting that *mer*R may be located elsewhere on the plasmid genes [130].
9. Gene position differs from one operon to another as *mer*B present between *mer*A and *mer*D in pDU1358 [133], while in broad spectrum part of *Pseudomonas stutzeri* pPB located between *mer*R and *mer*T [134].
10. In *Pseudomonas sp.* Tn502 and Tn512 *mer* operon are related to Tn5053 with exception of *mer*C and urf2M in newly recognized Tn502 compared to *mer*F in Tn512 and Tn5053 [129, 135].

7. CO-RESISTANCE OF MERCURY AND ANTIBIOTICS

mer operons are often part of group II transposons that carry integrons with multiple antibiotic resistance genes [11].

Antibiotics and metals resistance genes are located on the same plasmid [81, 136]. So, mercury exposure can also promote Horizontal Gene Transfer (HGT) of both mercury and antibiotic resistance [91, 125, 137, 138] that explain the increasing challenge in infectious bacteria treatment due to increasing of co-resistance.

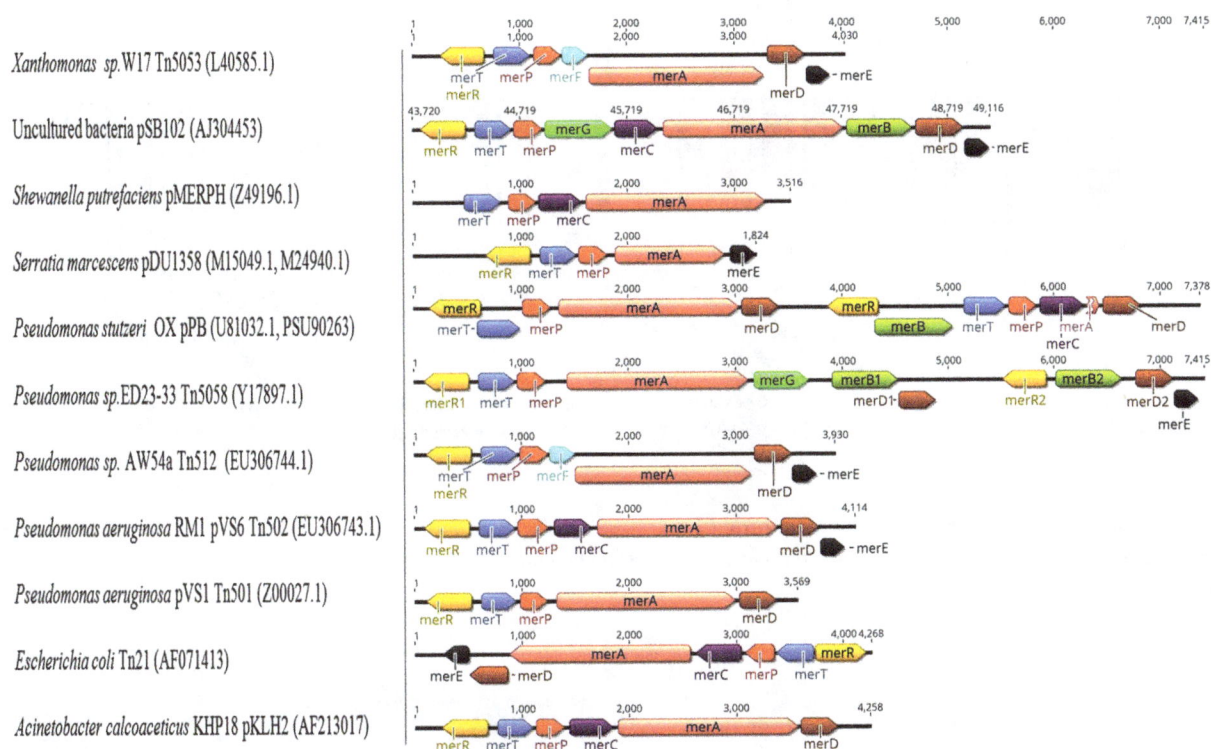

Fig. (6). Diversity of *mer* operons in Gram-negative bacteria. Arrows indicate the gene product translation direction.

Six adult monkeys were examined for both mercury and antibiotics resistances of different oral and intestinal bacterial strains before and during the installation, and after the replacement of the amalgam fillings. There was an increase in MRB during the 5 weeks after installation and during the 5 weeks after replacement. MRB was also resistant to one or more antibiotics as streptomycin, kanamycin, tetracycline, ampicillin, and chloramphenicol [81]. In a different study groups exposed to amalgam, MRB isolated from their fecal samples was found to be more resistant to antibiotics than MRB of those never exposed to amalgam [77], suggesting that both antibiotic and mercury resistance genes may be genetically linked [81] and contained within the same genetic mobile element [80] specially, the Tn*21* family of transposons in which the *mer* locus is linked to an antibiotic multi-resistance element [77].

A larger scale study by Pal *et al.*, for the co-occurrence of different metals and antibiotic resistance genes of fully sequenced 2522 bacterial genomes and 4582 plasmids as illustrated in Fig. (7), showed that although metal-antibiotic genes co-resistance was found to be rare on plasmid but, the only metal resistance genes commonly co-occurred with antibiotic-resistant genes on plasmids are mercury resistance genes as seen in Fig. (8). Moreover, about 86% of bacterial genomes contain different metals resistance genes of which 17% co-resistant with antibiotic resistance. Plasmids and genomes with different metals resistance genes were with high probability of carrying antibiotic resistant genes compared to those without metals resistance genes [138].

8. MRB DETECTION METHODS

8.1. Genotypic (Molecular) Techniques

8.1.1. Polymerase Chain Reaction

Simple PCR Specific DNA sequences of the *mer* system determinants could be amplified from the genomic DNA of all the isolates from environmental sources using *mer* determinants designed primers then product can be visualized under UV on electrophoresis agarose gel as bands by staining with ethidium bromide [101, 139 - 141].

Real-time Reverse Transcriptase-PCR is used to detect not only the presence but also the mRNA expression of the gene [87, 142].

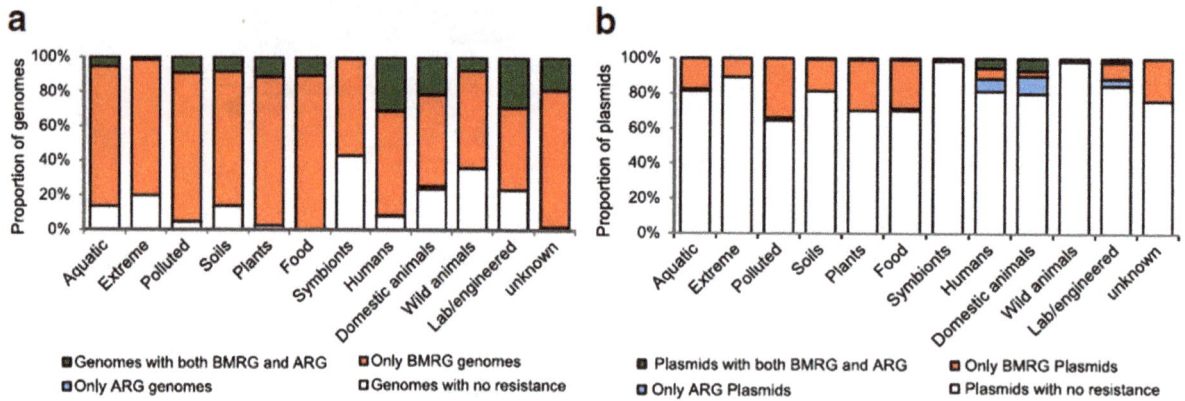

Fig. (7). Overview of the resistance information from (a) 2522completely sequenced bacterial genomes and (b) 1926 plasmids harboured by those genomes from different environments. (BMRGs) biocide and metal resistance genes. (ARGs) antibiotic resistant genes [138].

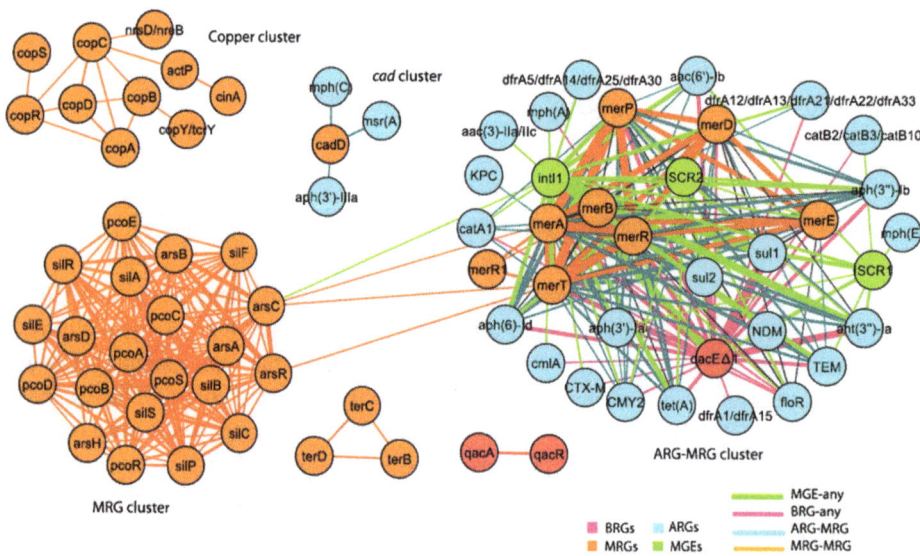

Fig. (8). Co-occurrence network of resistance genes on plasmids [138]. The thickness of each connection between two resistance genes is proportional to their co-occurrence times on the same plasmids. ARGs: Antibiotic Resistance Genes; MRGs: Metal ResistanceGenes; BRGs: Biocide Resistance Genes; MGEs: Mobile Genetic Elements.

8.1.2. DNA Hybridization Techniques

mer determinants encoding resistance to Hg^{2+} as *mer*A or other determinants are used as labeled probes in different hybridization procedures for detecting the presence of *mer* determinants that are complementary to probe nucleotide sequence in mercury resistant bacteria isolated from polluted environments through positive hybridization reaction [139 - 141]. Different hybridization techniques can be applied as Southern [139] and Northern blot hybridization [75], Colony blot hybridization [140, 143], Fluorescent *In Situ* Hybridization (FISH) [144], and microarray that could be used for detection of different multiple gene at once [145] also Restriction Fragment Length Polymorphism (RFLP) could be used [139, 146].

8.1.3. DNA Sequencing

PCR and/or hybridization products can be sequenced to detect its homology to *mer* determinants [139].

8.2. Phenotypic Techniques

8.2.1. Broth and Agar Methods (Direct Bacterial Growth)

Processed environmental samples are added into culture media as Luria-Bertani or nutrient agar (broth) supplemented with 10ppm or greater $HgCl_2$ to detect visible growth (turbidity or colonies) in media with 10ppm $HgCl_2$ or more. Grown bacteria are considered as MRB [104, 147, 148]. Optical Density (O.D.) in different time ranges at 600 nm can be measured to detect the growth pattern of tested microorganism in presence of mercury at constant conditions [80, 149].

8.2.2. Hg Utilization (Removal/Reduction) Rate

As only mercury resistant strains can utilize Hg and volatilize it, so the rate of Hg lose from the aqueous media was much higher when the cells are induced by growth in $HgCl_2$ [150]. Tested environmental samples are inoculated in aqueous growth medium supplemented with known concentration of $HgCl_2$ and incubated for 24 hrs at 37C. After centrifugation, the remaining concentration of mercury present in supernatant is measured using the Atomic Absorption Mercury Analyzer and compared to initial known concentration of Hg^{+2} to calculate its removal rate [83, 149].

Moreover, about 1ml samples are taken every 4 hrs and centrifuged. Then, remaining Hg concentration is measured in supernatants after processing. Hg^{+2} concentrations are calculated by measuring absorbance at OD600 using standard curve. Then, the initial and the final Hg^{+2} concentrations can be determined and reduction rates will be calculated [87].

8.2.3. Hg Volatilization Assay

The ability of bacteria to volatilize mercury from added mercuric chloride is a common mercury resistance mechanism by *mer* genes [150]. So, volatilization can be determined by using mercury vapor analyzer [75] or other atomic absorption spectrometry [78, 151], radioactive assay or using sensitive film containing Ag^+ as TEM photography film (X-ray film) which forms foggy areas due to reduction of the Ag^+ emulsion of the film by the mercury vapor [148, 151].

8.2.4. Mercuric Reductase (MR) Activity

MR activity measured depending on Hg^{+2} reduction mechanism by the enzyme in presence of NADPH. Hg^{2+} forms a complex with two cysteines of the enzyme active site. NADPH transfers a proton to FAD forming $FADH^-$. The resulting $FADH^-$ then reduces Hg^{2+} into Hg, and oxidizes back into a FAD. After reduction, the mercury is then released from the enzyme as a volatile vapor [152].

The MR activity was determined by different methods

1. Fluorimetric measurements by oxidation of fluorescent NADPH to the non-fluorescent $NADP^+$ due to the reduction of Hg^{+2} to Hg at specific wave lengths [144].
2. MR assays by measuring Mercury-dependent NADPH oxidation using a UV-visible spectrophotometer [153, 154] as NADPH increases the activity of MR while NADP+ has no effect on activity [154].
3. Determination of the remaining amount of NADPH by titration using phenazine methosulfate to produce visible formazan. The concentration of formed formazan is detected Spectrophotometric. As enzyme activity is related to the remaining amount of NADPH and the produced amount of formazan [155].

8.2.5. Biosorption Experiments

Analysis of control and inoculated samples for metal adsorption by inductively coupled plasma optical emission spectroscopy [156].

CONCLUSION

Industrialization has a great threat to the environmental mercury contamination already spread in the atmosphere, soil and water systems. Mercury causes more damage when reached to humans and animals. So, mercury usage should be restricted more to reduce its pollution. *mer* genes enable bacteria to convert the toxic organic or inorganic mercury forms to less toxic forms helping in mercury bioremediation as the most environmental friendly, safe and effective remediation technique.

The main *mer* operon genes are the *mer*B and *mer*A genes both can help bacteria to detoxify both organic mercurial compounds and inorganic mercury. *mer*A, is a mercuric ion reductase enzyme, catalyzes the reduction of Hg^{+2} into volatile Hg. *mer*B, codes for organomercurial lyase catalyzes demethylation of organic mercury compounds. *mer*T, *mer*P, *mer*C, *mer*E, *mer*F, and *mer*H express different transporter proteins. *mer*G helps in cellular effluxing of organomercurial compounds (Phenylmercury). *mer*R encodes for a metalloregulatory protein, it is Hg^{+2} dependent transcriptional repressor-activator that can sense metal concentration and control the expression of other functional *mer* operon genes. Evolution and diversity in structure and organization of *mer* genes between different Gram-negative and Gram-positive strains affects their resistance ability as the transporter *mer*P and *mer*T genes were more common to occur in earliest evolved less complex operons while, other transporter genes as *mer*C, *mer*F, *mer*E, *mer*H and effluxing *mer*G gene are commonly occurred in recently evolved, more complex operons [89].

Bacteria harboring the *mer* determinants can be genetically modified to code for resistance to other toxicants as toxic metals and be used during extreme environmental different conditions. *mer* determinants could be transferred into a new genetically engineered recipient suitable biological system (plant, bacteria or algae) to help in mercury remediation after removing undesired genes or adding genes to increase system harmony. Although *mer* system is highly understood mercury cycle in anaerobic conditions specially MeHg formation mechanism that prefers anoxic conditions is incompletely understood and requires more investigations as MeHg is the most bioaccumulative toxic form and understanding the anaerobic system will help in MeHg bioremediation.

ACKNOWLEDGEMENTS

The manuscript was drafted by Dr. Martha M. Naguib and revised and approved by Dr. Ahmed O. El-Gendy, Dr. Ahmed S. Khairalla respectively.

REFERENCES

[1]   Report IT. Heavy Metals "— A meaningless term ? (IUPAC Technical Report). 2002; 74(5): 793-807.

[2]   Montuelle B, Latour X, Volat B, Gounot AM. Toxicity of heavy metals to bacteria in sediments. Bull Environ Contam Toxicol 1994; 53(5): 753-8.
        [http://dx.doi.org/10.1007/BF00196950] [PMID: 7833613]

[3]   Nisa M, Coral Ü, Korkmaz H, Arikan B, Coral G. Plasmid mediated heavy metal resistances in Enterobacter spp. In: isolated from Sofulu landfill. in Adana, Turkey. Growth (Lakeland). 2005; 55: pp. (3)175-9.

[4]   Nies DH. Microbial heavy-metal resistance. Appl Microbiol Biotechnol 1999; 51(6): 730-50.
        [http://dx.doi.org/10.1007/s002530051457] [PMID: 10422221]

[5]   Wedepohl KH. The composition of the continental crust. Geochim Cosmochim Acta 1995; 59(7): 1217-32.
        [http://dx.doi.org/10.1016/0016-7037(95)00038-2]

[6]   Figueiredo NLL, Areias A, Mendes R, Canário J, Duarte A, Carvalho C. Mercury-resistant bacteria from salt marsh of Tagus Estuary: the influence of plants presence and mercury contamination levels. J Toxicol Environ Health A 2014; 77(14-16): 959-71.
        [http://dx.doi.org/10.1080/15287394.2014.911136] [PMID: 25072727]

[7]   Brooks S, Moore C, Lew D, Lefer B, Huey G, Tanner D. Temperature and sunlight controls of mercury oxidation and deposition atop the Greenland ice sheet. . Atmos Chem Phys 2011; 11(16): 8295-306.
        [http://dx.doi.org/10.5194/acp-11-8295-2011]

[8]   Dash HR, Das S. Bioremediation of mercury and the importance of bacterial mer genes. Int Biodeterior Biodegradation 2012; 75: 207-13.
        [http://dx.doi.org/10.1016/j.ibiod.2012.07.023]

[9]   Keating MH, Beauregard D, Benjey WG, *et al.* Mercury Study Report to Congress Volume II: An Inventory of Anthropogenic Mercury Emissions in the United States. United States Environmental Protection Agency 1998. EPA-452/R-:1-181.

[10]   Dash HR, Das S. International biodeterioration & biodegradation bioremediation of mercury and the importance of bacterial mer genes. Int Biodeterior Biodegradation 2012; 75: 207-13.
[http://dx.doi.org/10.1016/j.ibiod.2012.07.023]

[11]   Barkay T, Miller SM, Summers AO. Bacterial mercury resistance from atoms to ecosystems. FEMS Microbiol Rev 2003; 27(2-3): 355-84.
[http://dx.doi.org/10.1016/S0168-6445(03)00046-9] [PMID: 12829275]

[12]   Wang Q, Kim D, Dionysiou DD, Sorial Ga, Timberlake D. Sources and remediation for mercury contamination in aquatic systems - A literature review. Environ Pollut 2004; 131(2): 323-36.
[http://dx.doi.org/10.1016/j.envpol.2004.01.010] [PMID: 15234099]

[13]   Smith T, Pitts K, McGarvey JA, Summers AO. Bacterial oxidation of mercury metal vapor, Hg(0). Appl Environ Microbiol 1998; 64(4): 1328-32.
[PMID: 9546169]

[14]   Lindqvist O, Rodhe H. Atmospheric mercury—A review. Tellus B Chem Phys Meterol 1985; 37(3): 136-59.
[http://dx.doi.org/10.3402/tellusb.v37i3.15010]

[15]   Keeler G, Glinsorn G, Pirrone N. Particulate mercury in the atmosphere: Its significance, transport, transformation and sources. Water Air Soil Pollut 1995; 80(1-4): 159-68.
[http://dx.doi.org/10.1007/BF01189664]

[16]   Munthe J, McElroy WJ. Some aqueous reactions of potential importance in the atmospheric chemistry of mercury. Atmos Environ Part A 1992; 26(4): 553-7.
[http://dx.doi.org/10.1016/0960-1686(92)90168-K]

[17]   Matthew J, Colomboa JH, John R. Reinfeldera, Tamar Barkayb, Nathan Yeea. Anaerobic oxidation of Hg(0) and methylmercury formation by desulfovibrio desulfuricans ND132. Geochim Cosmochim Acta 2013; 112: 166-77.
[http://dx.doi.org/10.1016/j.gca.2013.03.001]

[18]   Allard B, Arsenie I. Abiotic reduction of mercury by humic substances in aquatic system — An important process for the mercury cycle. Water Air Soil Pollut 1991; 56(1): 457-64.
[http://dx.doi.org/10.1007/BF00342291]

[19]   Colombo MJ, Ha J, Reinfelder JR, Barkay T, Yee N. Oxidation of Hg(0) to Hg(II) by diverse anaerobic bacteria. Chem Geol 2014; 363: 334-40.
[http://dx.doi.org/10.1016/j.chemgeo.2013.11.020]

[20]   Zheng W, Liang L, Gu B. Mercury reduction and oxidation by reduced natural organic matter in anoxic environments. Environ Sci Technol 2012; 46(1): 292-9.
[http://dx.doi.org/10.1021/es203402p] [PMID: 22107154]

[21]   Gu B, Bian Y, Miller CL, Dong W, Jiang X, Liang L. Mercury reduction and complexation by natural organic matter in anoxic environments. Proc Natl Acad Sci USA 2011; 108(4): 1479-83.
[http://dx.doi.org/10.1073/pnas.1008747108] [PMID: 21220311]

[22]   Hamlett NV, Landale EC, Davis BH, Summers AO. Roles of the Tn21 merT, merP, and merC gene products in mercury resistance and mercury binding. J Bacteriol 1992; 174(20): 6377-85.
[http://dx.doi.org/10.1128/jb.174.20.6377-6385.1992] [PMID: 1328156]

[23]   Erie L. Mechanistic steps in the photoreduction of mercury in natural waters. 1994; 9697(1993).

[24]   Skogerboe RK, Wilson SA. Reduction of ionic species by fulvic acid. Anal Chem 1981; 53(2): 228-32.
[http://dx.doi.org/10.1021/ac00225a023]

[25]   2013.

[26]   Podar M, Gilmour CC, Brandt CC, *et al.* Global prevalence and distribution of genes and microorganisms involved in mercury methylation. Sci Adv 2015; 1(9): e1500675.
[http://dx.doi.org/10.1126/sciadv.1500675] [PMID: 26601305]

[27]   Parks JM, Johs A, Podar M, *et al.* The genetic basis for bacterial mercury methylation. Science 2013; 339(6125): 1332-5.
[http://dx.doi.org/10.1126/science.1230667] [PMID: 23393089]

[28]   Sedimentt AE. Principal methylators 1985; 50(2): 498-502.

[29]   Wood JM, Kennedy FS, Rosen CG. Synthesis of methyl-mercury compounds by extracts of a methanogenic bacterium. Nature 1968; 220(5163): 173-4.
[http://dx.doi.org/10.1038/220173a0] [PMID: 5693442]

[30]   Choi SC, Chase T, Bartha R. Metabolic pathways leading to mercury methylation in Desulfovibrio desulfuricans LS. Appl Environ Microbiol 1994; 60(11): 4072-7.
[PMID: 16349435]

[31]   Hamelin S, Amyot M, Barkay T, Wang Y, Planas D. Methanogens: Principal methylators of mercury in lake periphyton. Environ Sci Technol 2011; 45(18): 7693-700.
[http://dx.doi.org/10.1021/es2010072] [PMID: 21875053]

[32]   Kerin EJ, Gilmour CC, Roden E, Suzuki MT, Coates JD, Mason RP. Mercury methylation by dissimilatory iron-reducing bacteria. Appl Environ Microbiol 2006; 72(12): 7919-21.
[http://dx.doi.org/10.1128/AEM.01602-06] [PMID: 17056699]

[33]   Hu H, Lin H, Zheng W, et al. Oxidation and methylation of dissolved elemental mercury by anaerobic bacteria. Nat Geosci 2013; 6(9): 751-4.
[http://dx.doi.org/10.1038/ngeo1894]

[34]   Li Y, Cai Y. Progress in the study of mercury methylation and demethylation in aquatic environments. Chin Sci Bull 2013; 58(2): 177-85.
[http://dx.doi.org/10.1007/s11434-012-5416-4]

[35]   Cerrati G, Bernhard M, Weber JH. Model reactions for abiotic mercury (II) methylation: Kinetics of methylation of mercury (II) by mono□, di□, and tri□methyltin in seawater. Appl Organomet Chem 1992; 6(7): 587-95.
[http://dx.doi.org/10.1002/aoc.590060705]

[36]   Li Y, Yin Y, Liu G, et al. Estimation of the major source and sink of methylmercury in the Florida Everglades. Environ Sci Technol 2012; 46(11): 5885-93.
[http://dx.doi.org/10.1021/es204410x] [PMID: 22536798]

[37]   Begley TP, Walts AE, Walsh CT. Bacterial organomercurial Lyase: Over production, Isolation, and char acterization? 1986(1982); 7186-92.

[38]   Oremland RS, Culbertson CW, Winfrey MR. Methylmercury decomposition in sediments and bacterial cultures: Involvement of methanogens and sulfate reducers in oxidative demethylation. Appl Environ Microbiol 1991; 57(1): 130-7.
[PMID: 16348388]

[39]   Schaefer JK, Letowski J, Barkay T. mer-mediated resistance and volatilization of Hg (II) under anaerobic conditions. Geomicrobiol J 2002; 19(1): 87-102.
[http://dx.doi.org/10.1080/014904502317246192]

[40]   Summers AO, Silver S. Mercury resistance in a plasmid-bearing strain of Escherichia coli. J Bacteriol 1972; 112(3): 1228-36.
[PMID: 4565536]

[41]   Lu X, Liu Y, Johs A, et al. Anaerobic mercury methylation and demethylation by geobacter bemidjiensis Bem. Environ Sci Technol 2016; 50(8): 4366-73.
[http://dx.doi.org/10.1021/acs.est.6b00401] [PMID: 27019098]

[42]   Tjerngren I, Karlsson T, Björn E, Skyllberg U. Potential Hg methylation and MeHg demethylation rates related to the nutrient status of different boreal wetlands. Biogeochem 2012; 108(1): 335-50.
[http://dx.doi.org/10.1007/s10533-011-9603-1]

[43]   Suda I, Suda M, Hirayama K. Degradation of methyl and ethyl mercury by singlet oxygen generated from sea water exposed to sunlight or ultraviolet light. Arch Toxicol 1993; 67(5): 365-8.
[http://dx.doi.org/10.1007/BF01973709] [PMID: 8368946]

[44]   Seller P, Kelly CA, Rudd JWM, MacHutchon AR. Photodegradation of methylmercury in lakes. Nature 1996; 380(6576): 694-7.
[http://dx.doi.org/10.1038/380694a0]

[45]   Naftz DL, Cederberg JR, Krabbenhoft DP, Beisner KR, Whitehead J, Gardberg J. Diurnal trends in methylmercury concentration in a wetland adjacent to Great Salt Lake, Utah, USA. Chem Geol 2011; 283(1-2): 78-86.
[http://dx.doi.org/10.1016/j.chemgeo.2011.02.005]

[46]   Nascimento AM, Chartone-Souza E. Operon mer: Bacterial resistance to mercury and potential for bioremediation of contaminated environments. Genet Mol Res 2003; 2(1): 92-101.
[PMID: 12917805]

[47]   Braune B, Chételat J, Amyot M, et al. Mercury in the marine environment of the Canadian Arctic: review of recent findings. Sci Total Environ 2015; 509-510: 67-90.
[http://dx.doi.org/10.1016/j.scitotenv.2014.05.133] [PMID: 24953756]

[48]   Dixit R, Wasiullah EY, Malaviya D, et al. Bioremediation of heavy metals from soil and aquatic environment: An overview of principles and criteria of fundamental processes. Sustainability 2015; 7(2): 2189-212.
[http://dx.doi.org/10.3390/su7022189]

[49]   Deng X, Wang P. Isolation of marine bacteria highly resistant to mercury and their bioaccumulation process. Bioresour Technol 2012; 121: 342-7.
[http://dx.doi.org/10.1016/j.biortech.2012.07.017] [PMID: 22864169]

[50]   Rizwan M, Singh M, Mitra CK, Morve RK. Ecofriendly application of nanomaterials: Nanobioremediation. J Nanoparticles 2014; 2014: 7.
[http://dx.doi.org/10.1155/2014/431787]

[51]   McCarthy D, Edwards GC, Gustin MS, Care A, Miller MB, Sunna A. An innovative approach to bioremediation of mercury contaminated soils from industrial mining operations. Chemosphere 2017; 184: 694-9.
[http://dx.doi.org/10.1016/j.chemosphere.2017.06.051] [PMID: 28633064]

[52]   Elekes CC, Busuioc G. The mycoremediation of metals polluted soils using wild growing species of mushrooms. Jul 22; 2010; pp. In: In Proceedings of the 7th WSEAS international conference on Latest trends on Eng Educ, Corfu Island Greece 2010; 36-9.

[53]   Wang J, Feng X, Anderson CWN, Xing Y, Shang L. Remediation of mercury contaminated sites - A review. J Hazard Mater 2012; 221-222: 1-18.
[http://dx.doi.org/10.1016/j.jhazmat.2012.04.035] [PMID: 22579459]

[54]   Ruiz ON, Daniell H. Genetic engineering to enhance mercury phytoremediation. Curr Opin Biotechnol 2009; 20(2): 213-9.
[http://dx.doi.org/10.1016/j.copbio.2009.02.010] [PMID: 19328673]

[55]   Sone Y, Nakamura R, Pan-Hou H, Sato MH, Itoh T, Kiyono M. Increase methylmercury accumulation in *Arabidopsis thaliana* expressing bacterial broad-spectrum mercury transporter MerE. AMB Express 2013; 3(1): 52.
[http://dx.doi.org/10.1186/2191-0855-3-52] [PMID: 24004544]

[56]   Rugh CL, Wilde HD, Stack NM, Thompson DM, Summers AO, Meagher RB. Mercuric ion reduction and resistance in transgenic *Arabidopsis thaliana* plants expressing a modified bacterial merA gene. Proc Natl Acad Sci USA 1996; 93(8): 3182-7.
[http://dx.doi.org/10.1073/pnas.93.8.3182] [PMID: 8622910]

[57]   Bizily SP, Rugh CL, Summers AO, Meagher RB. Phytoremediation of methylmercury pollution: merB expression in *Arabidopsis thaliana* confers resistance to organomercurials. Proc Natl Acad Sci USA 1999; 96(12): 6808-13.
[http://dx.doi.org/10.1073/pnas.96.12.6808] [PMID: 10359794]

[58]   Rugh CL, Senecoff JF, Meagher RB, Merkle SA. Development of transgenic yellow poplar for mercury phytoremediation. Nat Biotechnol 1998; 16(10): 925-8.
[http://dx.doi.org/10.1038/nbt1098-925] [PMID: 9788347]

[59]   He YK, Sun JG, Feng XZ, Czakó M, Márton L. Differential mercury volatilization by tobacco organs expressing a modified bacterial merA gene. Cell Res 2001; 11(3): 231-6.
[http://dx.doi.org/10.1038/sj.cr.7290091] [PMID: 11642409]

[60]   Ruiz ON, Hussein HS, Terry N, Daniell H. Phytoremediation of organomercurial compounds *via* chloroplast genetic engineering. Plant Physiol 2003; 132(3): 1344-52.
[http://dx.doi.org/10.1104/pp.103.020958] [PMID: 12857816]

[61]   Yang H, Nairn J, Ozias-Akins P, Ozias-akins P. Transformation of peanut using a modified bacterial mercuric ion reductase gene driven by an actin promoter from *Arabidopsis thaliana*. J Plant Physiol 2003; 160(8): 945-52.
[http://dx.doi.org/10.1078/0176-1617-01087] [PMID: 12964870]

[62]   Czakó M, Feng X, He Y, Liang D, Márton L. Transgenic *Spartina alterniflora* for phytoremediation. Environ Geochem Health 2006; 28(1-2): 103-10.
[http://dx.doi.org/10.1007/s10653-005-9019-8] [PMID: 16528587]

[63]   Heaton AC, Rugh CL, Kim T, Wang NJ, Meagher RB. Toward detoxifying mercury-polluted aquatic sediments with rice genetically engineered for mercury resistance. Environ Toxicol Chem 2003; 22(12): 2940-7.
[http://dx.doi.org/10.1897/02-442] [PMID: 14713034]

[64]   Che D, Meagher RB, Heaton AC, Lima A, Rugh CL, Merkle SA. Expression of mercuric ion reductase in Eastern cottonwood (*Populus deltoides*) confers mercuric ion reduction and resistance. Plant Biotechnol J 2003; 1(4): 311-9.
[http://dx.doi.org/10.1046/j.1467-7652.2003.00031.x] [PMID: 17163907]

[65]   Lyyra S, Meagher RB, Kim T, *et al.* Coupling two mercury resistance genes in Eastern cottonwood enhances the processing of organomercury. Plant Biotechnol J 2007; 5(2): 254-62.
[http://dx.doi.org/10.1111/j.1467-7652.2006.00236.x] [PMID: 17309680]

[66]   Nagata T, Ishikawa C, Kiyono M, Pan-Hou H. Accumulation of mercury in transgenic tobacco expressing bacterial polyphosphate. Biol Pharm Bull 2006; 29(12): 2350-3.
[http://dx.doi.org/10.1248/bpb.29.2350] [PMID: 17142961]

[67]   Sinha A, Khare SK. Mercury bioaccumulation and simultaneous nanoparticle synthesis by *Enterobacter* sp. cells. Bioresour Technol 2011; 102(5): 4281-4.
[http://dx.doi.org/10.1016/j.biortech.2010.12.040] [PMID: 21216593]

[68]   Kretsinger RH, Uversky VN, Permyakov EA. Encyclopedia of Metalloproteins. Springer 2013.
[http://dx.doi.org/10.1007/978-1-4614-1533-6]

[69]   Baldi F, Pepi M, Filippelli M. Methylmercury resistance in desulfovibrio desulfuricans strains in relation to methylmercury degradation. Appl Environ Microbiol 1993; 59(8): 2479-85.
[PMID: 16349013]

[70]   Bhakta V, Balkrishna M, Thakuri C. Bacterial mer operon-mediated detoxification of mercurial compounds: Ashort review. Arch Microbiol 2011; 837-44.

[71]   Pan-Hou HS, Nishimoto M, Imura N. Possible role of membrane proteins in mercury resistance of Enterobacter aerogenes. Arch Microbiol 1981; 130(2): 93-5.
[http://dx.doi.org/10.1007/BF00411057] [PMID: 6459062]

[72]   Dash HR, Das S. Bioremediation of mercury and the importance of bacterial mer genes. Int Biodeterior Biodegradation 2012; 75: 207-13.
        [http://dx.doi.org/10.1016/j.ibiod.2012.07.023]

[73]   Sugio T, Komoda T, Okazaki Y, Takeda Y, Nakamura S, Takeuchi F. Volatilization of metal mercury from Organomercurials by highly
        mercury-resistant *Acidithiobacillus* ferrooxidans MON-1. Biosci Biotechnol Biochem 2010; 74(5): 1007-12.
        [http://dx.doi.org/10.1271/bbb.90888] [PMID: 20460735]

[74]   Zeng XX, Tagn JX, Jiang P, Liu H W, Dai Z-mM, Liu X-dD. Isolation, characterization and extraction of mer gene of Hg^{2+} resisting strain
        D2. Trans Nonferrous Met Soc China 2010; 20(50621063): 507-12. [English Edition].
        [http://dx.doi.org/10.1016/S1003-6326(09)60170-9]

[75]   Schelert J, Dixit V, Hoang V, Simbahan J, Drozda M, Blum P. Occurrence and characterization of mercury resistance in the
        hyperthermophilic archaeon Sulfolobus solfataricus by use of gene disruption. J Bacteriol 2004; 186(2): 427-37.
        [http://dx.doi.org/10.1128/JB.186.2.427-437.2004] [PMID: 14702312]

[76]   Liebert CA, Hall RM, Summers AO. Transposon Tn21, flagship of the floating genome. Microbiol Mol Biol Rev 1999; 63(3): 507-22.
        [PMID: 10477306]

[77]   Wireman J, Liebert CA, Smith T, *et al.* Association of mercury resistance with antibiotic resistance in the gram-negative fecal bacteria of
        primates. Association of Mercury Resistance with Antibiotic Resistance in the Gram-Negative Fecal Bacteria of Primates 1997; 63(11)

[78]   Pepi M, Gaggi C, Bernardini E, *et al.* Mercury-resistant bacterial strains *Pseudomonas* and *Psychrobacter* spp. isolated from sediments of
        Orbetello Lagoon (Italy) and their possible use in bioremediation processes. Int Biodeterior Biodegradation 2011; 65(1): 85-91.
        [http://dx.doi.org/10.1016/j.ibiod.2010.09.006]

[79]   Figueiredo NLL, Canário J, Duarte A, Serralheiro ML, Carvalho C. Isolation and characterization of mercury-resistant bacteria from
        sediments of Tagus Estuary (Portugal): Implications for environmental and human health risk assessment. J Toxicol Environ Health A 2014;
        77(1-3): 155-68.
        [http://dx.doi.org/10.1080/15287394.2014.867204] [PMID: 24555656]

[80]   D SR. Prevalence of Mercury-Resistant and Antibiotic-Resistant Bacteria found in Dental Amalgam. 2014; 3(4): 1-4.

[81]   Summers AO, Wireman J, Vimy MJ, *et al.* Mercury released from dental "silver" fillings provokes an increase in mercury- and antibiotic-
        resistant bacteria in oral and intestinal floras of primates. Antimicrob Agents Chemother 1993; 37(4): 825-34.
        [http://dx.doi.org/10.1128/AAC.37.4.825] [PMID: 8280208]

[82]   Ball MM, Carrero P, Castro D, Yarzábal LA. Mercury resistance in bacterial strains isolated from tailing ponds in a gold mining area near El
        Callao (Bolívar State, Venezuela). Curr Microbiol 2007; 54(2): 149-54.
        [http://dx.doi.org/10.1007/s00284-006-0347-4] [PMID: 17200804]

[83]   François F, Lombard C, Guigner JM, *et al.* Isolation and characterization of environmental bacteria capable of extracellular biosorption of
        mercury. Appl Environ Microbiol 2012; 78(4): 1097-106.
        [http://dx.doi.org/10.1128/AEM.06522-11] [PMID: 22156431]

[84]   De J, Ramaiah N. Characterization of marine bacteria highly resistant to mercury exhibiting multiple resistances to toxic chemicals. Ecol Indic
        2007; 7(3): 511-20.
        [http://dx.doi.org/10.1016/j.ecolind.2006.05.002]

[85]   Bogdanova ES, Bass IA, Minakhin LS, *et al.* Horizontal spread of mer operons among gram-positive bacteria in natural environments.
        Microbiology 1998; 144(Pt 3): 609-20.
        [http://dx.doi.org/10.1099/00221287-144-3-609] [PMID: 9534232]

[86]   Dash HR, Sahu M, Mallick B, Das S. Functional efficiency of MerA protein among diverse mercury resistant bacteria for efficient use in
        bioremediation of inorganic mercury. Biochimie 2017; 142: 207-15.
        [http://dx.doi.org/10.1016/j.biochi.2017.09.016] [PMID: 28966143]

[87]   Yu Z, Li J, Li Y, *et al.* A mer operon confers mercury reduction in a *Staphylococcus epidermidis* strain isolated from Lanzhou reach of the
        Yellow River. Int Biodeterior Biodegradation 1981; 90: 57-63.
        [http://dx.doi.org/10.1016/j.ibiod.2014.02.002]

[88]   Naik MM, Dubey S. Lead-and mercury-resistant marine bacteria and their application in lead and mercury bioremediation Marine Pollution
        and Microbial Remediation. Springer 2017; pp. 29-40.

[89]   Boyd ES, Barkay T. The mercury resistance operon: from an origin in a geothermal environment to an efficient detoxification machine. Front
        Microbiol 2012; 3: 349.
        [http://dx.doi.org/10.3389/fmicb.2012.00349] [PMID: 23087676]

[90]   Sone Y, Nakamura R, Pan-hou H, Itoh T, Kiyono M. MerP in resistance to mercurials and the transport of mercurials in Escherichia coli.
        2013; 36:1835-41.

[91]   Jan AT, Azam M, Ali A, Haq QMR. Molecular characterization of mercury resistant bacteria inhabiting polluted water bodies of different
        geographical locations in India. Curr Microbiol 2012; 65(1): 14-21.
        [http://dx.doi.org/10.1007/s00284-012-0118-3] [PMID: 22488489]

[92]   Wilson JR, Leang C, Morby AP, Hobman JL, Brown NL. MerF is a mercury transport protein: different structures but a common mechanism
        for mercuric ion transporters? FEBS Lett 2000; 472(1): 78-82.
        [http://dx.doi.org/10.1016/S0014-5793(00)01430-7] [PMID: 10781809]

[93]   Kiyono M, Oka Y, Sone Y, *et al.* Bacterial heavy metal transporter MerC increases mercury accumulation in *Arabidopsis thaliana*. Biochem Eng J 2013; 71: 19-24.
        [http://dx.doi.org/10.1016/j.bej.2012.11.007]

[94]   Schué M, Dover LG, Besra GS, Parkhill J, Brown NL. Sequence and analysis of a plasmid-encoded mercury resistance operon from *Mycobacterium marinum* identifies MerH, a new mercuric ion transporter. J Bacteriol 2009; 191(1): 439-44.
        [http://dx.doi.org/10.1128/JB.01063-08] [PMID: 18931130]

[95]   Schelert J, Rudrappa D, Johnson T, Blum P. Role of MerH in mercury resistance in the archaeon *Sulfolobus solfataricus*. Microbiology 2013; 159(Pt 6): 1198-208.
        [http://dx.doi.org/10.1099/mic.0.065854-0] [PMID: 23619003]

[96]   Kiyono M, Pan-Hou H. The merG gene product is involved in phenylmercury resistance in Pseudomonas strain K-62. J Bacteriol 1999; 181(3): 726-30.
        [PMID: 9922233]

[97]   Osborn AM, Bruce KD, Strike P, Ritchie DA. Distribution, diversity and evolution of the bacterial mercury resistance (mer) operon. FEMS Microbiol Rev 1997; 19(4): 239-62.
        [http://dx.doi.org/10.1111/j.1574-6976.1997.tb00300.x] [PMID: 9167257]

[98]   Das S, Dash HR, Chakraborty J. Genetic basis and importance of metal resistant genes in bacteria for bioremediation of contaminated environments with toxic metal pollutants. Appl Microbiol Biotechnol 2016; 100(7): 2967-84.
        [http://dx.doi.org/10.1007/s00253-016-7364-4] [PMID: 26860944]

[99]   Schottel JL. The mercuric and organomercurial detoxifying enzymes from a plasmid-bearing strain of Escherichia coli. J Biol Chem 1978; 253(12): 4341-9.
        [PMID: 350872]

[100]   Fermentation BK, Inage K, Chiba T, Metallic J, G S. Metallic in mercury-resistant pseudomonas has not and the electron described is containing group, and the enzyme for the decomposition mercuric chloride on bag, and subjected to a column (2.5 x 25 cm) of DEAE. 1971; 36(12): 217-6.

[101]   Essa AMM. The effect of a continuous mercury stress on mercury reducing community of some characterized bacterial strains. Afr J Microbiol Res 2012; 6(18): 4006-12.

[102]   Pan-Hou H. [Application of mercury-resistant genes in bioremediation of mercurials in environments]. Yakugaku Zasshi 2010; 130(9): 1143-56.
        [http://dx.doi.org/10.1248/yakushi.130.1143] [PMID: 20823672]

[103]   Parks JM, Guo H, Momany C, *et al.* Mechanism of Hg-C protonolysis in the organomercurial lyase MerB. J Am Chem Soc 2009; 131(37): 13278-85.
        [http://dx.doi.org/10.1021/ja9016123] [PMID: 19719173]

[104]   Mathema VB, Krishna B, Thakuri C, Sillanpää M, Amatya R. Study of mercury (II) chloride tolerant bacterial isolates from Baghmati River with estimation of plasmid size and growth variation for the high mercury (II) resistant Enterobacter spp. Culture 2011; 72-.

[105]   Sone Y, Mochizuki Y, Koizawa K, *et al.* Mercurial-resistance determinants in Pseudomonas strain K-62 plasmid pMR68. AMB Express 2013; 3(1): 41.
        [http://dx.doi.org/10.1186/2191-0855-3-41] [PMID: 23890172]

[106]   Guo H-B, Johs A, Parks JM, *et al.* Structure and conformational dynamics of the metalloregulator MerR upon binding of Hg(II). J Mol Biol 2010; 398(4): 555-68.
        [http://dx.doi.org/10.1016/j.jmb.2010.03.020] [PMID: 20303978]

[107]   Park S-j, Wireman JOY, Summers A. Genetic Analysis of the Tn2l operator-promoter 1992; 174(7): 2160-71.

[108]   Chang C-C, Lin L-Y, Zou X-W, Huang C-C, Chan N-L. Structural basis of the mercury(II)-mediated conformational switching of the dual-function transcriptional regulator MerR. Nucleic Acids Res 2015; 43(15): 7612-23.
        [http://dx.doi.org/10.1093/nar/gkv681] [PMID: 26150423]

[109]   Haberstroh L, Silver S. 1984.

[110]   Lee IW, Gambill BD, Summers AO. Translation of merD in Tn21. J Bacteriol 1989; 171(4): 2222-5.
        [http://dx.doi.org/10.1128/jb.171.4.2222-2225.1989] [PMID: 2539363]

[111]   Wireman J, Liebert CA, Smith T, Summers AO. Association of mercury resistance with antibiotic resistance in the gram-negative fecal bacteria of primates. Appl Environ Microbiol 1997; 63(11): 4494-503.
        [PMID: 9361435]

[112]   Zinder ND, Lederberg J. Genetic exchange in Salmonella. J Bacteriol 1952; 64(5): 679-99.
        [PMID: 12999698]

[113]   Lederbergi J. Gene recombination in the bacterium *Escherichia Coli*. Nature 1947; 673-84.

[114]   Frost LS, Leplae R, Summers AO, Toussaint A. Mobile genetic elements: the agents of open source evolution. Nat Rev Microbiol 2005; 3(9): 722-32.
        [http://dx.doi.org/10.1038/nrmicro1235] [PMID: 16138100]

188   Biotechnology: From Theories to Practice

[115]   Dash HR, Das S. Diversity, community structure, and bioremediation potential of mercury-resistant marine bacteria of estuarine and coastal environments of Odisha, India. Environ. Sci. Pollut Res. 2015.

[116]   Rankin DJ, Rocha EPC, Brown SP. What traits are carried on mobile genetic elements, and why? Heredity (Edinb) 2011; 106(1): 1-10.
        [http://dx.doi.org/10.1038/hdy.2010.24] [PMID: 20332804]

[117]   Dimitriu T, Misevic D, Lindner AB, Taddei F. Mobile genetic elements are involved in bacterial sociality. Mob Genet Elements 2015; 5(1): 7-11.
        [http://dx.doi.org/10.1080/2159256X.2015.1006110] [PMID: 26435881]

[118]   Martinez JL. The role of natural environments in the evolution of resistance traits in pathogenic bacteria. Proc Biol Sci 2009; 276(1667): 2521-30.
        [http://dx.doi.org/10.1098/rspb.2009.0320] [PMID: 19364732]

[119]   Martínez JL. Antibiotics and antibiotic resistance genes in natural environments. Science 2008; 321(5887): 365-7.
        [http://dx.doi.org/10.1126/science.1159483] [PMID: 18635792]

[120]   Bennett PM. Genome plasticity: insertion sequence elements, transposons and integrons, and DNA rearrangement. Methods Mol Biol 2004; 266: 71-113.
        [PMID: 15148416]

[121]   Mindlin S, Kholodii G, Gorlenko Z, et al. Mercury resistance transposons of gram-negative environmental bacteria and their classification. Res Microbiol 2001; 152(9): 811-22.
        [http://dx.doi.org/10.1016/S0923-2508(01)01265-7] [PMID: 11763242]

[122]   Bogdanova E, Minakhin L, Bass I, Volodin A, Hobman JL, Nikiforov V. Class II broad-spectrum mercury resistance transposons in Gram-positive bacteria from natural environments. Res Microbiol 2001; 152(5): 503-14.
        [http://dx.doi.org/10.1016/S0923-2508(01)01224-4] [PMID: 11446519]

[123]   Matsui K, Yoshinami S, Narita M, et al. Mercury resistance transposons in Bacilli strains from different geographical regions. FEMS Microbiol Lett 2016; 363(5): fnw013.
        [http://dx.doi.org/10.1093/femsle/fnw013] [PMID: 26802071]

[124]   Liebert CA, Hall RM, Summers AO. Transposon Tn 21. Flagship of the Floating Genome 1999; 63(3): 507-22.

[125]   Lal D, Lal R. Evolution of mercuric reductase (merA) gene: A case of horizontal gene transfer. Mikrobiologiia 2010; 79(4): 524-31.
        [PMID: 21058506]

[126]   Hall JPJ, Harrison E, Lilley AK, Paterson S, Spiers AJ, Brockhurst MA. Environmentally co-occurring mercury resistance plasmids are genetically and phenotypically diverse and confer variable context-dependent fitness effects. Environ Microbiol 2015; 17(12): 5008-22.
        [http://dx.doi.org/10.1111/1462-2920.12901] [PMID: 25969927]

[127]   Huang CC, Narita M, Yamagata T, Endo G. Identification of three merB genes and characterization of a broad-spectrum mercury resistance module encoded by a class II transposon of Bacillus megaterium strain MB1. Gene 1999; 239(2): 361-6.
        [http://dx.doi.org/10.1016/S0378-1119(99)00388-1] [PMID: 10548738]

[128]   Laddaga RA, Chu L, Misra TK, Silver S. Nucleotide sequence and expression of the mercurial-resistance operon from Staphylococcus aureus plasmid pI258. Proc Natl Acad Sci USA 1987; 84(15): 5106-10.
        [http://dx.doi.org/10.1073/pnas.84.15.5106] [PMID: 3037534]

[129]   Petrovski S, Blackmore DW, Jackson KL, Stanisich VA. Mercury(II)-resistance transposons Tn502 and Tn512, from Pseudomonas clinical strains, are structurally different members of the Tn5053 family. Plasmid 2011; 65(1): 58-64.
        [http://dx.doi.org/10.1016/j.plasmid.2010.08.003] [PMID: 20800080]

[130]   Osborn AM, Bruce KD, Ritchie DA, Strike P. The mercury resistance operon of the IncJ plasmid pMERPH exhibits structural and regulatory divergence from other Gram-negative mer operons. Microbiology 1996; 142(Pt 2): 337-45.
        [http://dx.doi.org/10.1099/13500872-142-2-337] [PMID: 8932707]

[131]   Narita M, Chiba K, Nishizawa H, et al. Diversity of mercury resistance determinants among Bacillus strains isolated from sediment of Minamata Bay. FEMS Microbiol Lett 2003; 223(1): 73-82.
        [http://dx.doi.org/10.1016/S0378-1097(03)00325-2] [PMID: 12799003]

[132]   Wang Y, Moore M, Levinson HS, Silver S, Walsh C, Mahler I. Nucleotide sequence of a chromosomal mercury resistance determinant from a Bacillus sp. with broad-spectrum mercury resistance. J Bacteriol 1989; 171(1): 83-92.
        [http://dx.doi.org/10.1128/jb.171.1.83-92.1989] [PMID: 2536669]

[133]   Griffin HG, Foster TJ, Silver S, Misra TK. Cloning and DNA sequence of the mercuric- and organomercurial-resistance determinants of plasmid pDU1358. Proc Natl Acad Sci USA 1987; 84(10): 3112-6.
        [http://dx.doi.org/10.1073/pnas.84.10.3112] [PMID: 3033633]

[134]   Reniero D, Galli E, Barbieri P. Cloning and comparison of mercury- and organomercurial-resistance determinants from a Pseudomonas stutzeri plasmid. Gene 1995; 166(1): 77-82.
        [http://dx.doi.org/10.1016/0378-1119(95)00546-4] [PMID: 8529897]

[135]   Kholodii GY, Mindlin SZ, Bass IA, Yurieva OV, Minakhina SV, Nikiforov VG. Four genes, two ends, and a res region are involved in transposition of Tn5053: A paradigm for a novel family of transposons carrying either a mer operon or an integron. Mol Microbiol 1995; 17(6): 1189-200.

[http://dx.doi.org/10.1111/j.1365-2958.1995.mmi_17061189.x] [PMID: 8594337]

[136]   Christopher M, Paul O, Hamadi B. Association of metal tolerance with multidrug resistance among Environmental Bacteria from wetlands of Lake Victoria Basin 2014.

[137]   Zeyaullah M, Islam B, Ali a. Isolation, identification and PCR amplification of merA gene from highly mercury polluted Yamuna river. Afr J Biotechnol 2010; 9(24): 3510-4.

[138]   Pal C, Bengtsson-Palme J, Kristiansson E, Larsson DGJ. Co-occurrence of resistance genes to antibiotics, biocides and metals reveals novel insights into their co-selection potential. BMC Genomics 2015; 16: 964.
        [http://dx.doi.org/10.1186/s12864-015-2153-5] [PMID: 26576951]

[139]   Trajanovska S, Britz ML, Bhave M. Detection of heavy metal ion resistance genes in gram-positive and gram-negative bacteria isolated from a lead-contaminated site. Biodegradation 1997; 8(2): 113-24.
        [http://dx.doi.org/10.1023/A:1008212614677] [PMID: 9342884]

[140]   Diels L, Mergeay M. DNA probe-mediated detection of resistant bacteria from soils highly polluted by heavy metals. Appl Environ Microbiol 1990; 56(5): 1485-91.
        [PMID: 16348196]

[141]   Bruce KD, Hiorns WD, Hobman JL, Osborn AM, Strike P, Ritchie DA. Amplification of DNA from native populations of soil bacteria by using the polymerase chain reaction. Appl Environ Microbiol 1992; 58(10): 3413-6.
        [PMID: 1444376]

[142]   Georgios M, Egki T. Phenotypic and Molecular Methods for the Detection of Antibiotic Resistance Mechanisms in Gram Negative Nosocomial Pathogens. Intech 2014.
        [http://dx.doi.org/10.5772/57582]

[143]   Barkay T, Fouts DL, Olson BH. Preparation of a DNA gene probe for detection of mercury resistance genes in gram-negative bacterial communities. Appl Environ Microbiol 1985; 49(3): 686-92.
        [PMID: 3994373]

[144]   Baldi F, Gallo M, Marchetto D, Faleri C, Maida I, Fani R. Manila clams from Hg polluted sediments of Marano and Grado lagoons (Italy) harbor detoxifying Hg resistant bacteria in soft tissues. Environ Res 2013; 125: 188-96.
        [http://dx.doi.org/10.1016/j.envres.2012.11.008] [PMID: 23398778]

[145]   Unc A, Zurek L, Peterson G, Narayanan S, Springthorpe SV, Sattar SA. Microarray assessment of virulence, antibiotic, and heavy metal resistance in an agricultural watershed creek. J Environ Qual 2012; 41(2): 534-43.
        [http://dx.doi.org/10.2134/jeq2011.0172] [PMID: 22370416]

[146]   Osborn AM, Bruce KD, Strike P, Ritchie DA. Polymerase Chain Reaction-Restriction Fragment Length Polymorphism Analysis Shows Divergence among mer Determinants from Gram-Negative Soil Bacteria Indistinguishable by DNA-DNA Hybridization 1993.

[147]   N. Mirzazei FK, Kargar M. Isolation and identification of mercury resistant bacteria from Kor river, Iran. 2008; pp. 935-9.

[148]   Taylor P, Figueiredo NLL, Areias A, Mendes R, Canário J, Duarte A. Mercury-Resistant Bacteria From Salt Marsh of Tagus Estuary: The Influence of Plants Presence and Mercury Contamination Levels. J Toxicol Environ Health 2014; •••: 37-41.

[149]   Cao D-j, Tian Z-f. Isolation and identification of a mercury resistant strain. Environ Prot Eng 2012; 38(4)

[150]   Summers AO, Lewis E. Volatilization of mercuric chloride by mercury-resistant plasmid-bearing strains of *Escherichia* coli, *Staphylococcus aureus*, and *Pseudomonas aeruginosa*. J Bacteriol 1973; 113(2): 1070-2.
        [PMID: 4632313]

[151]   Nakamura K, Nakahara H. Simplified X-ray film method for detection of bacterial volatilization of mercury chloride by *Escherichia* coli. Appl Environ Microbiol 1988; 54(11): 2871-3.
        [PMID: 3063210]

[152]   Lian P, Guo H-b, Riccardi D, Dong A, Parks JM, Xu Q. X-ray Structure of a Hg^{2+} Complex of Mercuric Reductase (MerA) and Quantum Mechanical/Molecular Mechanical Study of Hg^{2+} Transfer between the C-Terminal and Buried Catalytic Site Cysteine Pairs. 2014.

[153]   Freedman Z, Zhu C, Barkay T. Mercury resistance and mercuric reductase activities and expression among chemotrophic thermophilic Aquificae. Appl Environ Microbiol 2012; 78(18): 6568-75.
        [http://dx.doi.org/10.1128/AEM.01060-12] [PMID: 22773655]

[154]   Sandström A, Lindskog S. Activation of mercuric reductase by the substrate NADPH. Eur J Biochem 1987; 164(1): 243-9.
        [http://dx.doi.org/10.1111/j.1432-1033.1987.tb11017.x] [PMID: 3104042]

[155]   Ogunseitan OA. Protein method for investigating mercuric reductase gene expression in aquatic environments. Appl Environ Microbiol 1998; 64(2): 695-702.
        [PMID: 9464410]

[156]   Umrania VV. Bioremediation of toxic heavy metals using acidothermophilic autotrophes. Bioresour Technol 2006; 97(10): 1237-42.
        [http://dx.doi.org/10.1016/j.biortech.2005.04.048] [PMID: 16324838]

[157]   Bizily SP, Rugh CL, Summers AO, Meagher RB. Phytoremediation of methylmercury pollution: merB expression in *Arabidopsis thaliana* confers resistance to organomercurials. Proc Natl Acad Sci USA 1999; 96(12): 6808-13.
        [http://dx.doi.org/10.1073/pnas.96.12.6808] [PMID: 10359794]

[158]   Bizily SP, Rugh CL, Meagher RB. Phytodetoxification of hazardous organomercurials by genetically engineered plants. Nat Biotechnol 2000; 18(2): 213-7.
         [http://dx.doi.org/10.1038/72678] [PMID: 10657131]

[159]   Hsieh J-L, Chen C-Y, Chiu M-H, *et al.* Expressing a bacterial mercuric ion binding protein in plant for phytoremediation of heavy metals. J Hazard Mater 2009; 161(2-3): 920-5.
         [http://dx.doi.org/10.1016/j.jhazmat.2008.04.079] [PMID: 18538925]

[160]   Sasaki Y, Hayakawa T, Inoue C, Miyazaki A, Silver S, Kusano T. Generation of mercury-hyperaccumulating plants through transgenic expression of the bacterial mercury membrane transport protein MerC. Transgenic Res 2006; 15(5): 615-25.
         [http://dx.doi.org/10.1007/s11248-006-9008-4] [PMID: 16830224]

[161]   Schelert J, Drozda M, Dixit V, Dillman A, Blum P. Regulation of mercury resistance in the crenarchaeote Sulfolobus solfataricus. J Bacteriol 2006; 188(20): 7141-50.
         [http://dx.doi.org/10.1128/JB.00558-06] [PMID: 17015653]

17

A Review of Three-dimensional Printing for Biomedical and Tissue Engineering Applications

M. Gundhavi Devi[1], M. Amutheesan[2], R. Govindhan[3] and B. Karthikeyan[3,*]

[1]Centre for Bioscience and Nanoscience Research, Coimbatore, Tamil Nadu 641021, India

[2]Department of Aeronautical Engineering, Hindustan Institute of Technology & Science, Padur, Chennai, Tamil Nadu 603103, India

[3]Department of Chemistry, Annamalai University, Annamalai Nagar, Tamil Nadu 608 002, India

Abstract:

Background:

Various living organisms especially endangered species are affected due to the damaged body parts or organs. For organ replacement, finding the customized organs within the time by satisfying biomedical needs is the risk factor in the medicinal field.

Methods:

The production of living parts based on the highly sensitive biomedical demands can be done by the integration of technical knowledge of Chemistry, Biology and Engineering. The integration of highly porous Biomedical CAD design and 3D bioprinting technique by maintaining the suitable environment for living cells can be especially done through well-known techniques: Stereolithography, Fused Deposition Modeling, Selective Laser Sintering and Inkjet printing are majorly discussed to get final products.

Results:

Among the various techniques, Biomedical CAD design and 3D printing techniques provide highly precise and interconnected 3D structure based on patient customized needs in a short period of time with less consumption of work.

Conclusion:

In this review, biomedical development on complex design and highly interconnected production of 3D biomaterials through suitable printing technique are clearly reported.

Keywords: 3D bioprinting, 3D scaffold, Biomedical, Tissue engineering, Polymer, Rapid prototyping.

1. INTRODUCTION

Three-dimensional printing was first patented in 1986 by Charles Hull for Stereolithography Apparatus (SLA). Early researchers are known as Rapid prototyping technologies. Later, stereolithography is commonly known as 3D printing. 3D printing was initially used to create prototypes for product development within certain industries. Dr. Hideo, a Japanese lawyer was the first person to file a patent for rapid prototyping technology. Charles (chuck) Hull was the first person to invent the stereolithography machine (3D printer), which was the first ever device of its kind to print a real physical part from a digital (computer) generated file [1, 2]. Three-dimensional printing technology is one of the trending additive manufacturing methods. It is a process of making a 3D object by adding layer-by-layer of

* Address correspondence to this author at the Department of Chemistry, Annamalai University, Annamalai Nagar 608 002, Tamil Nadu, India; E-mail: bkarthi_au@yahoo.com

required material using a three-dimensional digital model [3, 4]. The most commonly used core material for additive manufacturing includes ceramic, metal, plastic and polymers (synthetic or natural polymers) [5 - 7]. 3D objects are mainly formed under the efficient control of digital computer, 3D modeling software (computer aided design or computer tomography scan images), machine equipment and layering materials [8]. Stereolithography is one of the commonly used software file types that is used for 3D printing [9]. After the given CAD model, 3D printing reads the input data from 3D modeling software. Finally, the highly sophisticated 3D objects are manufactured, which can easily produce tedious shapes and structures [10]. The three-dimensional printing process is shown step-by-step in Scheme **1**.

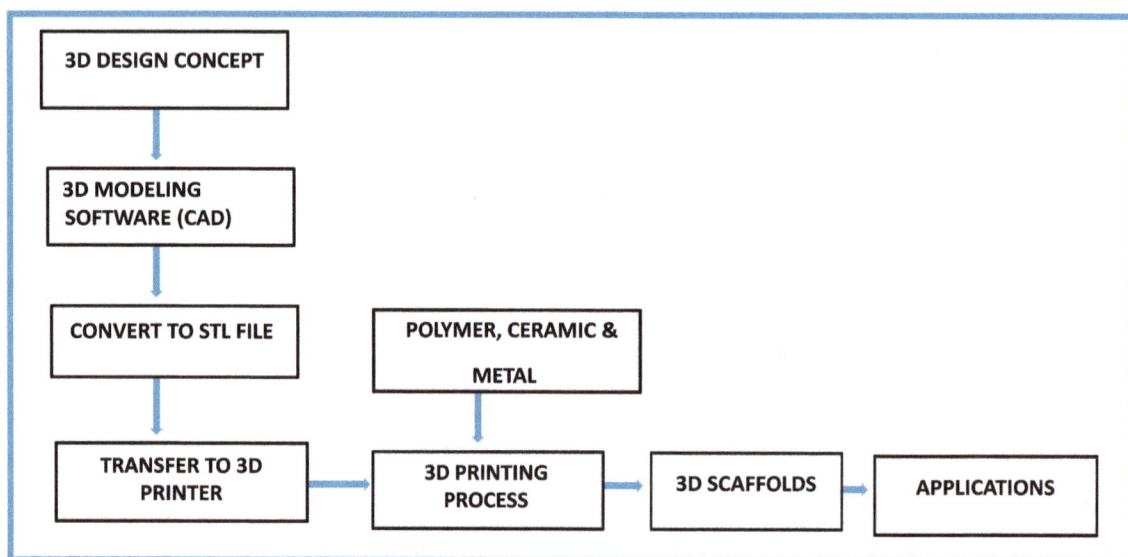

Scheme (1). Schematic process of three-dimensional printing.

Typically, many additive manufacturing processes are available in industries, laboratories, *etc*. It majorly integrated several fine parts such as vat photopolymerization, material extrusion, powder bed fusion, binder jetting, as shown in Table 1. Stereo-lithography (SLA) comes under the basics of vat photo-polymerization method [11, 12]. It is widely recognized as a first 3D printing method. SLA is a laser-based process, which mainly works with photopolymer resins to form a solid 3D object. In this process, photopolymer resin is finely placed in a VAT with a movable platform inside. A laser beam is sharply focused on the surface of the resin, and 3D structures are formed using CAD. Stereo-lithography is one of the most significant 3D printing processes with the good surface finish. Digital light processing is also depending on VAT photopolymerization. The huge difference between them is the light source, and it produces highly accurate parts with excellent resolution. The most widely used method in the material extrusion process is the Fused Deposition Modeling (FDM) or Fused Filament Fabrication (FFF) method. Fused Deposition Modeling uses continuous thermoplastic filament as the printing material to form fine structures and scaffolds [13 - 15]. In this process, it works by melting plastic filament in nozzle head and precisely deposit in build platform to form 3D structure according to the 3D data supplied to the printer. FDM process needs support structures for various bio-applications with overhanging geometries. FDM process is a highly accurate and reliable process that is studio-friendly. In the powder bed fusion process, the most widely used method is the Selective Laser Sintering (SLS) technique for scaffold fabrication. SLS is an additive manufacturing technique, which uses high power laser as a power source to sinter powdered material to fabricate various 3D scaffolds by 3D model [16, 17]. The significant material used in SLS are plastic, metal, ceramic, glass powders. SLS has the potential for creating prototypes, scaffolds, models and even final products, and it is mostly utilized in industry and all medical fields. It provides highly complex parts with adequate interior components and is the fastest Additive Manufacturing (AM) process for printing three-dimensional functional parts and organs with designed structural integrity. In Inkjet printing, the material being jetted is a binder, and it selectively drops into powdered bed of the part material to fuse for creating three-dimensional objects and scaffolds for various medical applications. A range of different materials can be used for three-dimensional printing for various fabrications [18, 19]. Different areas of the scientific community combined to form a 3D structure, which is shown in Scheme **2**.

Table 1. Additive manufacturing technology.

Methods	Classifications	Ref.
Vat photopolymerization	i) Stereolithography (SLA) ii) Digital Light Processing (DLP)	80-84
Material extrusion	i) Fused Deposition Modeling (FDM)	97-99
Powder bed fusion	i) Selective Laser Sintering	112-114
Binder jetting	i) Inkjet 3D printing	122

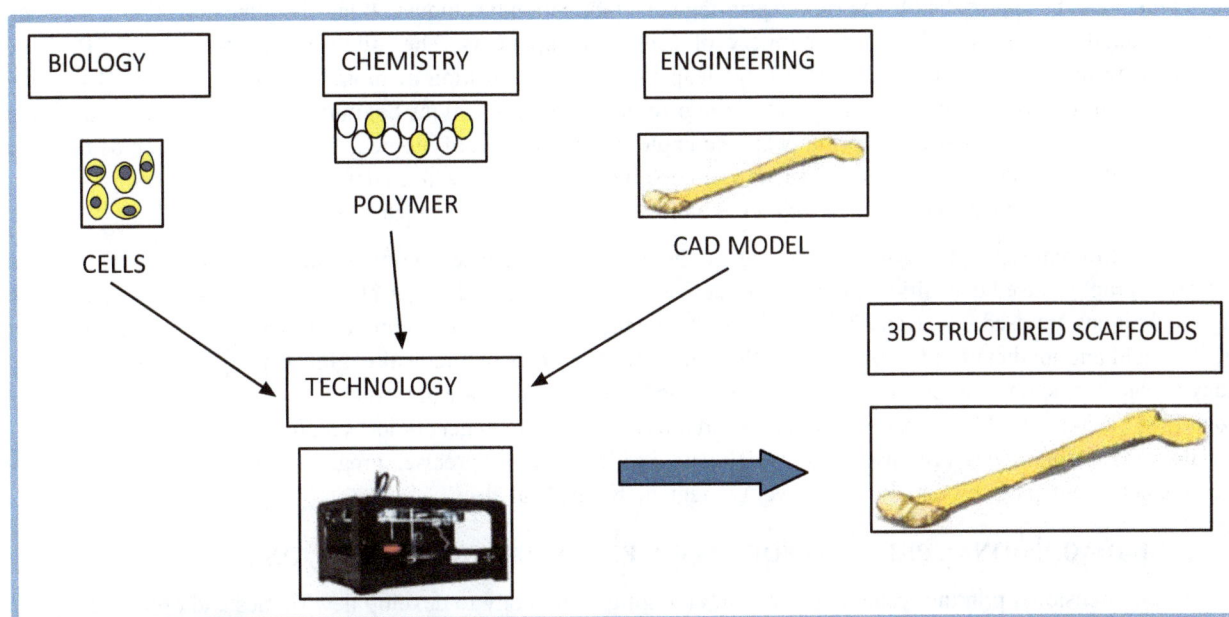

Scheme (2). A typical schematic workflow of multidisciplinary subjects used in the 3D printing method for tissue engineering applications.

3D file of the object can be created using Biomedical CAD software, with a 3D scanner. The variety of materials can be used for printing purposes such as plastics, alumide, ceramics, resins, metals, sand, textiles, biomaterials, glass, food and even lunar dust. Fused Deposition Modeling and Selective Laser Sintering use plastic and alumide for fabricating 3D scaffolds. Some FDM printers have two or more print heads to print multiple variable colors in scaffolds. Selective Laser Sintering consists of fabrication of 3D object by melting successive layers of powder together to form a scaffold. Stereolithography and Digital Light Processing use photopolymerization for developing highly sophisticated 3D products [20, 21]. 3D printing is excellent for developing healthcare products in many ways, including implantable and non-implantable medical devices along with cost-effective customizable devices, patient-specific products in orthopedics and maxillofacial surgery, fabricating human living tissue, prosthetics and accurate pre-op models for academic purpose. It is mainly helpful for fabricating different types of living tissues, recreating difficult bone disorders such as craniofacial disfigurement, hearing aids and dental delivery devices, which offer excellent visualization and great dimensional stability. 3D printing also creates medical fixtures, functional testing models, industrial design and end-use parts [22, 23].

Nowadays, the ultimate aim of tissue and organ engineering is to restore normal functions of living organs and tissues, regeneration, replacement of defective or injured organs and tissues using different technology. To achieve this aim amongst different technology, three-dimensional scaffolds are commonly used for biomedical, tissue engineering applications and all medical fields, which is made up of polymer (natural or synthetic polymers), metals, ceramics, *etc*. Three-dimensional Scaffolds provide several mandatory functions [24 - 26]. It creates an adequate internal pathway for the cell attachment and migration. It should transfer several growth factors, oxygen transport and waste product removal in 3D scaffolds [27, 28]. It must be biocompatible, good mechanical properties and keeps its shape and structure when the cells and tissue are growing [29 - 32]. Hence, 3D bioprinting technology is commonly used for controlling cell and tissue pattern to be retained viability and functionality of the cells inside the printed 3D structure using different biomaterials [33 - 35]. Many researchers have been studied development and improvement of the appropriate scaffold using 3D printing in tissue engineering applications. Advances developed by 3D printing enhance the ability to control pore volume, pore size, and pore interconnectivity in 3D scaffolds for patient specific applications. In addition, materials used in a 3D printing machine, which is very essential for fabricating 3D scaffolds using 3D modeling software and scan data. 3D bioprinting process uses living cells and bioactive molecules in biomaterials, which produces a 3D structure that does not affect the viability and functionality of the cells [36, 37].

In future, researchers should be considered in the biomaterials such as bio-ink for creating 3D objects in tissue engineering applications and all the medical fields. Recent advancement in the biomedical field of stem cell development can be approached to the Bio3D printing cells fabrication techniques. It has the huge potential of studying disease modelling, discover drugs and mimicry of cellular components. The microfluidic approach in 3D tissue fabrication printing has garnered to a significant leap in the vascularization of biomedical engineering.3D printing technology can be evolved to cover the entire range of biomedical applications beginning from diagnosis and ending with prognosis. The potential of 3D printers can be exploited in the field of biomedical engineering such as research works, drug delivery, lab testing, clinical practice and helping the surgeons with as detailed mock surgeries as possible. 3D printing is used as a one-step solution for all the biomedical engineering problems [38 - 40].

Among different rapid prototyping technologies, the widely used four technologies for biomedical applications are stereolithography, fused deposition modeling, selective laser sintering and inkjet 3D printing [41 - 43]. Even though these techniques are used in various fields such as architectural modeling, art, lightweight machines, aircraft industry, defense field and medical fields, but it is excellent in tissue and other biomedical engineering applications [44 - 48]. Many researchers keep on improving various methods and materials to create 3D structure by satisfying the mechanical properties, biocompatibility for regeneration of normal tissues and bone regeneration, *etc.* [49, 50]. In this review, only few ongoing technologies are discussed, which can produce highly precise, greatly customized and extremely interconnected bio parts by satisfying all requirement of biomedical needs.

2. THREE-DIMENSIONAL PRINTING FOR TISSUE ENGINEERING APPLICATION

Three-dimensional printing technologies are an emerging technology to develop new tissues and organs [51 - 55]. Many researchers are currently conducting a study for fabricating 3D structure, which is useful for tissue engineering fields [56 - 60]. Three-dimensional bioprinting creates unique 3D structure, which controls cell proliferation, attachment and migration within 3D printed structures [61 - 65]. Therefore, different types of three-dimensional bioprinting techniques are used for a variety of tissue and organ engineering applications [66 - 70]. Herein, we will discuss the four different types of three-dimensional bioprinting methods, which are most commonly used methods such as Stereolithography and Digital Light Processing in Vat Photopolymerization, Fused Deposition Modeling in material extrusion, Selective Laser Sintering in powder bed fusion and Inkjet printing in binder jetting methods. Table **2** represents some of the advantages and disadvantages of different 3D printing methods in tissue engineering applications.

Table 2. Additive manufacturing: Advantages and disadvantages.

Methods	Advantages	Disadvantages	Materials	Ref.
SLA, DLP	Simple and complex Fast and good resolution	Expensive equipment and materials	PEG, PCL, PEGDA	[79-84]
FDM	Easy to use good mechanical properties	Filament required Cannot used with cells	PCL, PLGA	[96-99]
SLS	No need for support materials Various of biomaterials	Rough surface Expensive equipment	PCL/HA, PCL	[111-114]
Inkjet	Cells and hydrogel printed, incorporation of drug and molecules	Low resolution Low mechanical properties	Fibrin, Gelatin	[121-122]

2.1. Stereolithography (SLA)

Stereolithography is also known as photo-solidification, which has been early and still widely used 3D printing method. SLA technique has commonly been used to fabricate 3D models, prototypes, patterns and production parts by using UV light in layer-by-layer. This technique has been obtained the patent by Charles (chuck) Hull in 1986 [71 - 73]. SLA method has the potential to create 3D scaffolds using photopolymerization. Similarly, the DLP technique has also been utilized to create 3D functional models and positive mold objects using visible light source [74, 75]. Both Stereolithography and DLP are formed on the vat photopolymerization [76 - 78]. In this process, UV light beam is directed onto the area of vat filled with a liquid photopolymer. UV light makes chains of molecules to bind and form polymers. And those polymers are essentially focused to fabricate three-dimensional objects. However, Photopolymerization method is formed free radicals that can affect cell membrane, protein and nucleic acids. Using computer Aided manufacturing or computer Aided design, Stereolithography can make any design that can be fast and expensive [79 - 81]. A typical schematic of stereolithography technique is shown in Scheme **3**. Many scientists obtained the SLA 3D product using various biomaterials in tissue engineering applications. Elomaa *et al.* fabricated cell-laden hydrogels constructs with biomimetic complexity for use in pharmaceutics, vascular and tissue engineering application. They reported that they used water-soluble methacrylated poly (ethylene glycol-co-depsipeptide) to synthesize and formed a biodegradable photocrosslinkable macromer for SLA [82]. Neiman's *et al.* created three-dimensional(3D) hydrogel scaffolds with open channels for post-seeding using photopolymerizable PEG in Stereolithography based method. They showed that structural and functional development of foster formation in 3D liver aggregates. The aim of this study was to develop a platform for drug toxicity study, liver pathophysiology and obtained micro perfusion flow within the open channels of this 3D hydrogel structure [83]. Justinas *et al.* utilized direct laser writing lithography to fabricate three-dimensional (3D) microstructured scaffolds for cartilage tissue engineering using ultrafast pulsed lasers. They reported that 3D microstructured scaffolds are excellent in spatial resolution, geometry complexity and hexagonal pore shaped hybrid organic and inorganic material micro-structured scaffold, which were fabricated using DLW technique in combining with Cho seeding [84]. Owen *et al.* created Polymerized High internal Phase Emulsions scaffolds using emulsion templating by combining with micro-stereolithography, which produces cell ingrowth, plasma penetration, tightly controlled and highly interconnected microporosity. Scaffolds constructed using two acrylate monomers with isobornyl acrylate and supported osteogenic differentiation of mesenchymal cells [85]. Du *et al.* created ceramic artificial bone scaffolds using stereolithography with acrylic resin, which produced correct external shape and internal architecture for bone tissue ingrowth [86]. Hang *et al.* produced a three-dimensional (3D) scaffold with desired architectures using Stereolithography (SLA) technique. They showed that they used two lentiviral gene constructs with human bone marrow-derived mesenchymal stem cells into a solution of photocrosslinkable gelatin, which was focused using visible light-based projection [87]. Main advantages of stereolithography in tissue engineering applications are fast speed, good resolution, easy to remove support materials, complex designs and fabrication of a simple, and the disadvantages in SLA are a limited range of photosensitive resin and polymers, expensive equipment and materials, cytotoxicity of uncured photoinitiator.

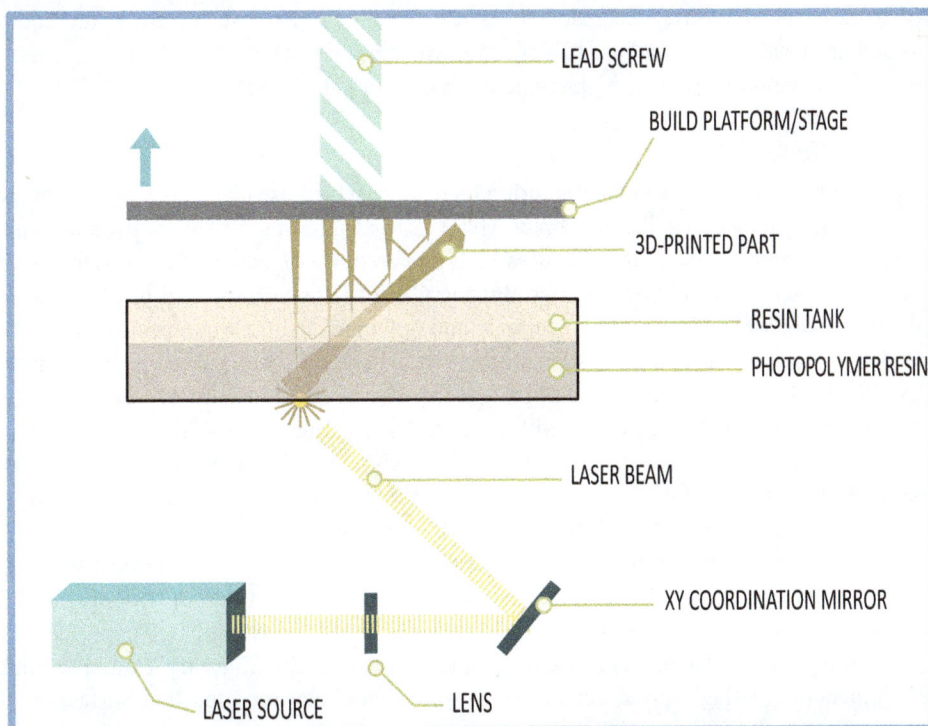

Scheme (3). A typical schematic representation of stereolithography.

2.2. Fused Deposition Modeling (FDM)

FDM is known as Solid-based AM technology. It is also an Additive Manufacturing technology (AM) which is mainly used for modeling, prototyping and production applications [88 - 90]. Fused Deposition Modeling was developed and commercialized by Scott Crump and Stratasys, which works under the controlling of Stereolithography (STL) file [91 - 93]. In this process, FDM printers use a continuous filament of a thermoplastic material in a material extrusion method. These filaments are heated to the melting point temperature, and molten material from the printhead nozzle is deposited on the surface of the growing workpiece to create 3D structures [94]. The nozzle and substrate are controlled by computer to print defined shape and structure, and nozzle can be travelled in both horizontal and vertical directions. Using computer-aided technology, FDM is very flexible to print 3D objects [95 - 97]. FDM used the thermoplastics PLA, ABS, ABSi, polyphenylsulfone, polycarbonate and among others material. Fused Deposition Modeling is a thermal heating technique, which is used for 3D scaffolds fabrication in tissue engineering applications. Many researchers were investigated using FDM method for tissue engineering applications. Pati *et al.* fabricated 3D printed scaffolds using a composite of polycaprolactone, polylactic-co-glycolic acid, β- tricalcium phosphate and mineralized ECM laid by human nasal inferior turbinate tissue-derived mesenchymal stromal cells. They studied that they improve the biological functionality of 3D printed synthetic scaffolds to mimics bony microenvironment using Fused Filament Fabrication, and they develop cellular responses and drive osteogenesis of stem cells. Jensen *et al.* created a polycaprolactone (PCL) scaffold using the combination of Fused Deposition Modeling and thermal induced phase separation to create nanoporous structure in polycaprolactone (PCL) scaffold. They studied PCL3D scaffold to be an excellent osteoconduction and osteointegration [98]. Xu *et al.* fabricated polycaprolactone nano-HA and polycaprolactone (PCL) 3D artificial bones to mimic natural goat femurs using Computed Tomography-guided Fused Deposition Modeling. They reported that polycaprolactone (PCL) /HA 3D artificial bones scaffolds are excellent in cell biocompatibility, biodegradation ability and bone formation ability, good biomechanical properties that reduce the stress shielding effect [99]. Idaszek *et al.* created a ternary polycaprolactone (PCL) scaffold, which consists of polycaprolactone (PCL), TCP and PLGA using Fused Deposition Modeling. They evaluated mechanical characteristics, degradation kinetics and surface properties through *in-vitro*. They resulted that the introduction of PLGA improved the degradation rate and surface roughness [100]. Li *et al.* used the Fused Deposition Modeling technique to make a comparison between Ti cage and PCL-TCP scaffold as a spinal fusion cage. They reported that inferior fusion performance of the PCL-TCP scaffold at 6 months is similar to Ti cage at 12 months. In addition, PCL-TCP scaffold is resulted in better bone ingrowth and distribution compared to Ti cage [101]. The main advantages of Fused Deposition Modeling are easy operation and easy use, adequate mechanical properties, low cost, solvent is not required and various lay- down patterns and the disadvantages in Fused Deposition Modeling are materials in filament form (thermoplastics), low speed, cannot used with cells and tissues, high temperature and smooth surface.

2.3. Selective Laser Sintering (SLS)

Selective Laser Sintering is an additive manufacturing technique, which uses laser as a power source to form solid 3D scaffolds [102 - 104]. It is very similar to Direct Metal Laser Sintering. SLS was patented and developed in mid1980's by Carl Deckard and Joe Beaman. SLS uses high powered lasers, which is too expensive. In this process, high powered lasers selectively fuse powdered material using CAD file or Scan data to form 3D objects [105]. This process is printed various materials; plastic, metal, ceramic and polymers and their composites [106 - 108]. Moreover, the SLS technique does not require a separate feeder for support material. SLS technique is the ability to make highly complex geometry directly from digital CAD data [109 - 111]. A typical schematic of Selective Laser Sintering is shown in Scheme **4**. Many researchers reported the SLS product using various biomaterials. In addition, SLS is used in tissue engineering applications as scaffolds from various biomaterials and their composites. Du *et al.* constructed three-dimensional bone scaffolds in Selective Laser Sintering technique with uniform multi-scaled porosity, moderate mechanical properties and good biocompatibility using PCL microspheres, and polycaprolactone /hydroxyapatite composite microspheres are used as the basic building materials. They showed that SLS derived scaffolds are excellent in multiple stem cells behavior, promoting cell adhesion, supporting cell proliferation, inducing cell differentiation, histocompatibility and adequate mechanical features [112]. Chen *et al.* fabricated polycaprolactone scaffolds for cartilage tissue engineering in craniofacial reconstruction using Selective Laser Sintering technique, which was surface modified through immersion coating with either gelatin or collagen. They reported that surface modification with collagen or gelatin improved the hydrophilicity, water uptake and good mechanical strength [113]. Roskies *et al.* created polyetheretherketone scaffolds using Selective Laser Sintering technique with a computer-aided design program. They evaluated that PEEK scaffolds maintain the viability of adipose and bone marrow-derived MSCs and induce the osteodifferentiation of the adipose-derived MSCs [114]. Feng *et al.* created highly interconnected porous scaffolds with β-TCP doping of zinc oxide powder using Selective Laser Sintering technique. They studied that porous scaffold resulted in excellent mechanical and biological properties by evaluating fracture toughness, compressive strength, osteoinduction and osteoconduction [115]. Shuai *et al.* developed poly (vinyl alcohol)/calcium silicate composite scaffolds with interconnected porous structures and customized shapes using Selective Laser Sintering. They found excellent compressive strength, good bioactivity and cytocompatibility in these scaffolds [116]. The main advantages of Selective Laser Sintering are a wide range of materials, good mechanical strength, relatively high precision, high porosity, support materials not required, and the disadvantages of selective laser sintering are materials in powder form, difficult to remove trapped materials, expensive equipment and rough surface.

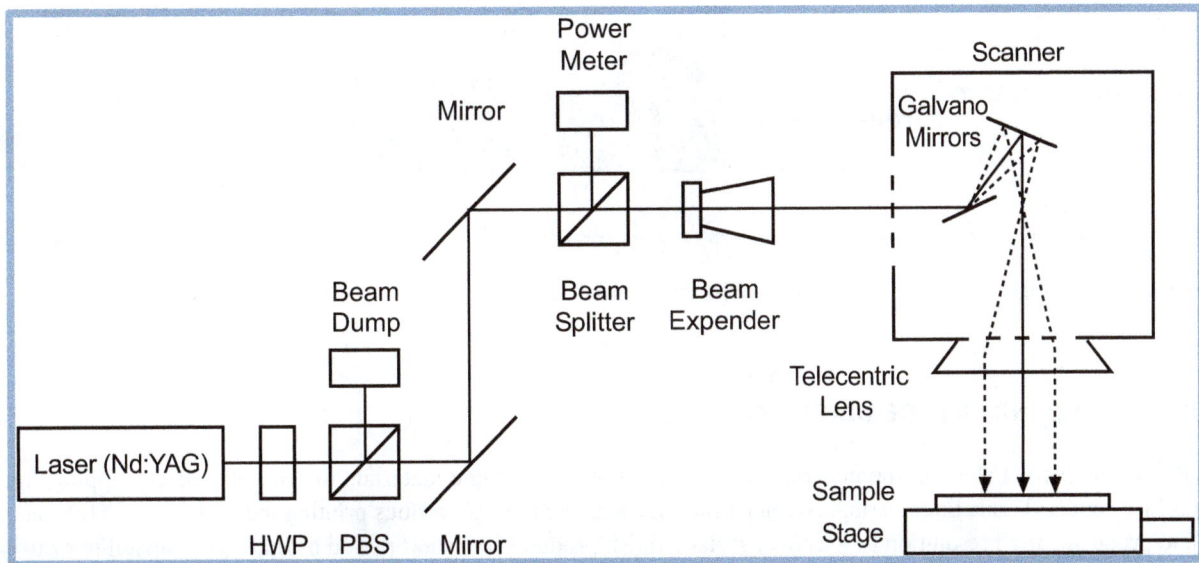

Scheme (4). A typical schematic representation of selective laser sintering.

2.4. Inkjet 3D Printing

Inkjet 3D printing technique is a rapid prototyping method, which is layered manufacturing technology for creating objects described by 3D modeling software and scan data [117]. Inkjet three-dimensional printing technique is similar to Inkjet head printing. In addition, Inkjet 3D printing method has the potential to use polymeric bio-inks for various applications such as Biomedical, Tissue engineering and all other Medical fields [118 - 120]. Nowadays, Inkjet bioprinter is a commonly used technology, which is useful for both non-biological and biological applications. Moreover, Inkjet bioprinter is a powerful technique for depositing cells, biomaterials and has become popular in creating cell-laden constructs, which can mimic the high complexity of native tissue and organ. A typical schematic of Inkjet bioprinting is shown in Scheme **5**. Many scientists were investigated using Inkjet 3D printing in tissue and organ engineering applications. Lorber *et al.* fabricated Retinal Ganglion Cells (RGC) and glia using piezoelectric inkjet printing. They found that the viability and survival/growth of the cells in culture were not affected by the inkjet printing process [121]. Pati *et al.* printed dome-shaped adipose tissue using human decellularized adipose tissue matrix bio-ink, which encapsulates human adipose tissue-derived mesenchymal stem cells from the biomimetic approach. They evaluated the efficacy of their printed tissue constructs for adipose tissue regeneration [122]. Irvine *et al.* created a patterned 3D structure using Inkjet bioprinter and used printable gelatin as an ideal material crosslinked with microbial transglutaminase to print cell bearing hydrogel for three-dimensional constructs. They confirmed excellent cell affinity [123]. The main advantages of Inkjet 3D printing in tissue engineering applications are cells, tissues and hydrogel that can be printed, patient-customized fabrication, incorporation of drug and biomolecules, low cost and rapid production, and the disadvantages in Inkjet printing are a limitation of size, low resolution, low mechanical properties, using limited biomaterials.

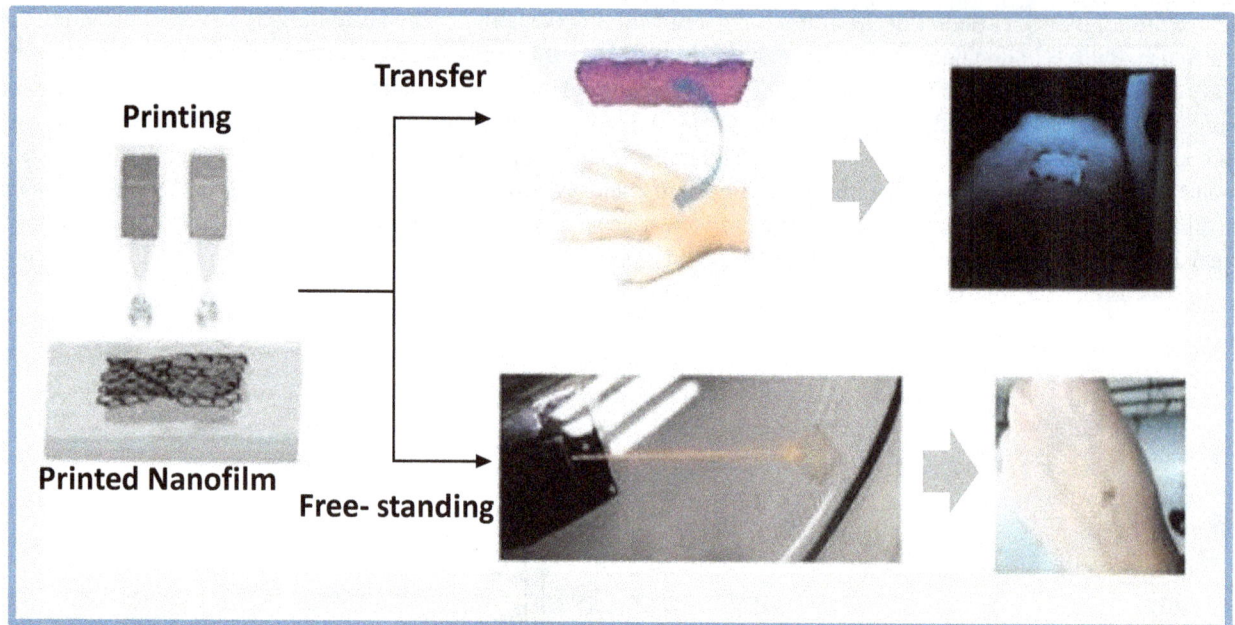

Scheme (5). A typical schematic representation of inkjet printing method.

CONCLUSION AND FUTURE DIRECTION

Rapid intellectual shine in the material science field uses nanoscopic materials for various societal applications: Nano Electronics, Tissue Engineering, Artificial Intelligence, *etc.* through various printing methodologies. SLA can be able to produce a high resolution of excellent surface finish products in a short period of time. The capability of using multiple prints heads provides choice to select different colors and materials on making the biomedical 3D products using FDM. SLS has the functionality to make highly complex geometry with more accuracy. Usage of bio-ink through ink jet printing open doors for a biomedical researcher for producing living cell products to satisfy the need of the rare species and other living organism. Biomedical researchers keep on working in Regenerative medicine. Self-healing property, natural Self-assembly of nanotubes and synergetic effect are the key factors in the artificial bio parts manufacturing. Natural Self-assembling property of the biocompatible peptide nanotubes can be able to reform its shape if the occurred damages in the 3D printed bio Parts are under the acceptable range of damages in biostructure like bone, tissues *etc*. This technique can be able to reduce the number of failure in the final products and minimize the requirement of design, production and overall cost. The precise designing using the Biomedical CAD software for customized design of tissue and organs for biomedical application, which can design the required 3D model. To manufacture the designed product, a unique combination of cells with the biopolymers can act as the core material for organs and tissues. The recent development of rapid prototyping process leads the manufacturing of living tissues and organs with highly porosity architecture. To obtain a highly efficient manufacturing model, the multifunctional 3D structure is undergoing recent development by combining more than two 3D printing technology or combination of 3D printing technology with other scaffold 3D printing technology. The quality of the functioning of the scaffold is mainly due to the integrity of design, materials and the manufacturing process. Improving biomaterials quality (Polymers and Bio-ink) is the most significant goal in Additive Manufacturing, which should be biocompatible, ease of processing, good mechanical properties for cell support and secure 3D structure. 3D printing also used for drug delivery, Chemical, Biological agents and Organ on-chip devices along with tissue engineering. The scientific community is now improving the resolution, speed of prototyping and quality of printing process by compatible with cells and tissues in 3D manufacturing. Patient customized 3D bioprinting is still a challengeable process of implementing for the whole global community. Moreover, the mimicking ability of the direct organ fabrication in the medical field is the goal of tissue engineering and regenerative medicine using 3D bioprinting. Fabrication of the scaffold for the mass bone defect is the crucial process in the 3D manufacturing due to unsatisfactory bone graft substitutes. The capability of integrating various disciplines: chemistry, materials science, computer-aided design, medical imaging and biomedical world should focus on to improve the availability of various models for different patient specific applications [124 - 126].

ACKNOWLEDGEMENTS

One of the authors, M. Gundhavi Devi would like to thank Prof. Insup Noh, Convergence institute of Biomedical and Biomaterials department, SEOUL TECH, Seoul, Korea for his valuable motivation to work on 3D-bioprinting.

REFERENCES

[1]   Sithole MN, Kumar P, du Toit LC, Marimuthu T, Choonara YE, Pillay V. A 3D bioprinted *in situ* conjugated-co-fabricated scaffold for potential bone tissue engineering applications. J Biomed Mater Res A 2018; 106(5): 1311-21.
        [http://dx.doi.org/10.1002/jbm.a.36333] [PMID: 29316290]

[2]   Wang LL, Highley CB, Yeh YC, Galarraga JH, Uman S, Burdick JA. Three□dimensional extrusion bioprinting of single□and double□ network hydrogels containing dynamic covalent crosslinks. J Biomed Mat Res A 2018; 106(4): 865-75.

[3]   Placone JK, Engler AJ. Recent advances in extrusion-based 3D printing for biomedical applications. Adv Healthc Mater 2017.
        [http://dx.doi.org/10.1002/adhm.201701161] [PMID: 29283220]

[4]   Kelly CN, Miller AT, Hollister SJ, Guldberg RE, Gall K. Design and structure-function characterization of 3d printed synthetic porous biomaterials for tissue engineering. Adv Healthc Mater 2018; 7(7): 1701095.
        [http://dx.doi.org/10.1002/adhm.201701095] [PMID: 29280325]

[5]   Govindhan R, Karthikeyan B. Nano Cu interaction with single amino acid tyrosine derived self-assemblies; study through XRD, AFM, confocal Raman microscopy, SERS and DFT methods. J Phy Chem Sol 2017; 111: 123.: 34.
        [http://dx.doi.org/111. 10.1016/j.jpcs.2017.07.025.]

[6]   Gao G, Huang Y, Schilling AF, Hubbell K, Cui X. Organ bioprinting: Are we there yet? Adv Healthc Mater. Adv Healthc Mater 2018; 7(1): 1701018.
        [http://dx.doi.org/10.1002/adhm.201701018] [PMID: 29193879]

[7]   Grémare A, Guduric V, Bareille R. Characterization of printed PLA scaffolds for bone tissue engineering. J Biomed Mat Res A 2018; 106(4): 887-7.
        [http://dx.doi.org/10.1002/jbm.a.36289.]

[8]   Palaganas NB, Mangadlao JD, de Leon ACC, *et al.* 3D printing of photocurable cellulose nanocrystal composite for fabrication of complex architectures *via* stereolithography. ACS Appl Mater Interfaces 2017; 9(39): 34314-24.
        [http://dx.doi.org/10.1021/acsami.7b09223] [PMID: 28876895]

[9]   Badea A, McCracken JM, Tillmaand EG, *et al.* 3D-printed phema materials for topographical and biochemical modulation of dorsal root ganglion cell response. ACS Appl Mater Interfaces 2017; 9(36): 30318-28.
        [http://dx.doi.org/10.1021/acsami.7b06742] [PMID: 28813592]

[10]   Cornelissen DJ, Faulkner-Jones A, Shu W. Current developments in 3D bioprinting for tissue engineering. Curr Opin Biomed Eng 2017; 2: 76-82.
        [http://dx.doi.org/10.1016/j.cobme.2017.05.004]

[11]   Billiet T, Gevaert E, De Schryver T, Cornelissen M, Dubruel P. The 3D printing of gelatin methacrylamide cell-laden tissue-engineered constructs with high cell viability. Biomaterials 2014; 35(1): 49-62.
        [http://dx.doi.org/10.1016/j.biomaterials.2013.09.078] [PMID: 24112804]

[12]   Pekkanen AM, Mondschein RJ, Williams CB, Long TE. 3D printing polymers with supramolecular functionality for biological applications. Biomacromolecules 2017; 18(9): 2669-87.
        [http://dx.doi.org/10.1021/acs.biomac.7b00671] [PMID: 28762718]

[13]   Bhargav A, Sanjairaj V, Rosa V, Feng LW, Fuh YH J. Applications of additive manufacturing in dentistry: A review. J Biomed Mater Res B Appl Biomater 2018; 106(5): 2058-64.
        [http://dx.doi.org/10.1002/jbm.b.33961] [PMID: 28736923]

[14]   Lim J, You M, Li J, Li Z. Emerging bone tissue engineering *via* Polyhydroxyalkanoate (PHA)-based scaffolds. Mater Sci Eng C 2017; 79: 917-29.
        [http://dx.doi.org/10.1016/j.msec.2017.05.132] [PMID: 28629097]

[15]   Gu BK, Choi DJ, Park SJ, Kim MS, Kang CM, Kim CH. 3-Dimensional bioprinting for tissue engineering applications. Biomater Res 2016; 20: 12.
        [http://dx.doi.org/10.1186/s40824-016-0058-2] [PMID: 27114828]

[16]   Sayyar S, Officer DL, Wallace GG. Fabrication of 3D structures from graphene-based biocomposites. J Mater Chem B Mater Biol Med 2017; 5: 3462-82.

[http://dx.doi.org/10.1039/C6TB02442D]

[17]   Dumanli AG. Nanocellulose and its composites for biomedical applications. Curr Med Chem 2017; 24(5): 512-28.
         [http://dx.doi.org/10.2174/0929867323666161014124008] [PMID: 27758719]

[18]   Gao G, Cui X. Three-dimensional bioprinting in tissue engineering and regenerative medicine. Biotechnol Lett 2016; 38(2): 203-11.
         [http://dx.doi.org/10.1007/s10529-015-1975-1] [PMID: 26466597]

[19]   Bendtsen ST, Quinnell SP, Wei M. Development of a novel alginate-polyvinyl alcohol-hydroxyapatite hydrogel for 3D bioprinting bone
         tissue engineered scaffolds. J Biomed Mater Res A 2017; 105(5): 1457-68.
         [http://dx.doi.org/10.1002/jbm.a.36036] [PMID: 28187519]

[20]   Chen C, Zhao ML, Zhang RK, et al. Collagen/heparin sulfate scaffolds fabricated by a 3D bioprinter improved mechanical properties and
         neurological function after spinal cord injury in rats. J Biomed Mater Res A 2017; 105(5): 1324-32.
         [http://dx.doi.org/10.1002/jbm.a.36011] [PMID: 28120511]

[21]   Souness A, Zamboni F, Walker GM, Collins MN. Influence of scaffold design on 3D printed cell constructs. J Biomed Mater Res B Appl
         Biomater 2018; 106(2): 533-45.
         [http://dx.doi.org/10.1002/jbm.b.33863] [PMID: 28194931]

[22]   Kim W, Lee H, Kim Y, et al. Versatile design of hydrogel-based scaffolds with manipulated pore structure for hard-tissue regeneration.
         Biomed Mater 2016; 11(5): 055002.
         [http://dx.doi.org/10.1088/1748-6041/11/5/055002] [PMID: 27586518]

[23]   Wei D, Sun J, Bolderson J, et al. Continuous fabrication and assembly of spatial cell-laden fibers for a tissue-like construct via a
         photolithographic-based microfluidic chip. ACS Appl Mater Interfaces 2017; 9(17): 14606-17.
         [http://dx.doi.org/10.1021/acsami.7b00078] [PMID: 28157291]

[24]   Muerza-Cascante ML, Shokoohmand A, Khosrotehrani K, et al. Endosteal-like extracellular matrix expression on melt electrospun written
         scaffolds. Acta Biomater 2017; 52: 145-58.
         [http://dx.doi.org/10.1016/j.actbio.2016.12.040] [PMID: 28017869]

[25]   Hong N, Yang GH, Lee J, Kim G. 3D bioprinting and its in vivo applications. J Biomed Mater Res B Appl Biomater 2018; 106(1): 444-59.
         [http://dx.doi.org/10.1002/jbm.b.33826] [PMID: 28106947]

[26]   Jo HH, Lee SJ, Park JS, et al. Characterization and preparation of three-dimensional-printed biocompatible scaffolds with highly porous
         strands. J Nanosci Nanotechnol 2016; 16: 11943-6.
         [http://dx.doi.org/10.1166/jnn.2016.13622]

[27]   Lind JU, Busbee TA, Valentine AD, et al. Instrumented cardiac microphysiological devices via multimaterial three-dimensional printing. Nat
         Mater 2017; 16(3): 303-8.
         [http://dx.doi.org/10.1038/nmat4782] [PMID: 27775708]

[28]   Jakus AE, Shah RN. Multi and mixed 3D-printing of graphene-hydroxyapatite hybrid materials for complex tissue engineering. J Biomed
         Mater Res A 2017; 105(1): 274-83.
         [http://dx.doi.org/10.1002/jbm.a.35684] [PMID: 26860782]

[29]   Ozler SB, Bakirci E, Kucukgul C, Koc B. Three-dimensional direct cell bioprinting for tissue engineering. J Biomed Mater Res B Appl
         Biomater 2017; 105(8): 2530-44.
         [http://dx.doi.org/10.1002/jbm.b.33768] [PMID: 27689939]

[30]   Zhu W, Ma X, Gou M, Mei D, Zhang K, Chen S. 3D printing of functional biomaterials for tissue engineering. Curr Opin Biotechnol 2016;
         40: 103-12.
         [http://dx.doi.org/10.1016/j.copbio.2016.03.014] [PMID: 27043763]

[31]   Datta P, Ayan B, Ozbolat IT. Bioprinting for vascular and vascularized tissue biofabrication. Acta Biomater 2017; 51: 1-20.
         [http://dx.doi.org/10.1016/j.actbio.2017.01.035] [PMID: 28087487]

[32]   Ouyang L, Highley CB, Rodell CB, Sun W, Burdick JA. 3D printing of shear-thinning hyaluronic acid hydrogels with secondary cross-
         linking. ACS Biomater Sci Eng 2016; 2: 1743-51.
         [http://dx.doi.org/10.1021/acsbiomaterials.6b00158]

[33]   Zhao J, Swartz LA, Lin WF, et al. Three-dimensional nanoprinting via scanning probe lithography-delivered layer-by-layer deposition. ACS
         Nano 2016; 10(6): 5656-62.
         [http://dx.doi.org/10.1021/acsnano.6b01145] [PMID: 27203853]

[34]   Morrison RJ, Kashlan KN, Flanangan CL, et al. Regulatory considerations in the design and manufacturing of implantable 3D-printed
         medical devices. Clin Transl Sci 2015; 8(5): 594-600.
         [http://dx.doi.org/10.1111/cts.12315] [PMID: 26243449]

[35]   Valverde I. Three-dimensional printed cardiac models: Applications in the field of medical education, cardiovascular surgery, and structural
         heart interventions. Rev Esp Cardiol (Engl Ed) 2017; 70(4): 282-91.
         [http://dx.doi.org/10.1016/j.rec.2017.01.012] [PMID: 28189544]

[36]   Das S, Pati F, Choi Y-J, et al. Bioprintable, cell-laden silk fibroin-gelatin hydrogel supporting multilineage differentiation of stem cells for
         fabrication of three-dimensional tissue constructs. Acta Biomater 2015; 11: 233-46.
         [http://dx.doi.org/10.1016/j.actbio.2014.09.023] [PMID: 25242654]

[37] Jose RR, Rodriguez MJ, Dixon TA, Omenetto F, Kaplan DL. Evolution of bioinks and additive manufacturing technologies for 3D bioprinting. ACS Biomater Sci Eng 2016; 2: 1662-78.
[http://dx.doi.org/10.1021/acsbiomaterials.6b00088]

[38] Xue C, Shi X, Fang X, et al. The "pure marriage" between 3D printing and well-ordered nanoarrays by using peald assisted hydrothermal surface engineering. ACS Appl Mater Interfaces 2016; 8(13): 8393-400.
[http://dx.doi.org/10.1021/acsami.6b01417] [PMID: 26974545]

[39] Park JY, Choi Y-J, Shim J-H, Park JH, Cho DW. Development of a 3D cell printed structure as an alternative to autologs cartilage for auricular reconstruction. J Biomed Mater Res B Appl Biomater 2017; 105(5): 1016-28.
[http://dx.doi.org/10.1002/jbm.b.33639] [PMID: 26922876]

[40] Kumar A, Nune KC, Misra RDK. Biological functionality of extracellular matrix-ornamented three-dimensional printed hydroxyapatite scaffolds. J Biomed Mater Res A 2016; 104(6): 1343-51.
[http://dx.doi.org/10.1002/jbm.a.35664] [PMID: 26799466]

[41] Zhang M, Vora A, Han W, et al. Dual-responsive hydrogels for direct-write 3D printing. Macromolecules 2015; 48: 6482-8.
[http://dx.doi.org/10.1021/acs.macromol.5b01550]

[42] Gonçalves EM, Oliveira FJ, Silva RF, et al. Three-dimensional printed PCL-hydroxyapatite scaffolds filled with CNTs for bone cell growth stimulation. J Biomed Mater Res B Appl Biomater 2016; 104(6): 1210-9.
[http://dx.doi.org/10.1002/jbm.b.33432] [PMID: 26089195]

[43] Xing JF, Zheng M-L, Duan X-M. Two-photon polymerization microfabrication of hydrogels: An advanced 3D printing technology for tissue engineering and drug delivery. Chem Soc Rev 2015; 44(15): 5031-9.
[http://dx.doi.org/10.1039/C5CS00278H] [PMID: 25992492]

[44] Jakus AE, Secor EB, Rutz AL, Jordan SW, Hersam MC, Shah RN. Three-dimensional printing of high-content graphene scaffolds for electronic and biomedical applications. ACS Nano 2015; 9(4): 4636-48.
[http://dx.doi.org/10.1021/acsnano.5b01179] [PMID: 25858670]

[45] Chia HN, Wu BM. Recent advances in 3D printing of biomaterials. J Biol Eng 2015; 9(1): 4.
[http://dx.doi.org/10.1186/s13036-015-0001-4] [PMID: 25866560]

[46] Bandyopadhyay A, Bose S, Das S. 3D printing of biomaterials. MRS Bull 2015; 40: 108-15.
[http://dx.doi.org/10.1557/mrs.2015.3]

[47] Nakayama Y, Takewa Y, Sumikura H, et al. In-body tissue-engineered aortic valve (Biovalve type VII) architecture based on 3D printer molding. J Biomed Mater Res B Appl Biomater 2015; 103(1): 1-11.
[http://dx.doi.org/10.1002/jbm.b.33186] [PMID: 24764308]

[48] Temple JP, Hutton DL, Hung BP, et al. Engineering anatomically shaped vascularized bone grafts with hASCs and 3D-printed PCL scaffolds. J Biomed Mater Res A 2014; 102(12): 4317-25.
[PMID: 24510413]

[49] Hung K-C, Tseng C-S, Hsu SH. Synthesis and 3D printing of biodegradable polyurethane elastomer by a water-based process for cartilage tissue engineering applications. Adv Healthc Mater 2014; 3(10): 1578-87.
[http://dx.doi.org/10.1002/adhm.201400018] [PMID: 24729580]

[50] Husár B, Hatzenbichler M, Mironov V, Liska R, Stampfl J, Ovsianikov A. Photopolymerization-based additive manufacturing for the development of 3D porous scaffolds. In Biomaterials for Bone Regeneration 2014; 149-201.
[http://dx.doi.org/10.1533/9780857098104.2.149.]

[51] Vaezi M, Yang S. Freeform fabrication of nanobiomaterials using 3D printing.Rapid Prototyping of Biomaterials. Elsevier 2014; pp. 16-74.
[http://dx.doi.org/10.1533/9780857097217.16]

[52] Gross BC, Erkal JL, Lockwood SY, Chen C, Spence DM. Evaluation of 3D printing and its potential impact on biotechnology and the chemical sciences. Anal Chem 2014; 86(7): 3240-53.
[http://dx.doi.org/10.1021/ac403397r] [PMID: 24432804]

[53] Hribar KC, Soman P, Warner J, Chung P, Chen S. Light-assisted direct-write of 3D functional biomaterials. Lab Chip 2014; 14(2): 268-75.
[http://dx.doi.org/10.1039/C3LC50634G] [PMID: 24257507]

[54] Duan B. State-of-the-Art review of 3D bioprinting for cardiovascular tissue engineering. Ann Biomed Eng 2017; 45(1): 195-209.
[http://dx.doi.org/10.1007/s10439-016-1607-5] [PMID: 27066785]

[55] Patra S, Young V. A review of 3D printing techniques and the future in biofabrication of bioprinted tissue. Cell Biochem Biophys 2016; 74(2): 93-8.
[http://dx.doi.org/10.1007/s12013-016-0730-0] [PMID: 27193609]

[56] Garreta E, Oria R, Tarantino C, et al. Tissue engineering by decellularization and 3D bioprinting. Mater Today 2017; 20: 166-78.
[http://dx.doi.org/10.1016/j.mattod.2016.12.005]

[57] Kelly CN, Miller AT, Hollister SJ, Guldberg RE, Gall K. Design and structure-function characterization of 3D printed synthetic porous biomaterials for tissue engineering. Adv Healthc Mater 2017; 7(7): 1701095.
[http://dx.doi.org/10.1002/adhm.201701095] [PMID: 29280325]

[58]   Alluri R, Jakus A, Bougioukli S, *et al.* 3D printed hyperelastic "bone" scaffolds and regional gene therapy: A novel approach to bone healing. J Biomed Mater Res A 2018; 106(4): 1104-10.
[http://dx.doi.org/10.1002/jbm.a.36310] [PMID: 29266747]

[59]   Richards D, Jia J, Yost M, Markwald R, Mei Y. 3D bioprinting for vascularized tissue fabrication. Ann Biomed Eng 2017; 45(1): 132-47.
[http://dx.doi.org/10.1007/s10439-016-1653-z] [PMID: 27230253]

[60]   Tan Z, Parisi C, Di Silvio L, Dini D, Forte AE. Cryogenic 3D printing of super soft hydrogels. Sci Rep 2017; 7(1): 16293.
[http://dx.doi.org/10.1038/s41598-017-16668-9] [PMID: 29176756]

[61]   Kim M, Kim W, Kim G. Topologically micropatterned collagen and poly(ε-caprolactone) struts fabricated using the poly(vinyl alcohol) fibrillation/leaching process to develop efficiently engineered skeletal muscle tissue. ACS Appl Mater Interfaces 2017; 9(50): 43459-69.
[http://dx.doi.org/10.1021/acsami.7b14192] [PMID: 29171953]

[62]   Blaeser A, Duarte Campos DF, Fischer H. 3D bioprinting of cell-laden hydrogels for advanced tissue engineering. Curr Opin Biomed Eng 2017; 2: 58-66.
[http://dx.doi.org/10.1016/j.cobme.2017.04.003]

[63]   Vella JB, Trombetta RP, Hoffman MD, Inzana J, Awad H, Benoit DSW. Three dimensional printed calcium phosphate and poly(caprolactone) composites with improved mechanical properties and preserved microstructure. J Biomed Mater Res A 2018; 106(3): 663-72.
[http://dx.doi.org/10.1002/jbm.a.36270] [PMID: 29044984]

[64]   McElheny C, Hayes D, Devireddy R. Design and fabrication of a low-cost three-dimensional bioprinter. J Med Device 2017; 11(4): 0410011-9.
[http://dx.doi.org/10.1115/1.4037259] [PMID: 29034057]

[65]   Huh J, Lee J, Kim W, Yeo M, Kim G. Preparation and characterization of gelatin/α-TCP/SF biocomposite scaffold for bone tissue regeneration. Int J Biol Macromol 2018; 110: 488-96.
[http://dx.doi.org/10.1016/j.ijbiomac.2017.09.030] [PMID: 28917939]

[66]   Kuss MA, Wu S, Wang Y. Prevascularization of 3D printed bone scaffolds by bioactive hydrogels and cell co-culture. J Biomed Mater Res B Appl Biomater 2017; 106(5): 1788-98.
[http://dx.doi.org/10.1002/jbm.b.33994] [PMID: 28901689]

[67]   Tan Z, Liu T, Zhong J, Yang Y, Tan W. Control of cell growth on 3D-printed cell culture platforms for tissue engineering. J Biomed Mater Res A 2017; 105(12): 3281-92.
[http://dx.doi.org/10.1002/jbm.a.36188] [PMID: 28865175]

[68]   Li L, Yu F, Shi J, *et al. In situ* repair of bone and cartilage defects using 3D scanning and 3D printing. Sci Rep 2017; 7(1): 9416.
[http://dx.doi.org/10.1038/s41598-017-10060-3] [PMID: 28842703]

[69]   Stephenson MK, Farris AL, Grayson WL. Recent advances in tissue engineering strategies for the treatment of joint damage. Curr Rheumatol Rep 2017; 19(8): 44.
[http://dx.doi.org/10.1007/s11926-017-0671-7] [PMID: 28718059]

[70]   Badea A, McCracken JM, Tillmaand EG, *et al.* 3D-printed pHEMA materials for topographical and biochemical modulation of dorsal root ganglion cell response. ACS Appl Mater Interfaces 2017; 9(36): 30318-28.
[http://dx.doi.org/10.1021/acsami.7b06742] [PMID: 28813592]

[71]   Li X, He J, Zhang W, Jiang N, Li D. Additive manufacturing of biomedical constructs with biomimetic structural organizations. Materials (Basel) 2016; 9(11): E909.
[http://dx.doi.org/10.3390/ma9110909] [PMID: 28774030]

[72]   Chua CK, Yeong WY, An J. Special issue: 3D printing for biomedical engineering. Materials (Basel) 2017; 10(3): E243.
[http://dx.doi.org/10.3390/ma10030243] [PMID: 28772604]

[73]   Gou M, Qu X, Zhu W, *et al.* Bio-inspired detoxification using 3D-printed hydrogel nanocomposites. Nat Commun 2014; 5: 3774.
[http://dx.doi.org/10.1038/ncomms4774] [PMID: 24805923]

[74]   De Santis R, D'Amora U, Russo T, Ronca A, Gloria A, Ambrosio L. 3D fibre deposition and stereolithography techniques for the design of multifunctional nanocomposite magnetic scaffolds. J Mater Sci Mater Med 2015; 26(10): 250.
[http://dx.doi.org/10.1007/s10856-015-5582-4] [PMID: 26420041]

[75]   Coelho RC, Marques AL, Oliveira SM. Extraction and characterization of collagen from Antarctic and Sub-Antarctic squid and its potential application in hybrid scaffolds for tissue engineering. 2017; 78: 787-95.

[76]   Lee VK, Lanzi AM, Haygan N, Yoo S-S, Vincent PA, Dai G. Generation of multi-scale vascular network system within 3D hydrogel using 3D bio-printing technology. Cell Mol Bioeng 2014; 7(3): 460-72.
[http://dx.doi.org/10.1007/s12195-014-0340-0] [PMID: 25484989]

[77]   Gao G, Yonezawa T, Hubbell K, Dai G, Cui X. Inkjet-bioprinted acrylated peptides and peg hydrogel with human mesenchymal stem cells promote robust bone and cartilage formation with minimal printhead clogging. Biotechnol J 2015; 10(10): 1568-77.
[http://dx.doi.org/10.1002/biot.201400635] [PMID: 25641582]

[78]   Kim T-H, Yun Y-P, Park Y-E, *et al. In vitro* and *in vivo* evaluation of bone formation using solid freeform fabrication-based bone morphogenic protein-2 releasing PCL/PLGA scaffolds. Biomed Mater 2014; 9(2): 025008.

[http://dx.doi.org/10.1088/1748-6041/9/2/025008] [PMID: 24518200]

[79]   Hong N, Yang G-H, Lee J, Kim G. 3D bioprinting and its *in vivo* applications. J Biomed Mater Res B Appl Biomater 2018; 106(1): 444-59.
         [http://dx.doi.org/10.1002/jbm.b.33826] [PMID: 28106947]

[80]   Walker V. Implementing a 3D printing service in a biomedical library. J Med Libr Assoc 2017; 105(1): 55-60.
         [PMID: 28096747]

[81]   Sochol RD, Gupta NR, Bonventre JV. A role for 3D printing in kidney-on-a-chip platforms. Curr Transplant Rep 2016; 3(1): 82-92.
         [http://dx.doi.org/10.1007/s40472-016-0085-x] [PMID: 28090431]

[82]   Elomaa L, Pan CC, Shanjani Y, Malkovskiy A, Seppälä JV, Yang Y. Three-dimensional fabrication of cell-laden biodegradable poly(ethylene glycol-co-depsipeptide) hydrogels by visible light stereolithography. J Mater Chem B Mater Biol Med 2015; 3(42): 8348-58.
         [http://dx.doi.org/10.1039/C5TB01468A] [PMID: 29057076]

[83]   Neiman JAS, Raman R, Chan V, *et al.* Photopatterning of hydrogel scaffolds coupled to filter materials using stereolithography for perfused 3D culture of hepatocytes. Biotechnol Bioeng 2015; 112(4): 777-87.
         [http://dx.doi.org/10.1002/bit.25494] [PMID: 25384798]

[84]   Mačiulaitis J, Deveikytė M, Rekštytė S, *et al.* Preclinical study of SZ2080 material 3D microstructured scaffolds for cartilage tissue engineering made by femtosecond direct laser writing lithography. Biofabrication 2015; 7(1): 015015.
         [http://dx.doi.org/10.1088/1758-5090/7/1/015015] [PMID: 25797444]

[85]   Owen R, Sherborne C, Paterson T, Green NH, Reilly GC, Claeyssens F. Emulsion templated scaffolds with tunable mechanical properties for bone tissue engineering. J Mech Behav Biomed Mater 2016; 54: 159-72.
         [http://dx.doi.org/10.1016/j.jmbbm.2015.09.019] [PMID: 26458114]

[86]   Du D, Asaoka T, Ushida T, Furukawa KS. Fabrication and perfusion culture of anatomically shaped artificial bone using stereolithography. Biofabrication 2014; 6(4): 045002.
         [http://dx.doi.org/10.1088/1758-5082/6/4/045002] [PMID: 25215543]

[87]   Lin H, Tang Y, Lozito TP, *et al.* Projection stereolithographic fabrication of BMP-2 gene-activated matrix for bone tissue engineering. Sci Rep 2017; 7(1): 11327.
         [http://dx.doi.org/10.1038/s41598-017-11051-0] [PMID: 28900122]

[88]   Zhang C, Bills BJ, Manicke NE. Rapid prototyping using 3D printing in bioanalytical research. Bioanalysis 2017; 9(4): 329-31.
         [http://dx.doi.org/10.4155/bio-2016-0293] [PMID: 28071134]

[89]   Tonsomboon K, Butcher AL, Oyen ML. Strong and tough nanofibrous hydrogel composites based on biomimetic principles. Mater Sci Eng C 2017; 72: 220-7.
         [http://dx.doi.org/10.1016/j.msec.2016.11.025] [PMID: 28024580]

[90]   Lin H, Tang Y, Lozito TP, *et al.* Projection stereolithographic fabrication of BMP-2 gene-activated matrix for bone tissue engineering. Sci Rep 2017; 7(1): 11327.
         [http://dx.doi.org/10.1038/s41598-017-11051-0] [PMID: 28900122]

[91]   Yeh YC, Highley CB, Ouyang L, Burdick JA. 3D printing of photocurable poly(glycerol sebacate) elastomers. Biofabrication 2016; 8(4): 045004.
         [http://dx.doi.org/10.1088/1758-5090/8/4/045004] [PMID: 27716633]

[92]   Zadpoor AA, Malda J. Additive manufacturing of biomaterials, tissues, and organs. Ann Biomed Eng 2017; 45(1): 1-11.
         [http://dx.doi.org/10.1007/s10439-016-1719-y] [PMID: 27632024]

[93]   Cox SC, Thornby JA, Gibbons GJ, Williams MA, Mallick KK. 3D printing of porous hydroxyapatite scaffolds intended for use in bone tissue engineering applications. Mater Sci Eng C 2015; 47: 237-47.
         [http://dx.doi.org/10.1016/j.msec.2014.11.024] [PMID: 25492194]

[94]   Mironov AV, Grigoryev AM, Krotova LI, Skaletsky NN, Popov VK, Sevastianov VI. 3D printing of PLGA scaffolds for tissue engineering. J Biomed Mater Res A 2017; 105(1): 104-9.
         [http://dx.doi.org/10.1002/jbm.a.35871] [PMID: 27543196]

[95]   Dababneh AB, Ozbolat IT. Bioprinting technology: A current state-of-the-art review. J Manuf Sci Eng 2014; 136(6)
         [http://dx.doi.org/10.1115/1.4028512]

[96]   Trombetta R, Inzana JA, Schwarz EM, Kates SL, Awad HA. 3D printing of calcium phosphate ceramics for bone tissue engineering and drug delivery. Ann Biomed Eng 2017; 45(1): 23-44.
         [http://dx.doi.org/10.1007/s10439-016-1678-3] [PMID: 27324800]

[97]   Panayotov IV, Orti V, Cuisinier F, Yachouh J. Polyetheretherketone (PEEK) for medical applications. J Mater Sci Mater Med 2016; 27(7): 118.
         [http://dx.doi.org/10.1007/s10856-016-5731-4] [PMID: 27259708]

[98]   Jensen J, Rölfing JHD, Le DQ, *et al.* Surface-modified functionalized polycaprolactone scaffolds for bone repair: *In vitro* and *in vivo* experiments. J Biomed Mater Res A 2014; 102(9): 2993-3003.
         [http://dx.doi.org/10.1002/jbm.a.34970] [PMID: 24123983]

[99]   Xu N, Ye X, Wei D, *et al.* 3D artificial bones for bone repair prepared by computed tomography-guided fused deposition modeling for bone repair. ACS Appl Mater Interfaces 2014; 6(17): 14952-63.

[http://dx.doi.org/10.1021/am502716t] [PMID: 25133309]

[100]   Idaszek J, Bruinink A, Święszkowski W. Ternary composite scaffolds with tailorable degradation rate and highly improved colonization by
        human bone marrow stromal cells. J Biomed Mater Res A 2015; 103(7): 2394-404.
        [http://dx.doi.org/10.1002/jbm.a.35377] [PMID: 25424876]

[101]   Li Y, Wu ZG, Li XK, et al. A polycaprolactone-tricalcium phosphate composite scaffold as an autograft-free spinal fusion cage in a sheep
        model. Biomaterials 2014; 35(22): 5647-59.
        [http://dx.doi.org/10.1016/j.biomaterials.2014.03.075] [PMID: 24743032]

[102]   Christensen K, Xu C, Chai W, Zhang Z, Fu J, Huang Y. Freeform inkjet printing of cellular structures with bifurcations. Biotechnol Bioeng
        2015; 112(5): 1047-55.
        [http://dx.doi.org/10.1002/bit.25501] [PMID: 25421556]

[103]   Wengerter BC, Emre G, Park JY, Geibel J. Three-dimensional printing in the intestine. Clin Gastroenterol Hepatol 2016; 14(8): 1081-5.
        [http://dx.doi.org/10.1016/j.cgh.2016.05.008] [PMID: 27189913]

[104]   Gu BK, Choi DJ, Park SJ, Kim MS, Kang CM, Kim C-H. 3-Dimensional bioprinting for tissue engineering applications. Biomater Res 2016;
        20: 12.
        [http://dx.doi.org/10.1186/s40824-016-0058-2] [PMID: 27114828]

[105]   Hernández-Córdova R, Mathew DA, Balint R, et al. Indirect three-dimensional printing: A method for fabricating polyurethane-urea based
        cardiac scaffolds. J Biomed Mater Res A 2016; 104(8): 1912-21.
        [http://dx.doi.org/10.1002/jbm.a.35721] [PMID: 26991636]

[106]   Hasan A, Memic A, Annabi N, et al. Electrospun scaffolds for tissue engineering of vascular grafts. Acta Biomater 2014; 10(1): 11-25.
        [http://dx.doi.org/10.1016/j.actbio.2013.08.022] [PMID: 23973391]

[107]   Shafiee A, Atala A. Printing technologies for medical applications. Trends Mol Med 2016; 22(3): 254-65.
        [http://dx.doi.org/10.1016/j.molmed.2016.01.003] [PMID: 26856235]

[108]   Hasan A, Ragaert K, Swieszkowski W, et al. Biomechanical properties of native and tissue engineered heart valve constructs. J Biomech
        2014; 47(9): 1949-63.
        [http://dx.doi.org/10.1016/j.jbiomech.2013.09.023] [PMID: 24290137]

[109]   Mandrycky C, Wang Z, Kim K, Kim DH. 3D bioprinting for engineering complex tissues. Biotechnol Adv 2016; 34(4): 422-34.
        [http://dx.doi.org/10.1016/j.biotechadv.2015.12.011] [PMID: 26724184]

[110]   Jakus AE, Rutz AL, Jordan SW, et al. Hyperelastic "bone": A highly versatile, growth factor-free, osteoregenerative, scalable, and surgically
        friendly biomaterial. Sci Transl Med 2016; 8(358): 358ra127.
        [http://dx.doi.org/10.1126/scitranslmed.aaf7704] [PMID: 27683552]

[111]   Hung BP, Naved BA, Nyberg EL, et al. Three-Dimensional printing of bone extracellular matrix for craniofacial regeneration. ACS Biomater
        Sci Eng 2016; 2(10): 1806-16.
        [http://dx.doi.org/10.1021/acsbiomaterials.6b00101] [PMID: 27942578]

[112]   Du Y, Liu H, Shuang J, Wang J, Ma J, Zhang S. Microsphere-based selective laser sintering for building macroporous bone scaffolds with
        controlled microstructure and excellent biocompatibility. Colloids Surf B Biointerfaces 2015; 135: 81-9.
        [http://dx.doi.org/10.1016/j.colsurfb.2015.06.074] [PMID: 26241919]

[113]   C-H. Chen, Ming-Yih Lee, Victor Bong-Hang Shyu, Yi-Chieh Chen, Chien-Tzung Chen, Jyh-Ping Chen. Surface modification of
        polycaprolactone scaffolds fabricated via selective laser sintering for cartilage tissue engineering. Mater Sci Eng C 2014; 40(1): 389-97.

[114]   Roskies M, Jordan JO, Fang D, et al. Improving PEEK bioactivity for craniofacial reconstruction using a 3D printed scaffold embedded with
        mesenchymal stem cells. J Biomater Appl 2016; 31(1): 132-9.
        [http://dx.doi.org/10.1177/0885328216638636] [PMID: 26980549]

[115]   Feng P, Wei P, Shuai C, Peng S. Characterization of mechanical and biological properties of 3-D scaffolds reinforced with zinc oxide for bone
        tissue engineering. PLoS One 2014; 9(1): e87755.
        [http://dx.doi.org/10.1371/journal.pone.0087755] [PMID: 24498185]

[116]   Shuai C, Mao Z, Han Z, Peng S. Preparation of complex porous scaffolds via selective laser sintering of poly(vinyl alcohol)/calcium silicate. J
        Bioact Compat Polym 2014; 29: 110-20.
        [http://dx.doi.org/10.1177/0883911514522570]

[117]   Muth JT, Dixon PG, Woish L, Gibson LJ, Lewis JA. Architected cellular ceramics with tailored stiffness via direct foam writing. Proc Natl
        Acad Sci USA 2017; 114(8): 1832-7.
        [http://dx.doi.org/10.1073/pnas.1616769114] [PMID: 28179570]

[118]   Hollister SJ, Flanagan CL, Morrison RJ, et al. Integrating image-based design and 3D biomaterial printing to create patient specific devices
        within a design control framework for clinical translation. ACS Biomater Sci Eng 2016; 2: 1827-36.
        [http://dx.doi.org/10.1021/acsbiomaterials.6b00332]

[119]   Shimomura K, Moriguchi Y, Ando W, et al. Osteochondral repair using a scaffold-free tissue-engineered construct derived from synovial
        mesenchymal stem cells and a hydroxyapatite-based artificial bone. Tissue Eng Part A 2014; 20(17-18): 2291-304.
        [http://dx.doi.org/10.1089/ten.tea.2013.0414] [PMID: 24655056]

[120] Alexander PG, Gottardi R, Lin H, Lozito TP, Tuan RS. Three-dimensional osteogenic and chondrogenic systems to model osteochondral physiology and degenerative joint diseases. Exp Biol Med (Maywood) 2014; 239(9): 1080-95.
[http://dx.doi.org/10.1177/1535370214539232] [PMID: 24994814]

[121] Lorber B, Hsiao W-K, Hutchings IM, Martin KR. Adult rat retinal ganglion cells and glia can be printed by piezoelectric inkjet printing. Biofabrication 2014; 6(1): 015001-10.
[http://dx.doi.org/10.1088/1758-5082/6/1/015001] [PMID: 24345926]

[122] Pati F, Ha DH, Jang J, Han HH, Rhie JW, Cho DW. Biomimetic 3D tissue printing for soft tissue regeneration. Biomaterials 2015; 62: 164-75.
[http://dx.doi.org/10.1016/j.biomaterials.2015.05.043] [PMID: 26056727]

[123] Irvine SA, Agrawal A, Lee BH, et al. Printing cell-laden gelatin constructs by free-form fabrication and enzymatic protein crosslinking. Biomed Microdevices 2015; 17(1): 16.
[http://dx.doi.org/10.1007/s10544-014-9915-8] [PMID: 25653062]

[124] Choi M, Heo J, Yang M, Hong J. Inkjet printing-based patchable multilayered biomoleculecontaining nanofilms for biomedical applications. ACS Biomater Sci Eng 2017.
[http://dx.doi.org/10.1021/acsbiomaterials.7b00138]

[125] Kwon J, Cho H, Eom H, et al. Low-temperature oxidation-free selective laser sintering of cu nanoparticle paste on a polymer substrate for the flexible touch panel applications. ACS Appl Mater Interfaces 2016; 8(18): 11575-82.
[http://dx.doi.org/10.1021/acsami.5b12714] [PMID: 27128365]

[126] Jill Z, Manapat , Joey Dacula Mangadlao, et al. High-strength stereolithographic 3D printed nanocomposites: Graphene oxide metastability. ACS Appl Mater Interfaces 2017.
[http://dx.doi.org/10.1021/acsami.6b16174]

18

Ochratoxin A Removal by *Lactobacillus Plantarum* V22 in Synthetic Substrates

Moncalvo A.[1], Dordoni R.[1], Silva A.[1], Fumi M.D.[1], Di Piazza S.[2] and Spigno G.[1,*]

[1]*DiSTAS Department for Sustainable Food Process, Università Cattolica del Sacro Cuore, Via Emilia Parmense 84, 29122 Piacenza, Italy*

[2]*Laboratory of Mycology, Department of Earth, Environmental and Life Sciences, Università degli Studi di Genova, Corso Europa 26, I, 16136 Genova, Italy*

Abstract:

Background:

Ochratoxin A is a nephrotoxin which may occur in wines characterised by higher pH than the average. In the last decades the mechanisms responsible for ochratoxin A reduction by lactic acid bacteria have been investigated and identified as mainly cell walls adsorption and / or enzymatic conversion to ochratoxin-α, a non-toxic metabolite. Since lactic acid bacteria are involved in the malolactic fermentation during the wine-making process, selected starter cultures could be exploited to guarantee safe ochratoxin A level in wines also from contaminated grapes. A lactic acid bacteria strain (*Lactobacillus plantarum* V22) was previously selected for its ability of both degrading ochratoxin A and carrying out malolactic fermentation at high pH.

Objective:

This study was aimed at assessing if the selected *L. plantarum* strain, can reduce ochratoxin A because it can use it as a carbon source.

Methods:

L. plantarum V22 was grown in the presence of ochratoxin A in two different synthetic substrates, with or without malic acid, monitoring the reduction of ochratoxin A and the presence of ochratoxin α as an indicator for a toxin enzymatic hydrolysis. The presence of residual not hydrolysed ochratoxin A bound to the bacteria cell walls was also evaluated to quantify the ochratoxin A removal due to simple adsorption.

Result:

A significant reduction of 19.5 ± 2.0% in ochratoxin A concentration was observed only in the presence of malic acid. The quantified fraction of ochratoxin A adsorbed on cell walls was irrelevant and the metabolite ochratoxin α could not be detected.

Conclusion:

There is a possibility that *L. plantarum* V22 can degrade ochratoxin A through a not yet identified metabolic pathway.

Keywords: Bio-decontamination, *Lactobacillus plantarum*, Malolactic fermentation, Ochratoxin A, Wine, Synthetic Substrates.

[*] Address correspondence to this author at the DiSTAS Department for Sustainable Food Process, Università Cattolica del Sacro Cuore, Via Emilia Parmense 84, 29122 Piacenza, Italy; E-mail: giorgia.spigno@unicatt.it

1. INTRODUCTION

Ochratoxin A (OTA) is a secondary metabolite of moulds and one of the most common contaminating agents in raw materials and food products. It is a well-known nephrotoxin for different species, classified as possible human carcinogen by IARC [1].

OTA was identified in grape juices and wines for the first time in 1996 [2]. After that, several approaches have been tested to reduce this toxin in grape and to remove it during winemaking [3, 4]. A maximum residual contamination level in wine of two parts-per-billion (ppb, ~2 µg/L) was established by the European Union (Commission Regulation No. 1881/2006). In red winemaking process, OTA concentration increases during maceration as an effect of prolonged contact between skins and must. However, operations such as clarification and filtration, and processes as Malolactic Fermentation (MLF), can allow for a relevant decrease [5]. MLF is carried out by lactic acid bacteria and OTA degradation by LAB has also been investigated [6], showing that LAB can hydrolyse (through a carboxypeptidase) OTA in L-phenylalanine and OTα, a non-toxic metabolite [7].

OTA can be particularly found in musts and wines from specific regions of Europe as southern Italy, Greece and Spain [8]. Furthermore, wines from these regions are characterized by higher pH values which can negatively affect MLF [9].

Several approaches have been tested to reduce the occurrence of OTA in grape and to remove the toxin during winemaking [10 - 12]. During red winemaking process the OTA concentration increases in maceration probably due to prolonged contact between skins and must, but the balance of OTA in wine during winemaking is overall negative because operations as clarification and filtration, and processes as MLF cause a relevant decrease of this mycotoxin. The use of selected bacterial strains for OTA reduction in winemaking requires further studies because the available literature is limited. In fact, although many studies were carried out on adsorption process by yeasts during winemaking [13 - 15], the degradation process by LAB and its mechanisms have been little investigated. Rodriguez et al. [16] investigated the capability to hydrolyse OTA in some Brevibacterium species finding that different strains of B. casei, B. linens, B. iodinum and B. epidermidis were able to completely degrade OTA in OTα (in basal salt medium with 40 µg/L OTA). Abrunhosa et al. [17] showed that in Man Rogosa Sharpe (MRS) with 1 µg/mL OTA, Pediococcus parvulus strains were able to degrade OTA in OTα, while L. plantarum reduced OTA concentration by 10-14% (without forming OTα).

A selection program of malolactic starter culture capable of degrading OTA in wine, driven by the former Institute of Oenology and Agro-Food Engineering of the Università Cattolica del Sacro Cuore of Piacenza (Italy), led to the identification of a Lactobacillus plantarum strain (V22™Lactobacillus plantarum, Lallemand) with demonstrated ability to carry out MLF especially in wine with high pH.

The aim of this work was to get some more insight into the mechanisms behind the reduction of OTA levels in wine by L. plantarum V22 to better exploit its use for bio-decontamination of wines with high OTA contents. We wanted to evaluate whether the strain can use OTA as carbon source at the end of MLF (when the L-malic acid is completely decarboxylated in L-lactic acid) or if the reduction can occur only during MLF.

2. MATERIALS AND METHODS

2.1. Standards, Chemicals and Bacteria

All solvent used for chromatographic analysis were of HPLC grade, OTA and OTα standards were from Biopure (Waterlooville, UK), MRS broth was purchased from Oxoid (Basingstoke, UK) and Agar was obtained from Bacto™ (Sparks, MD, USA). Yeast Nitrogen Base (YNB) without amino acids and ammonium sulfate was purchased from Difco™ (Sparks, MD, USA). L-malic acid was from Sigma-Aldrich (Merck).

The L. plantarum V22 was stored in glycerol at -80°C in the laboratories of the DiSTAS of the Università Cattolica del Sacro Cuore of Piacenza (Italy). For the preparation of the inoculum for the planned trials, the strain was reactivated through growth in MRS broth for 48 h at 25 °C under aerobic conditions. The cells were harvested by centrifugation at 6000xg for 15 min at 4°C, washed with and resuspended in sterilized physiological solution (NaCl 8.5 g/L). The cell content of this suspension was determined by turbidimetric measurement (optical density at 630 nm, OD_{630}, using a spectrophotometer UV-1601 UV-visible, Shimadzu) and using a calibration curve for OD_{630} versus cell number concentration (determined by counting the cells at the optical microscope). The correspondence of total cell content with the content in colony forming units (CFU/mL) was verified by plate count on MRS-Agar, 2 days of incubation at

25°C.

2.2. Ochratoxin A Removal Assay

OTA removal was assessed in YNB containing all essential nutrients except amino acids, nitrogen and carbohydrate.

The synthetic substrate without L-malic acid (L-malic-) consisted in YNB 6.7 g/L in water, pH adjusted at 3.8 using phosphoric acid 1 M. The substrate with L-malic acid (L-malic+) was obtained as the previous one adding 2 g/L L-malic acid.

OTA was added to both the substrates at a concentration of 25 μg/L (actual starting level was measured by HPLC analysis). Aliquots of 30 mL of different synthetic substrates were poured into 50 mL Falcon tubes and then inoculated with adequate volumes of the inoculum suspension of L. plantarum V22 inoculum to have an initial concentration of 10^9 CFU/mL.

Not-inoculated samples of L-malic+ substrate with OTA were prepared as negative control samples.

As positive control, B. linens able to degrade OTA without any other carbon source was tested in L-malic- [16].

All the prepared samples were kept at 25°C in a thermostatic incubator under anaerobic (closed Falcon tubes) and static condition. OTα, OTA, and L-malic acid contents and total cell content (as OD_{630}) were evaluated at inoculation time (time 0) and after 3, 6 and 10 days. Total viable cell concentration (CFU/mL) was determined by plate count on MRS-Agar only at time 0 and after 10 days. OD_{630} and CFU/mL were not evaluated for the positive control.

For the analysis of OTα, OTA and L-malic acid, the biomass and supernatant were separated by centrifugation at 6000xg for 15 min at 4°C. The supernatant was filtered (0.45 μm) and directly analysed for OTα, OTA (by HPLC) and for L-malic acid (by L-malic acid assay kit, Megazyme, Wicklow, Ireland).

The biomass pellet was rinsed two times with water, dried and then suspended in 2 mL of absolute methanol for 1 h. After centrifugation at 6000xg for 15 min at 4°C, methanol was separated, collected in 5 mL vials and evaporated to dryness under dry nitrogen gas stream. For evaluation of OTA and OTα contents by HPLC, the dry residue was reconstituted with the HPLC mobile phase immediately before analysis.

2.3. Chromatographic Analysis

Chromatographic analysis was performed on a HPLC system consisting of a Perkin Elmer (Norwalk, CT, USA) instrument equipped with a 200 Series pump, a Perkin-Elmer 650-10S fluorescence detector, a Jasco LC-Net II/ADC (Oklahoma City, OK, USA) communication module and operated by ChromNAV Control Center software. OTA and OTα were separated as described in Muñoz et al. [7] using a Gemini C18 column 250 mm×4.6 mm fitted with 5 μm (Phenomenex). The analysis was carried out at room temperature with an injection volume of 20 μL. The mobile phases consisted of water/acetic acid (98:2 v/v)(phase A) and acetonitrile/acetic acid (98:2 v/v) (phase B) eluted at a flow rate of 1.0 mL/min. Stepwise gradient was: 0-5 min 25% B, 5-10 min 25-30% B, 10-18 min 30-52% B, 18-23 min 52% B, 23-31 min 52-75% B, 31-36 min 75-100% B, 36-41 min 100% B, 41-44 min 100-25% B, 44-54 min 25% B. Fluorescence detector was set at 333 nm excitation and 460 nm emission wavelengths.

2.4. Statistical Analysis

Experiments were carried out in duplicate, and the analyses were performed in triplicate. Results were expressed as means ± standard deviation. The influence of the substrate and time on OD_{630}, OTA and malic acid concentration was evaluated by one-way Analysis of Variance (ANOVA) using statistical software SPSS® (version 19.0, SPSS Inc., Chicago, IL, USA). Differences at p ≤ 0.05 were considered significant and in case of significant difference, the means were discriminated applying the post-hoc Tukey's test (p ≤ 0.05).

3. RESULTS AND DISCUSSION

There was no significant influence of malic acid presence on neither cell concentration (OD_{630}) nor its trend over time (Table 1).

Table 1. Bacterial cell count (OD$_{630}$) into L-malic- and L-malic+ substrates. Data represent mean values ± standard deviations. Different letters indicate significant differences between samples, as assessed by ANOVA ($p \leq 0.05$).

Days	OD$_{630}$	
	L-malic+	L-malic-
0	0.9255 ± 0.0021^a	0.9235 ± 0.0064^a
3	0.8025 ± 0.0078^b	0.7895 ± 0.0078^b
6	0.7705 ± 0.0092^b	0.7675 ± 0.0276^b
10	0.6705 ± 0.0403^c	0.6445 ± 0.0290^c

Cell concentration significantly decreased since the beginning indicating, as expectable, that the synthetic medium represents stressful conditions due to the lack of nutrients. However, evaluation of cell viability after 10 days, showed a higher level in L-malic+ (1.3×10^6 CFU/mL) than in L-malic- (1.2×10^3 CFU/mL), confirming that L-malic acid is used as a carbon source. L-malic acid was degraded almost completely in 3 days by *L. plantarum* V22 (Fig. 1).

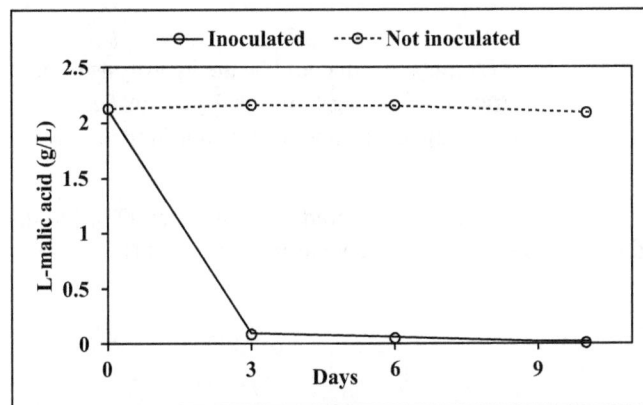

Fig. (1). Trends in L-malic acid concentration into inoculated and not inoculated substrates. Each point represents mean value; error bars indicate ± standard deviations.

The OTA content (Fig. 2) decreased only in the L-malic+ medium and only in the first 3 days after inoculum. OTA levels in L-malic- and in the control samples were statistically comparable and did not significantly change through the time, indicating that OTA reduction was linked to both bacteria and L-malic acid presence.

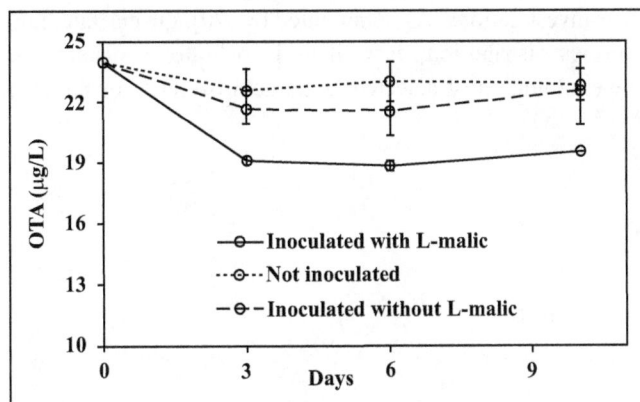

Fig. (2). Trends in OTA concentration into inoculated substrates with and without L-malic acid, and into not inoculated substrate. Each point represents mean value, error bars indicate ± standard deviations.

OTA reduction in the L-malic+ medium was $19.5 \pm 2.0\%$. This limited decrease could be related both to the unfavourable YNB composition and to the high applied initial OTA concentration (25 µg/L), although the positive control *B. linens* showed a total degradation of OTA after three days confirming the results of Rodriguez *et al.* [16] (data not showed). Our results agree with Piotrowska & Żakowska [6] who proved that high OTA levels have a negative effect on the LAB growth. However, Abrunhosa *et al.* [17] observed an OTA degradation by *Pediococcus*, *Lactobacillus* and *Oenococcus* on Man Rogosa Sharpe (MRS) with an initial OTA level of 1000 µg/L, showing that LAB can grow also in the presence of high OTA concentration.

OTA distribution between supernatant and pellet was also investigated, showing that just a little amount of OTA ($0.7 \pm 0.2\%$) remained bound to the bacteria walls in the L-malic+ samples. This value was comparable with the $1.3 \pm 1.0\%$ of adsorbtion on cells of *P. parvulus* described by Abrunhosa *et al.* [17] and confirm that in our trials the adsorption mechanism cannot explain the observed OTA reduction. Regarding this point, it could also be assumed that high stress conditions, caused by lack of nutrients, led to cell lysis processes (as indicated by the decrease in OD_{630}) with potential reduction of available surface for the adsorption and release in the media of the already adsorbed toxin [3, 18].

Since OTA biodegradation process was evident only in the presence of L-malic acid, it can be said that the OTA cannot be used as a carbon source unless another carbon source is present and that its metabolization is related to bacteria viability.

One hypothetical metabolic pathway for OTA metabolism could be the hydrolysis of peptide bond to form OTα and L-β-phenylalanine. The initial OTA concentration of 25 µg/L was selected to better detect the hypothetical OTα production. However, OTα could not be detected neither in supernatants nor in the biomass pellets, in agreement with [17].

Based on this finding other mechanisms of OTA degradation, without OTα production, as opening of the isocoumarin ring in a ring-opened lactone form of OTA, or dechlorination from OTA to OTB [19] must be identified.

CONCLUSION

L. plantarum V22 strain proved to reduce OTA into synthetic medium only when L-malic acid was present even though as the only carbon source. This underlines the potential of exploiting this strain to combine malolactic fermentation with OTA reduction in wine-making. OTA showed a reduction of $19.5 \pm 2.0\%$ with a negligible amount of OTA adsorbed on bacterial cell wall ($0.7 \pm 0.2\%$) indicating a biological degradation. Lack of OTα in supernatant and in pellet led to suppose that OTA reduction is not associated with the presence of enzymes able to hydrolyse the amine bond between L-β-phenylalanine and isocoumarinic part (*i.e.* carboxipeptidase), therefore a different metabolic pathway must be involved which needs further investigations. As commented by [20], the current climate change scenarios may lead to modifications in fungal species distribution, as well as in ochratoxin occurrence in wine, therefore bio-detoxification methods, such as the exploitation of selected lactic acid bacteria for malolactic fermentation, will be important.

ABBREVIATIONS

OTA	=	OchraToxin A
LAB	=	Lactic Acid Bacteria
OTα	=	Ochratoxin α
MLF	=	Malolactic Fermentation
YNB	=	Yeast Nitrogen Base

HUMAN AND ANIMAL RIGHTS

No animals/humans were used for studies that are the basis of this research.

REFERENCES

[1]   IARC (International Agency for Research on Cancer). Some Naturally Occurring Substances, Food Items and Constituents, Heterocyclic Aromatic Amines and Mycotoxins. Monographs on the Evaluation of Carcinogenic Risks to Humans 2003;56, Lyon, France.

[2]   Zimmerli B, Dick R. Ochratoxin A in table wine and grape-juice: Occurrence and risk assessment. Food Addit Contam 1996; 13(6): 655-68.
        [http://dx.doi.org/10.1080/02652039609374451] [PMID: 8871123]

[3]   Del Prete V, Rodriguez H, Carrascosa AV, de las Rivas B, Garcia-Moruno E, Muñoz R. *In vitro* removal of ochratoxin A by wine lactic acid bacteria. J Food Prot 2007; 70(9): 2155-60.
        [http://dx.doi.org/10.4315/0362-028X-70.9.2155] [PMID: 17900096]

[4]   Garcia-Moruno E, Sanlorenzo C, Beccaccino B, Di Stefano R. Treatment with yeast to reduce the concentration of ochratoxin A in red wine. Am J Enol Vitic 2005; 56: 73-6.

[5]   Grazioli B, Fumi MD, Silva A. The role of processing on ochratoxin A content in Italian must and wine: A study on naturally contaminated grapes. Int J Food Microbiol 2006; 111(Suppl. 1): S93-6.
        [http://dx.doi.org/10.1016/j.ijfoodmicro.2006.01.045] [PMID: 16714068]

[6]   Piotrowska M, Żakowska Z. The elimination of ochratoxin A by lactic acid bacteria strains. Pol J Microbiol 2005; 54(4): 279-86.
        [PMID: 16599298]

[7]   Muñoz K, Blaszkewicz M, Degen GH. Simultaneous analysis of ochratoxin A and its major metabolite ochratoxin alpha in plasma and urine for an advanced biomonitoring of the mycotoxin. J Chromatogr B Analyt Technol Biomed Life Sci 2010; 878(27): 2623-9.
        [http://dx.doi.org/10.1016/j.jchromb.2009.11.044] [PMID: 20031488]

[8]   Moncalvo A, Marinoni L, Dordoni R, Duserm Garrido G, Lavelli V, Spigno G. Waste grape skins: Evaluation of safety aspects for the production of functional powders and extracts for the food sector. Food Addit Contam Part A Chem Anal Control Expo Risk Assess 2016; 33(7): 1116-26.
        [http://dx.doi.org/10.1080/19440049.2016.1191320] [PMID: 27295010]

[9]   Fumi MD, Silva A, Krieger S, Vagnoli P, Domeneghetti D. Applicabilità di preparati liofilizzati di *Lactobacillus plantarum* per il controllo della fermentazione malolattica in mosti/vini con alti valori di pH. Paper presented at the Enoforum 2009, Piacenza.

[10]   Bornet A, Teissedre PL. Reduction of toxins and contaminants with biological tools. Bull OIV 2007; 80: 471-81.

[11]   Castellari M, Versari A, Fabiani A, Parpinello GP, Galassi S. Removal of ochratoxin A in red wines by means of adsorption treatments with commercial fining agents. J Agric Food Chem 2001; 49(8): 3917-21.
        [http://dx.doi.org/10.1021/jf010137o] [PMID: 11513689]

[12]   Silva A, Galli R, Grazioli B, Fumi MD. Metodi di riduzione di residui di ocratossina A nei vini. Ind Bevande 2003; 32: 467-72.

[13]   Bejaoui H, Mathieu F, Taillandier P, Lebrihi A. Ochratoxin A removal in synthetic and natural grape juices by selected oenological *Saccharomyces* strains. J Appl Microbiol 2004; 97(5): 1038-44.
        [http://dx.doi.org/10.1111/j.1365-2672.2004.02385.x] [PMID: 15479420]

[14]   Shetty PH, Jespersen L. *Saccharomyces cerevisiae* and lactic acid bacteria as potential mycotoxin decontaminating agents. Trends Food Sci Technol 2006; 17: 48-55.
        [http://dx.doi.org/10.1016/j.tifs.2005.10.004]

[15]   Nunez YP, Pueyo E, Carrascosa AV, Martínez-Rodríguez AJ. Effects of aging and heat treatment on whole yeast cells and yeast cell walls and on adsorption of ochratoxin A in a wine model system. J Food Prot 2008; 71(7): 1496-9.
        [http://dx.doi.org/10.4315/0362-028X-71.7.1496] [PMID: 18680954]

[16]   Rodriguez H, Reveron I, Doria F, *et al.* Degradation of ochratoxin a by *Brevibacterium* species. J Agric Food Chem 2011; 59(19): 10755-60.
        [http://dx.doi.org/10.1021/jf203061p] [PMID: 21892825]

[17]   Abrunhosa L, Inês A, Rodrigues AI, *et al.* Biodegradation of ochratoxin A by *Pediococcus parvulus* isolated from Douro wines. Int J Food Microbiol 2014; 188: 45-52.
        [http://dx.doi.org/10.1016/j.ijfoodmicro.2014.07.019] [PMID: 25087204]

[18]   Haskard CA, El-Nezami HS, Kankaanpää PE, Salminen S, Ahokas JT. Surface binding of aflatoxin B(1) by lactic acid bacteria. Appl Environ Microbiol 2001; 67(7): 3086-91.
        [http://dx.doi.org/10.1128/AEM.67.7.3086-3091.2001] [PMID: 11425726]

[19]   Li S, Marquardt RR, Frohlich AA. Identification of ochratoxins and some of their metabolites in bile and urine of rats. Food Chem Toxicol 2000; 38(2-3): 141-52.
        [http://dx.doi.org/10.1016/S0278-6915(99)00153-2] [PMID: 10717354]

[20]   Gil-Serna J, Vázquez C, González-Jaén MT, Patiño B. Wine contamination with Ochratoxins: A Review. Beverages 2018; 4(6): 1-21.

19

Optimization of Cellulase Production by *Aspergillus niger* Isolated from Forest Soil

Srilakshmi Akula and Narasimha Golla[*]

Applied Microbiology Laboratory, Department of Virology, Sri Venkateswara University, Tirupati-517502 Andhra Pradesh, India

Abstract:

Background:

An impressive increase in the application of cellulases in various fields over the last few decades demands extensive research in improving its quality and large-scale production. Therefore, the current investigation focuses on factors relevant for optimal production of cellulase by *Aspergillus niger* isolated from forest soil.

Method:

Throughout this study, the fungal strain *Aspergillus niger* was maintained under the submerged condition for a period of 7 days at 120 rpm rotational speed. Various physical and chemical conditions were employed in examining their influence on cellulase production by the selected fungal strain. After appropriate incubation, culture filtrates were withdrawn and checked for FPase, CMCase, and β-D-glucosidase activities.

Results:

The optimum pH and temperature for cellulase production were found to be 5.0 and 32°C, respectively. Among the various carbon sources tested in the present study, amendment of lactose in the medium yielded peak values of FPase (filter paperase) and CMCase (Carboxy-methyl cellulase) whereas fructose supported the higher titers of β-glucosidase. Among the nitrogen sources, profound FPase and CMCase activity were recorded when urea was used but higher β-glucosidase activity was noticed when yeast extract was added. Various natural lignocellulosic substrates like bagasse, coir, corncob, groundnut shells, litter, rice bran, rice husk, sawdust and wheat bran were tested to find out the induction of cellulase. Among the lignocelluloses, sawdust and litter served as good substrates for cellulase production by *Aspergillus niger*.

Conclusion:

In gist, the outcome of this study sheds light on the cellulolytic potentiality of the fungal strain *Aspergillus niger* promising in its future commercial applications which may be economically feasible.

Keywords: Forest soil, *Aspergillus niger*, Cellulase, Lignocelluloses, Optimization conditions, Fungal strain.

1. INTRODUCTION

Cellulose is the major skeletal component of the cell wall of green biomass found together with hemicelluloses, pectin and lignin [1, 2]. Cellulose is the fibrous, insoluble, crystalline polysaccharide composed of repeating units of D-glucose that is cemented by β-1, 4-glucosidic bonds [3]. Cellulose-rich plant biomass is one of the expected renewable sources of fuel, animal feedstock and feed for chemical synthesis [4]. Nowadays, huge quantities of municipal, industrial and agriculture cellulose wastes have been aggregated or exploited inefficiently due to the drastically lifted

[*] Address correspondence to this author at the Department of Virology, Sri Venkateswara University, Tirupati-517502 Andhra Pradesh, India, E-mail: gnsimha123@rediffmail.com

cost of their utilisation process [5]. In order to diminish food-feed-fuel conflicts, it is essential to assimilate all these kinds of bio-waste into biomass economy [6]. The rate of utilization of plant biomass is largely dependent on active soil mycoflora. Fungi effectively degrade cellulose, hemicellulose and lignin in plants by secreting a multifarious set of hydrolytic and oxidative enzymes [7, 8] being the hydrolyses (especially cellulase) the most abundant. Thus, scientists are studying fungi at the molecular level [9, 10] trying to discover cellulolytic fungi [11] and are developing mutant strains to enhance the production of cellulases [12].

At present, the production of cellulase has widely been studied in submerged culture processes, but the relatively high costs of enzyme production have hindered the industrial application of cellulose bioconversion [13]. Thus, the economics of cellulase production needs to be improved by reducing production cost or increasing the enzyme activities. Several kinds of research have shown that the production costs of cellulase are tightly associated with the productivity of enzyme producing microbial strain, the final activity in the fermentation broth and type of substrate used in the fermentation production of the enzyme [14 - 17]. Keeping the above observations into consideration, the present work focused on the optimization of process parameters for enhanced cellulose production by *Aspergillus niger* isolated from the forest soil.

2. MATERIALS AND METHODS

2.1. Microbial Strain

The fungal culture *Aspergillus niger* used in the present study was isolated from the forest soil [18].

2.2. Fungus Cultivation for Spore Production and Inoculums Preparation

The fungal inoculum density of $2x10^6$ spores [19] was used for cellulase production by submerged fermentation method and quantification was carried out by enzyme assays such as Filter paper Endoglucanase and β-D-Glucosidase.

2.3. Optimization and Experimental Design

2.3.1. Effect of Medium pH on Cellulose Production by Aspergillus niger

Czapek-Dox medium (HIMEDIA) used in this method contained (g/l); sucrose – 30, $NaNO_3$ – 2, K_2HPO_4 . 1, $MgSO_4$ – 0.05, KCl – 0.5, $FeSO_4$ – 0.01. Czapek-Dox liquid medium amended with 1% cellulose was adjusted to different pH ranges (3, 4, 5, 6, 7 & 8) and distributed in 250 ml Erlenmeyer conical flasks. The flasks were sterilized, cooled and inoculated with the fungal spore suspension. The flasks were incubated at 32°C on a rotary shaker at 120 rpm for 7 days. After incubation, dry weight of fungal mass, protein content, reducing sugars and individual enzyme components of cellulase activity was determined.

2.3.2. Influence of Temperature on Cellulase Production by Aspergillus niger

To check the influence of temperature on cellulase production, the conical flasks containing fermentation medium supplemented with 1% cellulose (pH 5.0) were incubated at different temperatures (25,32 and 40°C). The flasks were inoculated with fungal spores. After incubation, dry mass of fungal mat, the extracellular protein, total soluble sugar, and activity of total cellulase was determined.

2.3.3. Effect of Carbon Source on Cellulase Production by Aspergillus niger

To determine the effect of various carbon sources on cellulase production in the Czapek-Dox medium, various carbon sources at 1% (W/V) level were used. Glucose, fructose, lactose, galactose, maltose, and CMC were separately added to different Erlenmeyer flasks containing 100 mL of Czapek-Dox medium, the pH was adjusted to 5.0. The flasks with cellulose amended medium without a carbon source served as a control. The flasks were autoclaved at 121°C, 15 lb pressure for 15 minutes. After sterilization, the flasks were cooled and inoculated with the spore suspension of *Aspergillus niger*. The flasks were incubated at 32°C on the rotary shaker at 120 rpm for 7 days. Then the contents of flasks were filtered through Whatman No. 1 filter paper. Biomass, the extracellular protein content, total soluble sugar, and the total cellulase activity were determined in the collected culture filtrate.

2.3.4. Effect of Nitrogen Sources on Cellulose Production by Aspergillus niger

To know the impact of nitrogen sources on cellulase production, Erlenmeyer conical flasks (250ml) containing 100 ml of Czapek-Dox medium with 1% (W/V) cellulose were separately added with various nitrogen sources like peptone, yeast extract, KNO_3, $(NH_4)_2SO_4$ and urea at a concentration of 0.03%. The flasks with cellulose amended medium without nitrogen source served as control. All the flasks were aseptically inoculated with spores of *Aspergillus niger* and the flasks were incubated at 32°C on a rotary shaker at 120 rpm for 7 days. Biomass, extracellular protein content, glucose content and total cellulase activity were determined in the collected culture filtrate.

2.3.5. Effect of Lignocellulosic Substrates on Cellulase Production by Aspergillus niger

To detect the supportive native lignocellulosic substrates for the production of cellulase complex, different natural lignocellulosic substrates which include sawdust, rice bran, rice husk, wheat bran, bagasse, litter, groundnut shells, and coir and corn cob were used in the present study. 100 ml of Czapek-Dox broth medium was distributed into separate Erlenmeyer conical flasks. Each flask was provided with one gram of finely powdered untreated lignocelluloses. The pH of the medium was adjusted to 5.0. After sterilization, the flasks were cooled and inoculated with the spore suspension of *Aspergillus niger*. The flasks were incubated at 32°C for 7 days on a rotary shaker at 120 rpm. After 7 days of incubation, the contents of the flasks were filtered through Whatman filter paper No. 1. The culture filtrates were used for the estimation of biomass, the extracellular protein content, total soluble sugar and total cellulase activities.

2.4. Analytical Methods

2.4.1. Estimation of Protein Content

The extracellular protein content in the fungal filtrate was determined [20]. Suitable aliquots of filtrate were mixed with 5 ml of alkaline solution. After 30 min, 0.5 ml Folin-Ciocalteu's - reagent was added. The color developed was read at 550 nm by using the spectrophotometer (Spectronic- 20D). Bovine serum albumin was used as a protein standard.

2.4.2. Estimation of Sugar Content

The total soluble sugar content in the culture filtrates was determined [21]. Glucose was used as a standard. Suitable aliquots of culture filtrates were mixed with 3 ml of DNS reagent. The contents were boiled vigorously in a boiling water bath for exactly five minutes and the colour developed was read at 540 nm by using the spectrophotometer (Spectronic-20D).

2.4.3. Determination of Fungal Biomass

The fungal biomass was estimated by incubating the fungal culture for 7 days. After incubation, the contents of the flasks were aseptically passed through a pre-weighed filter paper (Whatman No.1) to separate mycelial mat from culture filtrate. The filter paper along with mycelial mat was dried at 70°C in an oven until constant weight and the weight was recorded. Difference between the weights of the filter paper bearing mycelial mat and weight of pre-weighed filter paper represented fungal biomass, which was expressed in terms of dry weight of mycelial mat (mg/100 ml of Czapek-Dox medium).

2.4.4. Determination of Cellulase Activity

The filtrates obtained after removal of mycelia mat by filtration through filter paper was used as an enzyme source. Flasks containing the growing culture of *Aspergillus niger* were withdrawn at every 7- day interval for processing.

2.4.5. Determination of Filter Paper Assay (FPA)

Filter paper activity of the culture filtrates was determined according to the method of Mandels and Weber (1969) [22]. Whatman filter paper strips containing 50 mg weight was suspended in one ml of 0.05 M sodium citrate buffer (pH 4.8) at 50°C in a water bath. Suitable aliquots of enzyme source were added to the above mixture and incubated for 60 minutes at 50°C. After incubation, the liberated reducing sugars were estimated by the addition of 3, 5- Dinitrosalicylic acid [21]. After cooling, colour developed in tubes was read at 540 nm in a spectrophotometer (Spectronic-20D). Appropriate control without enzyme was simultaneously made to run. The activity of cellulase was expressed in filter paper units. One unit of Filter Paper Unit (FPU) was defined as the amount of enzyme releasing one micromole of reducing sugar from filter paper /ml /h.

2.4.6. Determination of Endoglucanase Activity

The activity of endoglucanase in the culture filtrate was quantified by carboxymethylcellulose method [23]. The reaction mixture with 1.0 ml of 1% carboxymethyl cellulose in 0.2 M acetate buffer (pH 5.0) was pre-incubated at 50°C in a water bath for 20 minutes. An aliquot of 0.5 ml of culture filtrate with appropriate dilution was added to the reaction mixture and incubated at 50°C in a water bath for an hour. Appropriate control without enzyme was simultaneously made to run. The reducing sugar produced in the reaction mixture was determined by Dinitrosalicylic acid (DNS) method [21]. 3, 5-Dinitro-salicylic acid reagent was added to aliquots of the reaction mixture and the colour developed was read at a wavelength of 540 nm by using the spectrophotometer (Spectronic-20D). One unit of endoglucanase activity was defined as the amount of enzyme releasing one micromole of reducing sugar /ml /h.

2.4.7. Estimation of β-D-Glucosidase Activity

The activity of β-glucosidase in the culture filtrates was determined based on the method of Herr (1979) [24]. 200 microlitres of 5mM *p*-nitro phenyl β-D-glucopyranoside (PNPG,) in 0.05 M citrate buffer pH 4.8 was added to 0.2 ml of diluted enzyme solution with appropriate controls. After incubation for 30 min at 50°C, the reaction was stopped by adding 4 ml of 0.05 M NaOH-Glycine buffer (pH 10.6) and the liberated yellow colored *p*-nitrophenol was determined at 450 nm by using the spectrophotometer (Spectronic-20D). One unit of β-glucosidase activity was defined as the amount of enzyme liberating one micromole of þ-nitro phenol /ml /h under standard assay conditions.

3. RESULTS

3.1. Fungal Culture

The fungal culture *Aspergillus niger* used in this study was isolated from forest soil [18]. Similarly, Narasimha *et al.*, [25] isolated and identified a potent cellulolytic fungi *Aspergillus niger* from soil contaminated with cotton ginning industry effluents.

3.2. Optimization Conditions for Cellulase Production by Aspergillus niger

A series of experiments were carried out to determine the factors for enhanced production of cellulase by *Aspergillus niger*.

3.2.1. Effect of Medium pH

The effect of different pH ranges (3-8) on cellulase production by *Aspergillus niger* was tested and listed (Figs. 1, 2 and Table 1). Though maximum growth (1800mg/100ml of Czapek-Dox medium) of tested organism was obtained at pH 4.0, maximum activities of FPase (14.16 U/ml), CMCase (64.00 U/ml), extracellular protein (1.65 mg/ml), total soluble sugar (7.11 mg/ml) were recorded at pH 5.0. Relatively lower activities of FPase (0.38-1.6 U/ml), CMCase (8.11-10.2 U/ml) and β-glucosidase (0.006 U/ml), lower amounts of total soluble sugar (0.07-0.42 mg/ml), the extracellular protein (0.22-0.41 mg/ml) was found when the culture was grown at pH 3.0, 4.0 and 8.0. The less vegetative growth of 1460 mg/100ml was found at pH value of 8.0. β-glucosidase activity was not detected when pH of the medium was set above 5.0. Thus, it was clear from the above results that pH 5.0 was found to optimum pH value for cellulase production in case of *Aspergillus niger*.

Table 1. Effect of medium pH ranges on cellulase production.

Sr. No.	Medium pH Range	FPase[a] (U/ml/h)	CMCase[b] (U/ml/h)	β-Glucosidase[c] (U/ml/h)
1.	3	0.38	8.11	ND
2.	4	0.444	10.22	0.006
3.	5	14.16	64.00	0.014
4.	6	13.56	29.78	ND
5	7	13.33	28.88	ND
6.	8	1.611	9.22	ND

ND= Not detected, The results are the mean of three different experiments.

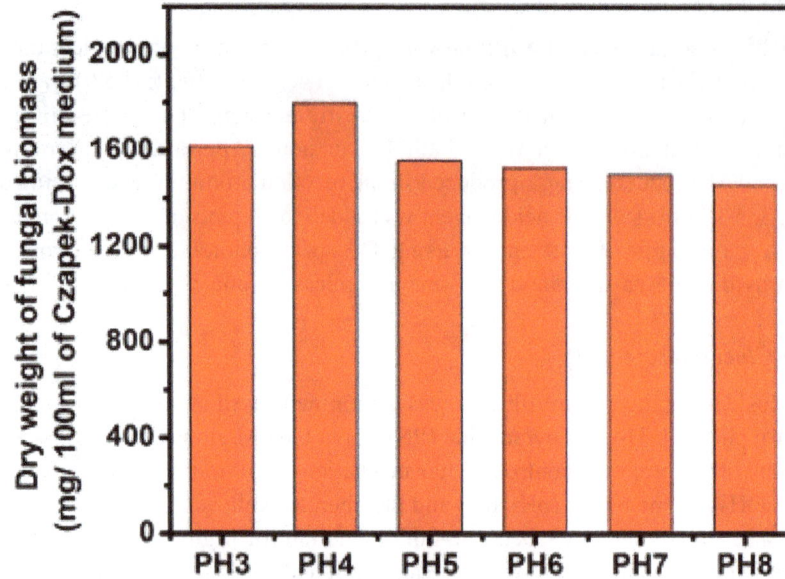

Fig. (1). Effect of different medium pH on *A. niger* biomass production. The results are the mean of three different experiments.

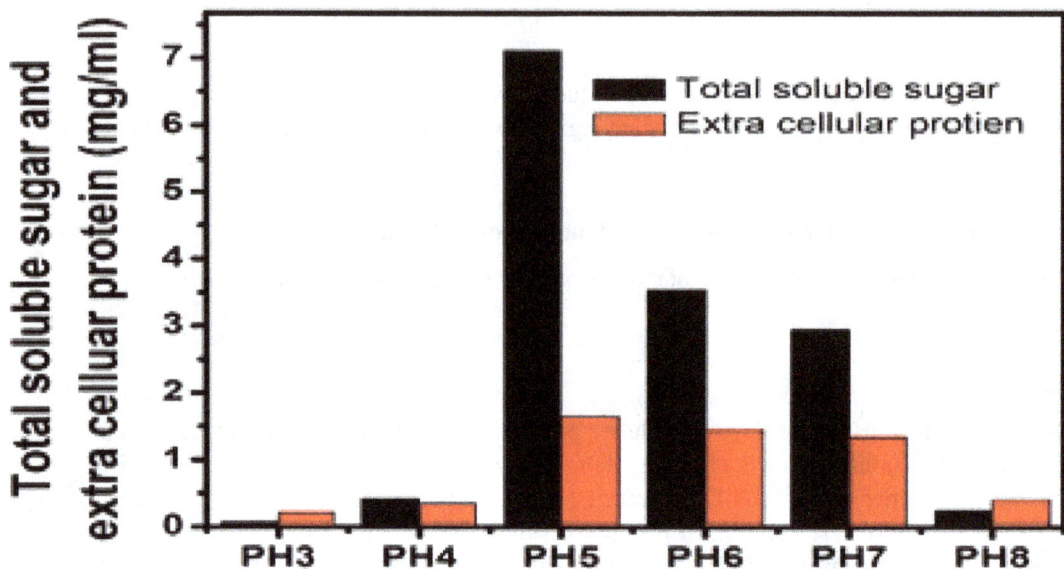

Fig. (2). Effect of medium pH on soluble sugar and protein production. The results are the mean of three different experiments.

3.2.2. Effect of Temperature

Temperature highly influences growth and enzymatic activities of microorganisms. The effect of different temperatures on cellulase production was studied and results were depicted in Figs. (3, 4 and Table 2). The culture filtrates obtained from the flasks incubated at 32°C exhibited higher of FPase activity (14.66 U/ml), CMCase (64.00 U/ml) and β-glucosidase (0.014 U/ml), fungal biomass of 1670 mg/100ml of Czapek-Dox medium, the extracellular protein of 1.65 mg/ml and total soluble sugar of 7.11 mg/ml (Figs. 10). Comparatively lower activities of FPase (2.66 U/ml), CMCase (9.32 U/ml) and β-glucosidase (0.006 U/ml) as well as lower contents of extracellular protein (0.71 mg/ml), total soluble sugar (0.55 mg/ml) and biomass of (1480 mg/100ml) were recorded at 40°C. The culture filtrate obtained from the flask incubated at 25°C shows FPase of 7.62 U/ml, CMCase of 30.88 U/ml and β-glucosidase of 0.006 U/ml. The dry weight of fungal biomass was 1580 mg/100ml, the extracellular protein was 1.58 mg/ml and total soluble sugar was 6.88 mg/ml. These values lie in between the activity ranges of the above-discussed temperatures (32°C, 40°C).

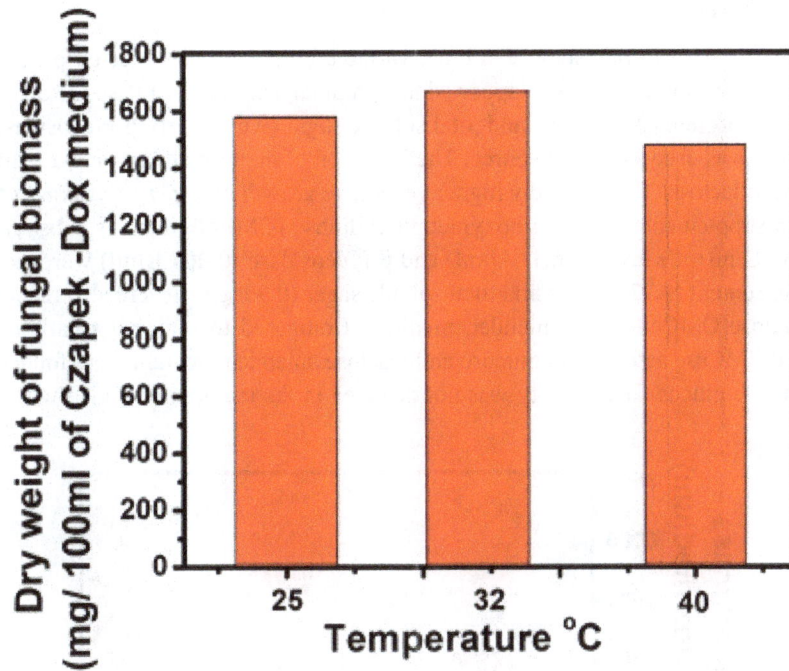

Fig. (3). Effects of temperature on biomass production.

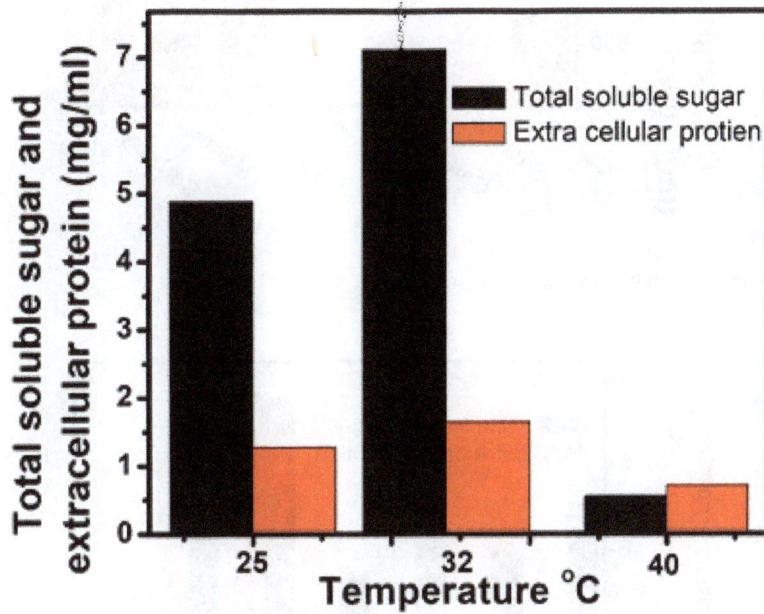

Fig. (4). Effects of temperature on total sugar and protein content. The results are the mean of three different experiments.

Table 2. Effect of various temperatures on cellulase production.

Serial Number	Temperature (°C)	FPase[a] (U/ml/h)	CMCase[b] (U/ml/h)	β-Glucosidase[c] (U/ml/h)
1.	25	7.62	30.88	0.006
2.	32	14.66	64.00	0.014
3.	40	2.66	9.32	0.006

ND= Not Detected. The results are the mean of three different experiments.

3.2.3. Effect of Carbon Sources

Different carbon sources used in this study exhibited variable effects on the selected parameters (Figs. **5**, **6** and Table **3**). Of the carbon sources applied, lactose supported the maximal activities of FPase (38.33 U/ml) CMCase (85.54 U/ml), higher extracellular protein (9.6 mg/ml) and total soluble sugar (9.6 mg/ml). β-glucosidase activity was higher (0.02 U/ml) when fructose was used as carbon source. The inclusion of maltose and galactose exhibited higher titres of FPase and CMCase next to lactose. The relatively higher vegetative growth of *Aspergillus niger* of 3330mg/100ml was obtained in the medium supplemented with carboxymethyl cellulose (CMCellulose) as substrate. The three enzyme components FPase (4.33 U/ml), CMCase (16.11 U/ml) and β-glucosidase (0.007 U/ml) were exhibited at low levels. The extracellular protein content (0.77 mg/ml) and total soluble sugar (0.9 mg/ml) were also recorded at lower levels in the Carboxymethylcellulose(CMCellulose) amended medium. Compared to remaining carbon sources used in the present study, FPase and CMCase activities in glucose and fructose amended medium were found to be intermediate to the above-described ones. β-glucosidase activity was not detected in the medium amended with glucose, lactose, and maltose.

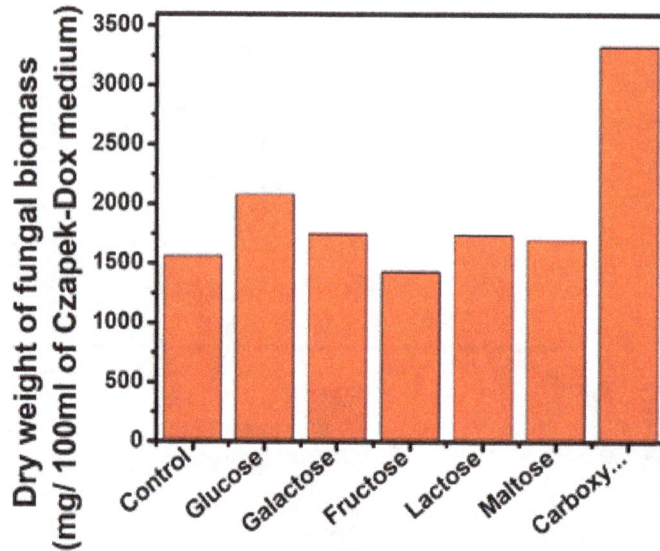

Fig. (5). Effect of carbon source on soluble sugar and protein content.

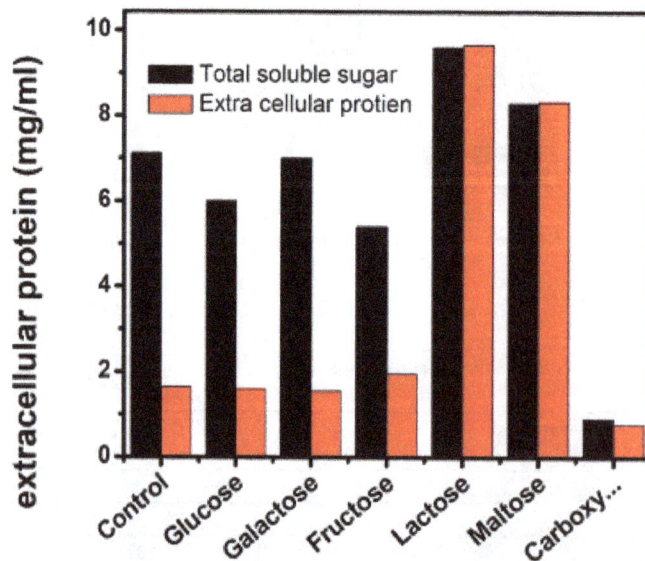

Fig. (6). Effect of carbon sources on biomass production. The results are the mean of three different experiments.

Table 3. Effect of carbon sources on cellulase production.

Serial Number	Carbon Source	FPase[a] (U/ml/h)	CMCase[b] (U/ml/h)	β-Glucosidase[c] (U/ml/h)
1.	Glucose	26.38	47.76	ND
2.	Galactose	32.77	63.32	0.007
3.	Fructose	23.33	52.22	0.020
4.	Lactose	38.33	85.54	ND
5.	Maltose	37.77	77.76	ND
6.	CMC	4.33	16.11	0.007
7.	Control[d]	14.16	64.00	0.014

ND= Not Detected. The results are the mean of three different experiments.

The results are the mean of three different experiments.

3.2.4. Effect of Nitrogen Sources

The effects of various nitrogen sources on cellulase production were studied and compared with control (Figs. **7, 8** and Table **4**). Among the list of nitrogen sources, urea appeared to be the best as reflected by the highest production of extracellular protein (210mg/ml). 38.88 U/ml of FPase, 68.88 U/ml of CMCase, and total soluble sugar of 8.70 mg/ml. β-glucosidase was undetected not only in urea but also in peptone and KNO_3 amended medium. Though higher β-glucosidase activity (0.230 U/ml) was observed in yeast extract supplemented medium lower activities of FPase (0.77 U/ml), CMCase (7.32 U/ml) was recorded. Maximum (2080 mg/100ml) and minimum (1490 mg/100ml) fungal growth were observed in $(NH_4)_2$, SO_4 and KNO_3 amended medium, respectively. Considerably lower activities of FPase and CMCase along with lower amounts of extracellular protein and total soluble sugar yeast extract and $(NH_4)_2$, SO_4 supplemented medium.

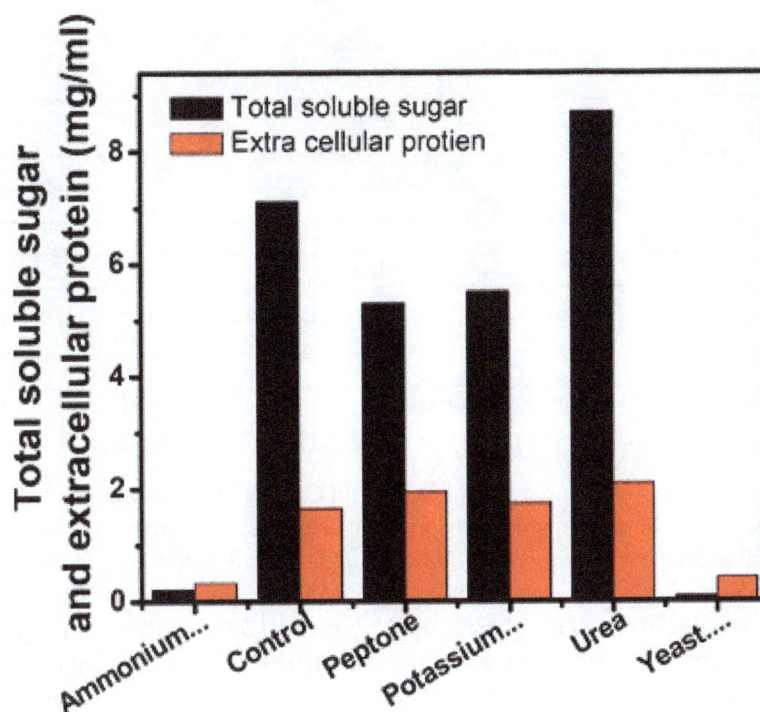

Fig. (7). Effect of nitrogen sources on soluble sugar and protein content.

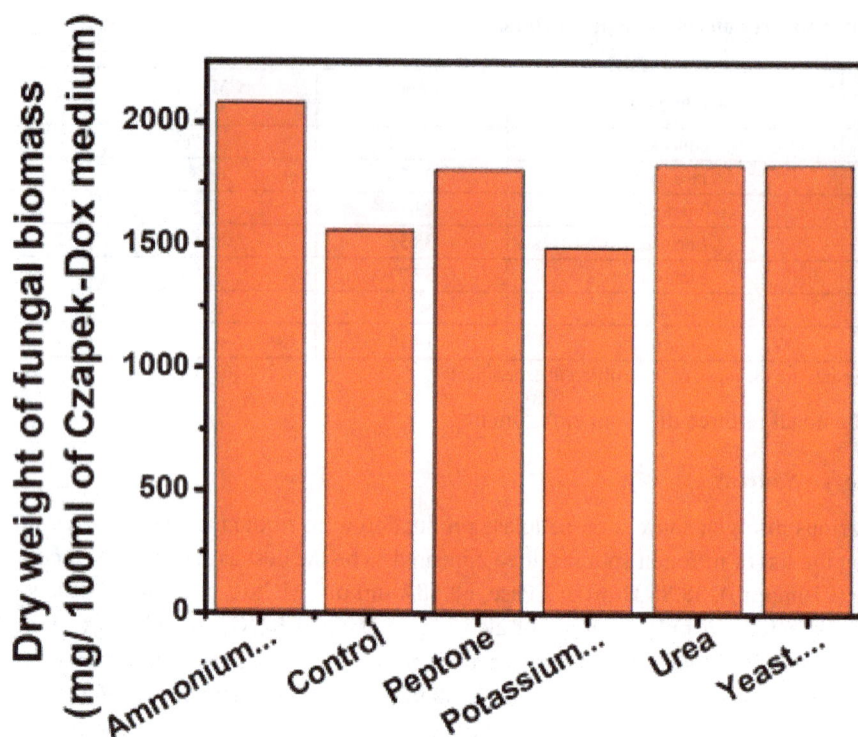

Fig. (8). Effect of nitrogen source on biomass production. The results are the mean of three different experiments.

Table 4. Effect of carbon sources on cellulase production.

Serial Number	Carbon Source	FPase[a] (U/ml/h)	CMCase[b] (U/ml/h)	β-glucosidase[c] (U/ml/h)
1.	Glucose	26.38	47.76	ND
2.	Galactose	32.77	63.32	0.007
3.	Fructose	23.33	52.22	0.020
4.	Lactose	38.33	85.54	ND
	Maltose	37.77	77.76	ND
6.	CMC	4.33	16.11	0.007
7.	Control[d]	14.16	64.00	0.014

ND= Not detected .The results are the mean of three different experiments.

3.3. Effect of Lignocellulosic Substrates

Various natural lignocelluloses were employed at 1% level in Czapek-Dox medium to induce cellulase production by *Aspergillus niger*. The results were compared to the yields of cellulase synthesized by the same microorganism on cellulose (Figs. **9**, **10**, Table **5**). Among the lignocelluloses supplemented in the present study, the sawdust induced maximum biomass of 1650 mg/100ml, the total soluble sugar of 7.90 mg/ml and the highest titres of FPase (31.11 U/ml) CMCase (67.54 U/ml) and lower yields of extracellular protein (1.71 mg/ml) were noticed. Higher quantities of β-glucosidase (0.43 U/ml) were obtained with bagasse. Supplementation of forest litter induced the yields of FPase (29 U/ml) CMCase (60.44 U/ml), total soluble sugar (5.0 mg/ml) which proved it as a good inducer of cellulase production next to sawdust. The secretion of extracellular protein was maximum (1710 mg/ml) when groundnut shells were added as substrate. The decreased activities of FPase and CMCase were noticed with rice bran, groundnut shells, wheat bran, and bagasse. The lowest β-glucosidase activity (0.02 U/ml) was recorded in wheat bran and groundnut shells. Least values of total soluble sugar (0.87 mg/ml), extracellular protein (0.75 mg/ml) and dry weight (690 mg/100ml) were noticed in the experiments which used bagasse as a substrate.

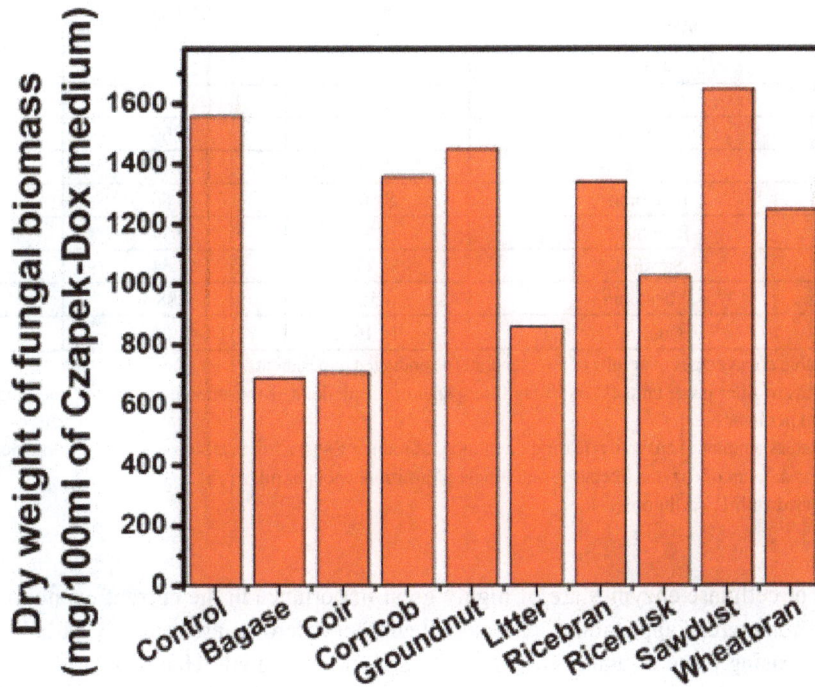

Fig. (9). Effect of lignocelluloses on biomass production.

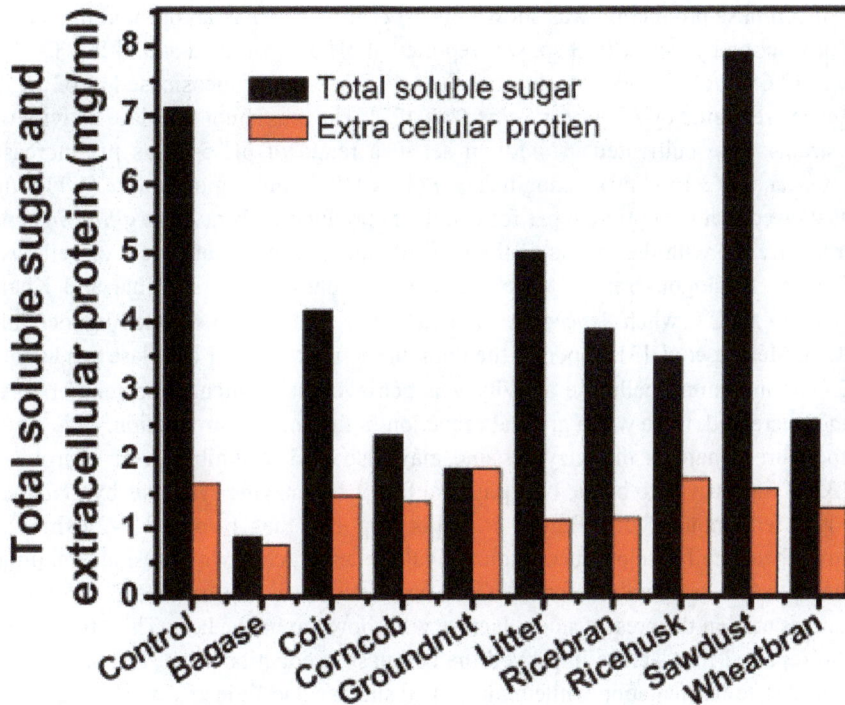

Fig. (10). Effect of lignocelluloses on soluble sugar and protein content. The results are the mean of three different experiments.

Table 5. Influence of Lignocellulosic substrates for cellulase production.

Serial Number	Lignocellulose	FPase[a] (U/ml/h)	CMCase[b] (U/ml/h)	β-Glucosidase[c] (U/ml/h)
1.	Bagasse	3.5	8.88	0.43

Serial Number	Lignocellulose	FPase[a] (U/ml/h)	CMCase[b] (U/ml/h)	β-Glucosidase[c] (U/ml/h)
2.	Coir	19.55	21.32	0.05
3.	Corncob	15.33	28.44	0.32
4.	Groundnut shells	8.50	11.66	0.02
5.	Litter	29.0	60.44	0.06
6.	Rice bran	8.66	16.66	0.13
7.	Rice husk	23.33	40.44	0.16
8.	Sawdust	31.11	67.54	0.32
9.	Wheatbran	4.55	3.88	0.02
10.	Control[d]	14.16	64.00	0.014

*Values represented in the table are averages of results of two separately conducted experiments.

[a] FPase was expressed in terms of filter paper units (U/ml). One filter paper unit is defined as the amount of enzyme releasing one µmole of reducing sugar from filter paper per ml per hour.

[b] CMCase activity was defined as amount of enzyme releasing one µmole of reducing sugar from carboxy-methyl cellulose per ml per hour.

[c] β-glucosidase was defined as amount of enzyme liberating one µmole of þ-nitrophenol per ml per /h

[d] Czapek-Dox medium amended with 1% cellulose.

4. DISCUSSION

Microbial sources of cellulase enzymes are of highly great importance in the current economic value due to their versatile industrial and commercial applications. Cellulose degrading microbes are basically carbohydrate degraders and are usually incapable of using proteins and lipids as energy sources of growth. Hence, a large number of filamentous fungi, bacteria and actinomycetes produce cellulase for the fragmentation of cellulose biopolymer. The characteristic of certain fungi to secrete, abundant amounts of extracellular protein makes them most suited for the production of higher levels of extracellular cellulases. Therefore, in the present study, a cellulolytic fungus *Aspergillus niger* was isolated from forest litter soil and was grown under different physical and chemical conditions for the optimal production of cellulase. High titter of cellulase production was shown by *Aspergillus niger* at an optimal pH 5.0. Similarly, optimal growth and cellulase production by *Penicillium* sp. was reported at pH 5.0 Prasanna *et al.* [26]. The highest activities of Endo-β-1, 4-glucanase (17.65 U/g), Exo-β-1, 4-glucanase (13.49 U/g) and β-glucosidase (14.62 U/g) were obtained at pH of 5.0 with sugar cane press mud by *Pleurotus Sajor-Caju* [27]. The maximum cellulase activity was achieved when *Trichoderma viride* strains were cultivated in medium set to a range of pH 5–6; as pH increased up to 5.5, the hyperactivities of exoglucanase (2.16 U/ml), endoglucanase (1.94 U/ml) and β-glucosidase (1.71 U/ml) were observed [28]. Pham *et al.* [29] showed that the optimum pH for cellulases production from *Aspergillus niger* VTCC-F021 strain was 5.0. These reports correlate with the results of the present study. A maximum yield of cellulase was obtained at 32°C in this investigation. Analogous results were achieved by many workers. Gilna and Khaleel [30] reported maximum cellulase activity at 32°C when *Aspergillus fumigatus* was cultured on selected lignocellulosic wastes under liquid state fermentation. Mekala *et al.* [31] reported the optimum temperature for cellulase production by *Trichoderma* RUT C30 was 33°C. The maximum cellulase activity was achieved at an incubation temperature of 32°C. As the temperature was further increased, there was a gradual reduction in the enzyme production. This may be because of the fact that higher temperature denatures the enzymes and may also lead to inhibition of microbial growth [32]. The temperature level of 32°C was proved to be the best physical factor for enzyme synthesis by *Trichoderma* sp. on apple pomace under solid state fermentation [33]. Bansal *et al.* [34] reported that *A. niger* NS-2 exhibited a wide range of temperatures for its growth and cellulase production on agriculture and kitchen waste residues. Similarly, El-Hadi *et al.* [35] reported that the optimum temperature for CMCase production was observed at 37°C for *Aspergillus hortai*. Among the carbon sources used in the present study, lactose was shown to be the best. This study substantiates the work of Sun *et al.* [33] who reported that lactose improved the cellulase production when *Trichoderma* sp. was grown on apple pomace under solid state fermentation while maltose and sucrose had little effect. The maximal cellulase activity (1.18 U/ml) was observed when lactose used as a carbon source for *A. hortai* [35]. Kathiresan and Manivannan [36] and Devanathan *et al.* [37] proved lactose as the best inducer of *Aspergillus* sp. Muthuvelayudham and Viruthagiri [38] reported maximum growth and cellulase enzyme production by *T. reesei* C5 with the provision of lactose as a sole carbon source. These reports were found to be in agreement with the results of the current study. The present study proves that supplementation of urea-induced maximum enzyme production. Identical reports were made by Jyotsna *et al.* [39]. Urea appeared as the best nitrogen source for the highest activity of exo-β-glucanase with 4.8 IU/ml, endo-β-glucanase with 5.3 IU/ml and β-glucosidase with 1.6 IU/ml by *Streptomyces albaduncus*. Dharmdutt and Alokkumar [40] gave a report that there was a substantial increase in the cellulase activity when the medium was supplemented with complex nitrogen sources like yeast extract and urea by *Aspergillus flavus* AT-2 and *Aspergillus niger* AT-3.The culture filtrate of *Aspergillus niger* exhibited relatively higher cellulolytic activity (1.682 U/ml) grown on Czapek-Dox medium containing 0.03% urea followed by peptone and $NaNO_3$ [19]. These findings corroborate with the results of the present study that organic nitrogen sources such as urea and peptone served as the best compared to inorganic nitrogen sources such as $(NH_4)_2 SO_4$ and KNO_3. Of all the various lignocelluloses amended to the medium, sawdust supported higher yields of cellulase by *Aspergillus niger*. These findings were in agreement with the previous reports of Praveen Kumar *et al.* [41] that the wild strain of *Penicillium chrysogenum* PCL501 produces significant cellulase activity (100.0 U/ml and 92.2 U/ml) and extracellular proteins when sawdust was used as the sole carbon source. The highest amount of cellulase activity was exhibited by *Aspergillus niger* followed by *Trichoderma viride* by solid-state fermentation with sawdust as a substrate [42]. Supplementation of sawdust at 1% level resulted in higher activities of FPase (2.412 U/ml), CMCase (0.775 U/ml) and β-glucosidase (1.322 U/ml) by *Aspergillus niger* [19].

CONCLUSION

The potent cellulolytic fungi, *Aspergillus niger* was grown in Czapek-Dox medium under optimal physical and chemical conditions. A temperature of 32°C and pH value of 5.0 were found to be the optimal condition for cellulase production. Among various carbon and nitrogen sources tested in this study, lactose and urea yielded peak values of cellulase. Natural lignocellulosic materials like sawdust and forest litter served as good substrates for enhanced production of cellulase.

HUMAN AND ANIMAL RIGHTS

No animals/humans were used for studies that are the basis of this research.

ACKNOWLEDGEMENTS

The first author was highly thankful to Prof. D.V.R. Sai Gopal, Department of Virology, and S.V. University for providing laboratory facility for carrying out this work.

REFERENCES

[1]   Brijwani K, Oberoi HS, Vadlani PV. Production of cellulolytic enzyme system in mixed culture solid-state fermentation of soybean hulls supplemented with wheat bran. Process Biochem 2010; 45: 120-8. [http://dx.doi.org/10.4061/2011/860134]. [http://dx.doi.org/10.1016/j.procbio.2009.08.015]

[2]   Ram L, Kuldeep K, Sandeep S. Screening isolation and characterization of cellulase producing microorganisms from soil. Int J PharmSci Invent 2014; 3(3): 12-8.

[3]   Jagtap S, Rao M. Purification and properties of a low molecular weight 1,4-beta-d-glucan glucohydrolase having one active site for carboxymethyl cellulose and xylan from an alkalothermophilic *Thermomonospora* sp. Biochem Biophys Res Commun 2005; 329(1): 111-6. [http://dx.doi.org/10.1016/j.bbrc.2005.01.102] [PMID: 15721281]

[4]   Bhat MK. Cellulases and related enzymes in biotechnology. Biotechnol Adv 2000; 18(5): 355-83. [http://dx.doi.org/10.1016/S0734-9750(00)00041-0] [PMID: 14538100]

[5]   Kim KC, Seung-Soo Y, Oh Young A, Seong-Jun K. Isolation and characterization of *Trichoderma harzianum* FJ1 producing cellulase and xylanase. J Microbiol Biotechnol 2003; 13: 1-8.

[6]   Mahro B, Timm M. Potential of bio-waste from the food industry as a biomass resource. Eng Life Sci 2007; 7: 457-68. [http://dx.doi.org/10.1002/elsc.200620206]

[7]   Abd-Elzaher FH, Fadel M. Production of bioethanol *via* enzymatic saccharification of rice straw by cellulase produced by *Trichoderma reesei* under solid state fermentation. Int J Appl Microbiol Biotechnol Res 2010; 43(3): 72-8.

[8]   Amir I, Zahid A, Yusuf Z, *et al.* Optimization of cellulase enzyme production from corn cobs using *Alternaria alternata* by solid state

fermentation. J Cell Mol Biol 2011; 9(2): 51-6.

[9] Nevalainen H, Penttila M. Kuck U. Molecular Biology of Cellulolytic Fungi. In: Berlin, Heidelberg: Springer Verlag: The Mycota, Vol. II, Genetics and Biotechnology 2003; pp. 303-19.

[10] Shimosaka M, Kumehara M, Zhang XY, *et al.* Cloning and characterization of a chitosanase gene from the plant pathogenic fungus *Fusarium solani*. J Ferment Bioeng 1996; 82: 426-31.
[http://dx.doi.org/10.1016/S0922-338X(97)86977-2]

[11] Ariunaa J, Temuulen G. Distribution of Cellulolytic Fungi in Soil of Mongolia. Scientific Journal "Biology" Mongolian National University, 2001, 10, 71-3.

[12] Chand P, Aruna A, Maqsood A M, Rao LV. Novel mutation method for increased cellulase production. J Appl Microbiol 2005; 98(2): 318-23.

[13] Pandey A, Soccol CR, Poonam Nigam P, Saccol VT. Biotechnological potential of agro-Industrial residues.1: Sugarcane bagasse. Bioresour Technol 2000; 7(1): 69-80.
[http://dx.doi.org/10.1016/S0960-8524(99)00142-X]

[14] Duff SJB, Murray WD. Bioconversion of forest products and industry waste cellulosics to fuel ethanol: A review. Bioresour Technol 1996; 55: 1-33.
[http://dx.doi.org/10.1016/0960-8524(95)00122-0]

[15] Nieves RA, Ehrman CL, Adney WS, Elander RT, Himmel ME. Technical communication: Survey and analysis of commercial cellulase preparations suitable for biomass conversion to ethanol. World J Microbiol Biotechnol 1998; 14: 301-4.
[http://dx.doi.org/10.1023/A:1008871205580]

[16] Chahal PS, Chahal DS, Andre G. Cellulase production profile of *Trichoderma reesei* on different cellulosic substrates at various pH levels'. Ferment Bioeng 1992; 74: 126-8.
[http://dx.doi.org/10.1016/0922-338X(92)80015-B]

[17] Reczey K, Szengyel Z, Eklund RC, Zacchi G. Cellulase production by *T. reesei*. Bioresour Technol 1996; 57: 25-30.
[http://dx.doi.org/10.1016/0960-8524(96)00038-7]

[18] Srilakshmi A, Narasimha G. Production of cellulases by fungal cultures isolated from forest litter soil. Ann For Res 2012; 55(1): 85-92.

[19] Narasimha G, Sridevi A, Viswanath B, Subhosh Chandra M, Rajasekhar Reddy B. Nutrient effects on production of cellulolytic enzymes by *Aspergillus niger*. Afr J Biotechnol 2006; 5(5): 472-6.

[20] Lowry OH, Rosebrough NJ, Farr AL, Randall RJ. Protein measurement with the folin phenol reagent. J Biol Chem 1951; 193(1): 265-75.
[PMID: 14907713]

[21] Miller GL. Use of dinitrosalicylic acid reagent for determination of reducing sugars. Anal Chem 1959; 31: 426-9. [DOI: 10.1021/ac60147a030].
[http://dx.doi.org/10.1021/ac60147a030]

[22] Mandels M, Weber J. Cellulases and its application. In: Am Chem Soc. Washington, DC: Advances in chemistry Series (R.F. Gould. Ed) 1969; 95: pp. 391-414.

[23] Ghosh TK. Measurement of cellulase activities. Pure Appl Chem 1987; 59(2): 257-68. [http://dx.doi.org/10.1351/pac198759020257].
[http://dx.doi.org/10.1351/pac198759020257]

[24] Herr D. Secretion of cellulase and beta-glucosidase by *Trichoderma viride* ITCC-1433 in submerged culture on different substrates. Biotechnol Bioeng 1979; 21(8): 1361-71. [DOI: 10.1002/bit.260210805]. [PMID: 110377].
[http://dx.doi.org/10.1002/bit.260210805] [PMID: 110377]

[25] Narasimha G, Babu GVAK, Rajasekhar Reddy B. Cellulolytic activity of fungal cultures isolated from soil contaminated with effluents of cotton ginning industry. J Sci Ind Res (India) 1998; 57: 617-20.

[26] Prasanna HN, Ramanjaneyulu G, Rajasekhar Reddy B. Optimization of cellulase production by *Penicillium sp.* 3 Biotech. 2016; 6: 162.

[27] Pandit Nitin Prakash, Maheshwari Sanjiv Kumar. Optimization of cellulase enzyme production from sugarcane Press mud using oyster mushroom – *pleurotus sajor-Caju* by solid state fermentation. J Bioremediat Biodegrad 2012; 3(3): 1-5.

[28] Gautam SP, Bundela PS, Pandey AK, Awasthi MK, Sarsaiya S. Screening of cellulolytic fungi for management of municipal solid waste. J Appl Sci in Environ Sanita 2010; 4(3): 391-5.

[29] Pham TH, Quyen DT, Nghiem NM. Optimization of endoglucanase production by *Aspergillus niger* VTCCF021. Aust J Basic Appl Sci 2010; 4(9): 4151-7.

[30] Gilna VV, Khaleel KM. Biochemistry of cellulase enzyme activity of *Aspergillus fumigatus* from mangrove soil on lignocellulosics substrate. Recent ResSciTechnol 2011; 3(1): 132-4.

[31] Mekala NK, Singhania RR, Sukumaran RK, Pandey A. Cellulase production under solid-state fermentation by *Trichoderma reesei* RUT C30: Statistical optimization of process parameters. Appl Biochem Biotechnol 2008; 151(2-3): 122-31.
[http://dx.doi.org/10.1007/s12010-008-8156-9] [PMID: 18975142]

[32] Shazia KM, Hamid M, Ammad AF, Haq IU. Optimization of process parameters for the biosynthesis of cellulose by *Trichoderma viride*. Pak J Bot 2010; 42(6): 4243-51.

[33]   Sun H, Xiangyang Ge, ZhikuiHao Ming P. Cellulase production by *Trichoderma sp.* on apple pomace under solid state fermentation. Afr J Biotechnol 2010; 9(2): 163-6. [DOI: 10.4314/ajb.v9i2.].

[34]   Bansal N, Tewari R, Soni R, Soni SK. Production of cellulases from *Aspergillus niger* NS-2 in solid state fermentation on agricultural and kitchen waste residues. Waste Manag 2012; 32(7): 1341-6.
[http://dx.doi.org/10.1016/j.wasman.2012.03.006] [PMID: 22503148]

[35]   El-Hadi A. Anwar.optimization of cultural and nutritional conditions for carboxymethylcellulase production by *Aspergillus hortai*. J Radiat Res Appl Sci 2014; 7(23): 28-F36.

[36]   Kathiresan K, Manivannan S. Cellulase production by *Penicillium fellutanum* isolated from coastal mangrove rhizosphere soil. Res J Microbial 2006; 1(5): 438-42. [http://dx.doi.org/10.3923/jm.2006.438.442].
[http://dx.doi.org/10.3923/jm.2006.438.442]

[37]   Devanathan G, Shanmugan A, Balasubramanian T, *et al.* Cellulase production by *Aspergillus niger* isolated from coastal mangrove debris. Trends Appl Sci Res 2007; 2: 23-7. [http://dx.doi.org/10.3923/tasr.2007.23.27].
[http://dx.doi.org/10.3923/tasr.2007.23.27]

[38]   Muthuvelayudham R, Viruthagiri T. Fermentative production and kinetics of cellulase protein on *Trichoderma reesei* using sugarcane bagasse and rice straw. Afr J Biotechnol 2006; 5(20): 1873-81.

[39]   Jyotsna PK, Ramakrishna Rao A, Devaki K. Effect of nutritional factors on cellulase production by *Streptomyces albaduncus* from the gut of earthworm, *Eiseniafoetida*. Pest Manage Hortic Ecosyst 2015; 21(1): 75-80.

[40]   Dharm Dutt and AlokAumar. Optimization of cellulase production under solid-state fermentation by *Aspergillus flavus* (AT-2) and *Aspergillus niger* (AT-3) and its impact on stickies and ink particle size of sorted office paper. Cellulose ChemTechnol 2014; 48(3-4): 285-98.

[41]   Praveen Kumar Reddy G, Narasimha G, Dileep Kumar K, *et al.* Cellulase production by *Aspergillus niger* on different natural lignocellulosic substrates. Int J Curr Microbial Appl Sci 2014; 4(1): 835-45.

[42]   Bhoos Reddy GL. Comparative study of cellulase production by *Aspergillus niger* and *Trichoderma viride* using solid state fermentation on cellulosic substrates corncob, cane bagasse and sawdust. Int J Sci Res (Ahmedabad) 2014; 3(5): 324-6.

Permissions

All chapters in this book were first published in TOBJ, by BENTHAM Open; hereby published with permission under the Creative Commons Attribution License or equivalent. Every chapter published in this book has been scrutinized by our experts. Their significance has been extensively debated. The topics covered herein carry significant findings which will fuel the growth of the discipline. They may even be implemented as practical applications or may be referred to as a beginning point for another development.

The contributors of this book come from diverse backgrounds, making this book a truly international effort. This book will bring forth new frontiers with its revolutionizing research information and detailed analysis of the nascent developments around the world.

We would like to thank all the contributing authors for lending their expertise to make the book truly unique. They have played a crucial role in the development of this book. Without their invaluable contributions this book wouldn't have been possible. They have made vital efforts to compile up to date information on the varied aspects of this subject to make this book a valuable addition to the collection of many professionals and students.

This book was conceptualized with the vision of imparting up-to-date information and advanced data in this field. To ensure the same, a matchless editorial board was set up. Every individual on the board went through rigorous rounds of assessment to prove their worth. After which they invested a large part of their time researching and compiling the most relevant data for our readers.

The editorial board has been involved in producing this book since its inception. They have spent rigorous hours researching and exploring the diverse topics which have resulted in the successful publishing of this book. They have passed on their knowledge of decades through this book. To expedite this challenging task, the publisher supported the team at every step. A small team of assistant editors was also appointed to further simplify the editing procedure and attain best results for the readers.

Apart from the editorial board, the designing team has also invested a significant amount of their time in understanding the subject and creating the most relevant covers. They scrutinized every image to scout for the most suitable representation of the subject and create an appropriate cover for the book.

The publishing team has been an ardent support to the editorial, designing and production team. Their endless efforts to recruit the best for this project, has resulted in the accomplishment of this book. They are a veteran in the field of academics and their pool of knowledge is as vast as their experience in printing. Their expertise and guidance has proved useful at every step. Their uncompromising quality standards have made this book an exceptional effort. Their encouragement from time to time has been an inspiration for everyone.

The publisher and the editorial board hope that this book will prove to be a valuable piece of knowledge for researchers, students, practitioners and scholars across the globe.

List of Contributors

Kehinde Odelade
Department of Pure and Applied Biology, Ladoke Akintola University of Technology, P.M.B 4000, Ogbomoso, Oyo State, Nigeria

Adeyemi Ojutalayo Adeeyo
Department of Pure and Applied Biology, Ladoke Akintola University of Technology, P.M.B 4000, Ogbomoso, Oyo State, Nigeria
School of Environmental Sciences, University of Venda, Thohoyandou 0950, South Africa

John Odiyo
School of Environmental Sciences, University of Venda, Thohoyandou 0950, South Africa

Georgi Dobrev, Hristina Strinska, Anelia Hambarliiska and Boriana Zhekova
Department of Biochemistry and Molecular Biology, University of Food Technologies, 26 Maritza Blvd., 4002 Plovdiv, Bulgaria

Valentina Dobreva
Department of Engineering Ecology, University of Food Technologies, 26 Maritza Blvd., 4002 Plovdiv, Bulgaria

Akiyoshi Taniguchi
Cellular Functional Nanomaterials Group, Research Center for Functional Materials, National Institute for Materials Science, 1-1 Namiki, Tsukuba, Ibaraki 305-0044, Japan
Graduate School of Advanced Science and Engineering, Waseda University, 3-4-1 Okubo, Shinjuku-ku, Tokyo 169-8555, Japan

Le Thi Minh Phuc
Cellular Functional Nanomaterials Group, Research Center for Functional Materials, National Institute for Materials Science, 1-1 Namiki, Tsukuba, Ibaraki 305-0044, Japan
Graduate School of Advanced Science and Engineering, Waseda University, 3-4-1 Okubo, Shinjuku-ku, Tokyo 169-8555, Japan
Institute of Biotechnology, Vietnam Academy of Science and Technology, 18 Hoang Quoc Viet, Cau Giay, Hanoi, Vietnam

Nguyen Thi Minh Huyen, Nguyen Thi Thu Thuy and Le Quang Huan
Institute of Biotechnology, Vietnam Academy of Science and Technology, 18 Hoang Quoc Viet, Cau Giay, Hanoi, Vietnam

Tetsuji Sasaki and Hisayo Shimizu
Kyokuto Pharmaceutical Industrial Co., Ltd. 3333-26 Aza-Asayama, Kamitezuna, Takahagi-shi, Ibaraki 318-0004, Japan

E. Calabrò
Department of Mathematical and Informatics Sciences, Physical Sciences and Earth Sciences of Messina University, Viale Ferdinando Stagno D' Alcontres 31, 98166 Messina, Italy

S. Magazù and M. T. Caccamo
Department of Mathematical and Informatics Sciences, Physical Sciences and Earth Sciences of Messina University, Viale Ferdinando Stagno D' Alcontres 31, 98166 Messina, Italy
Istituto Nazionale di Alta Matematica "F. Severi" – INDAM - Gruppo Nazionale per la Fisica Matematica – GNFM, Messina, Italy

Anika Guliani and Amitabha Acharya
Biotechnology Division, CSIR-Institute of Himalayan Bioresource Technology, Palampur (H.P.) 176061, India
Academy of Scientific & Innovative Research (AcSIR), CSIR-Institute of Himalayan Bioresource Technology, Palampur (H.P.) 176061, India

Rajni Hatti-Kaul and Bo Mattiasson
Department of Biotechnology, Center for Chemistry and Chemical Engineering, Lund University, Lund, Sweden

Kevin Raymond Oluoch
Department of Biotechnology, Center for Chemistry and Chemical Engineering, Lund University, Lund, Sweden
Department of Biochemistry, University of Nairobi, Nairobi, Kenya

Patrick Wafula Okanya and Francis Jakim Mulaa
Department of Biochemistry, University of Nairobi, Nairobi, Kenya

Ravi Danielsson, Tove Sandberg and Håkan Eriksson
Department of Biomedical Science, Faculty of Health and Society, Malmö University, SE-205 06 Malmö, Sweden

Dasha Mihaylova and Albert Krastanov
Department of Biotechnology, University of Food Technologies, 26 Maritza Blvd., Plovdiv 4002, Bulgaria

Aneta Popova and Iordanka Alexieva
Department of Catering and Tourism, University of Food Technologies, 26 Maritza Blvd., Plovdiv 4002, Bulgaria

Anna Lante
Department of Agronomy, Food, Natural Resources, Animals and Environment – DAFNAE, University of Padova, Viale Università 16, Agripolis 35020, Italy

Narin Kijkriengkraikul
Technopreneurship and Innovation Management Program, Graduate School, Chulalongkorn University, Bangkok 10330, Thailand

Issarang Nuchprayoon
Department of Pediatrics, Faculty of Medicine, Chulalongkorn University, Bangkok 10330, Thailand

Wei Tang
College of Horticulture and Gardening, Yangtze University, Jingzhou, Hubei 434025, China

Alejandro Ruiz-Marin, Yunuen Canedo-López, Asteria Narvaez-García and Juan Carlos Robles-Heredia
Faculty of Chemical Engineering, University Autónoma de Ciudad del Carmen, Concordia, 24180 Ciudad del Carmen, Campeche, México

Jose del Carmen Zavala-Loria
Department of Industrial Engineering, University International Hiberoamericana, IMI III, Mexico

Sridevi Ayla and Suvarnalathadevi Pallipati
Department of Applied Microbiology, Sri Padmavati Mahila University, Tirupati, A.P, India

Narasimha Golla
Applied Microbiology Laboratory, Department of Virology, Sri Venkateswara University, Tirupati -517502, A. P, India

Daryoush Asgarpoor
Student Research Center, Zanjan University of Medical Sciences, Zanjan, Iran

Department of Microbiology, Zanjan University of Medical Sciences, Zanjan, Iran

Fakhri Haghi and Habib Zeighami
Department of Microbiology, Zanjan University of Medical Sciences, Zanjan, Iran

Ebrahim Zohourvahid Karimi and Mohammad Ansari
Department of Metallurgy and Ceramic, Mashhad Branch, Islamic Azad University, Mashhad, Iran

Martha M. Naguib
Department of Biotechnology and Life Sciences, Faculty of Post Graduate Studies for Advanced Sciences, Beni-Suef University, Beni-Suef 62511, Egypt

Ahmed O. El-Gendy and Ahmed S. Khairalla
Department of Microbiology and Immunology, Faculty of Pharmacy, Beni-Suef University, Beni-Suef 62511, Egypt

M. Gundhavi Devi
Centre for Bioscience and Nanoscience Research, Coimbatore, Tamil Nadu 641021, India

M. Amutheesan
Department of Aeronautical Engineering, Hindustan Institute of Technology & Science, Padur, Chennai, Tamil Nadu 603103, India

R. Govindhan and B. Karthikeyan
Department of Chemistry, Annamalai University, Annamalai Nagar, Tamil Nadu 608 002, India

Moncalvo A., Dordoni R., Silva A., Fumi M. D. and Spigno G.
DiSTAS Department for Sustainable Food Process, Università Cattolica del Sacro Cuore, Via Emilia Parmense 84, 29122 Piacenza, Italy

Di Piazza S.
Laboratory of Mycology, Department of Earth, Environmental and Life Sciences, Università degli Studi di Genova, Corso Europa 26, I, 16136 Genova, Italy

Srilakshmi Akula and Narasimha Golla
Applied Microbiology Laboratory, Department of Virology, Sri Venkateswara University, Tirupati-517502 Andhra Pradesh, India

Index